**Vehicle Scanning Method for Bridges**

# Vehicle Scanning Method for Bridges

*Yeong-Bin Yang*
School of Civil Engineering
Chongqing University
China

*Judy P. Yang*
Department of Civil Engineering
National Chiao Tung University
Taiwan

*Bin Zhang*
School of Civil Engineering
Chongqing University
China

*Yuntian Wu*
School of Civil Engineering
Chongqing University
China

This edition first published 2020


*Registered Offices*
John Wiley & Sons, Inc., 111 River Street, Hoboken, NJ 07030, USA
John Wiley & Sons Ltd, The Atrium, Southern Gate, Chichester, West Sussex, PO19 8SQ, UK

*Editorial Office*
The Atrium, Southern Gate, Chichester, West Sussex, PO19 8SQ, UK

For details of our global editorial offices, customer services, and more information about Wiley products visit us at www.wiley.com.

Wiley also publishes its books in a variety of electronic formats and by print-on-demand. Some content that appears in standard print versions of this book may not be available in other formats.



*Library of Congress Cataloging-in-Publication Data*

Names: Yang, Yeong-Bin, 1954– author.
Title: Vehicle scanning method for bridges / Yeong-Bin Yang, School of Civil Engineering, Chongqing University.
Description: First edition. | Hoboken : Wiley, 2020. | Includes bibliographical references and index.
Identifiers: LCCN 2019024957 (print) | LCCN 2019024958 (ebook) | ISBN 9781119539582 (hardback) | ISBN 9781119539490 (adobe pdf) | ISBN 9781119539612 (epub)
Subjects: LCSH: Bridges–Testing. | Bridges–Live loads. | Structural health monitoring. | Automatic data collection systems. | Dynamic testing.
Classification: LCC TG305 .Y36 2019 (print) | LCC TG305 (ebook) | DDC 624.2/5–dc23
LC record available at https://lccn.loc.gov/2019024957
LC ebook record available at https://lccn.loc.gov/2019024958

Cover Design: Wiley
Cover Image: © gaspr13/Getty Images

Set in 10/12pt Warnock by SPi Global, Pondicherry, India
Printed and bound in Singapore by Markono Print Media Pte Ltd

10 9 8 7 6 5 4 3 2 1

# Contents

# Preface

Bridges constitute an essential part of transportation systems such as highways, railways, city rail systems, and high-speed railways. Regardless of their irreplaceable role in ensuring the free and safe passage of passengers and cargoes, bridges often suffer from varying degrees of damage due to degradation in stiffness of structural members, connections, supports, or material strength, caused by vehicles' overloading, weathering, or natural disasters, such as earthquakes, typhoons, or deluges. The number of bridges that have been built in the past three decades has increased tremendously. For example, in China there is a total of some 800 000 highway bridges built in this period. Many of them have been ranked among the top in the world in terms of span length, bridge type, and column height. Perhaps the most fantastic is the newly built record-breaking Hong Kong-Zhuhai-Macao Bridge system that connects Hong Kong, Zhuhai, and Macao, totaling a length of 55 km, including seabed tunnels of some 35 km. From the global picture, there is clearly an urgent need to develop efficient and mobile techniques to detect bridge damage so as to enhance the quality of maintenance and possibly rehabilitation.

To monitor the operational and/or damage conditions of bridges, vibration-based methods have been adopted for half a century or longer. Most of the methods require the installation of quite a number of sensors on the bridge for detecting the modal properties, such as frequencies, mode shapes, and damping coefficients. They were referred to as the direct approach, in that the modal properties were retrieved from the vibration data taken directly from the bridge. An enormous volume of research has been carried out along these lines using the ambient vibration, traffic vibration, forced vibration, impact vibration, etc. One drawback with the direct approach is that it usually requires numerous sensors to be installed on the bridge, along with data acquisition systems, for which the deployment and maintenance cost is generally high. Another drawback is that the vast amount of data generated, the so-called sea-like data, may not be effectively digested. It should be added that the monitoring system tailored for one bridge can hardly be transferred to another bridge and work there, a problem known as the lack of mobility.

The vehicle scanning method (VSM) for bridge measurement was proposed by the senior author and coworkers in 2004 mainly to circumvent the drawbacks of the direct approach. This method was known in the early days of development as the indirect approach. However, the term *indirect approach* is not self-explanatory, since it can only be explained along with the direct approach. Recently, we started to use the term *vehicle scanning method for bridges* instead, for its better conveyance of the meaning implied. With this technique, the vibration data collected by one or few sensors installed on the moving test vehicle are used to retrieve the modal properties of the sustaining bridge. No sensors are needed on the bridge. Compared with the direct approach, the VSM shows great potential in economy, mobility, and efficiency, although further research in software and hardware is required to enhance its robustness in field applications.

To our knowledge, this book is the first one on the subject of VSM for bridges. After some 15 years of research on the VSM, we believe it is timely, and indeed necessary, to present an in-depth coverage of the technique, at least based on the works by the senior author and coworkers. The contents of the book have been arranged such that they are reflective of the progressive advancement of the technique, which is also good for pedagogical reasons. By and large, each chapter can be comprehended by readers with little reference to the previous chapters, since a minimum amount of background information is provided in the introductory section. The following is a summary of the content in each of the 11 chapters.

In Chapter 1, a state-of-the-art review is given of the works known to the authors up to roughly 2018 on the subject of the VSM. Among these, a substantial part is the series of papers published by the senior author and coworkers. It can be seen that research has been extended from the original goal of bridge frequency extraction to a variety of applications, including damage detection, modal identification, and damping estimation of the bridges. Aside from the theoretical explorations, small-scale lab experiments and field tests have also been attempted.

In Chapter 2, the vehicle-bridge interaction (VBI) model used for extracting the bridge frequencies is introduced. For the first time, the feasibility of extracting bridge frequencies from the passing vehicle's response is theoretically investigated. For simplicity, only the first mode of vibration of the bridge is considered. From the closed-form solution derived for the passing vehicle, the key parameters involved in the VMS technique are unveiled.

Chapter 3 differs from Chapter 2 in that all the modes of vibration of the bridge are included in the formulation. The general theory presented in this chapter confirms the validity of the simplified theory presented in Chapter 2. In addition, the parameters involved in the VBI are evaluated with potential applications identified.

The first field test of the technique is presented in Chapter 4 for scanning the frequencies of vibration of a bridge in northern Taiwan. The device used is a single-axle test cart towed by a light tractor. This test confirms that the bridge frequencies can be successfully retrieved from the response recorded of the test cart during its passage over the bridge by the fast Fourier transformation (FFT).

Chapter 5 is aimed at enhancing the visibility of bridge frequencies from the vehicle's response. First, the vehicle response is processed by the empirical mode decomposition (EMD) to yield the intrinsic mode functions (IMFs). Then the IMFs are processed by the FFT to yield bridge frequencies not restricted to the first mode.

Chapter 6 deals with road roughness of the bridge, a polluting factor that may render bridge frequencies unidentifiable from the vehicle's response. Both numerical and closed-form solutions are used to physically interpret the effect of road roughness on the retrieval of bridge frequencies. Then a dual vehicle model is proposed for reducing such an effect by deducting the response of one vehicle from the other.

In Chapter 7, three filtering techniques are assessed for removing the (undesired) vehicle frequency from the vehicle's spectrum, so as to enhance the visibility of the (desired) bridge frequencies. The singular spectrum analysis with band-pass filter is demonstrated to be most effective among the three schemes.

Chapter 8 is aimed at tuning the various parameters of the test vehicle for field use. As such, a hand-drawn single-axle cart is extensively tested in the lab and in the field. The qualitative guidelines drawn from this part of study using the handy test cart serve as a useful reference for the design of practical test vehicles.

Chapter 9 presents a theoretical framework for retrieving the mode shapes of a bridge from the passing vehicle's dynamic response. By the Hilbert transform, the mode shape is recognized as the envelope of the instantaneous amplitude of the component response of the moving test vehicle. Factors that may affect such a procedure are studied.

In Chapter 10, the contact point of the vehicle with the bridge, rather than the vehicle body itself, is proposed as a better parameter for use in the VSM technique. The contact-point response, back calculated from the vehicle response, is free of the vehicle frequency that may overshadow the bridge frequencies. The relatively better performance of the contact-point response is demonstrated in the numerical simulations.

As a sequel to Chapter 10, the capability of the contact-point response for damage detection of the bridge is presented in Chapter 11. By the Hilbert transform, the instantaneous amplitude squared (IAS) calculated of the driving component of the contact-point response is demonstrated to be effective for detecting bridge damages for scenarios, including the presence of ongoing traffic.

In the Appendix, the derivation of the VBI element is given in detail based on Chang et al. (2010), a modification from Yang and Yau (1997). Also given is the procedure for assembling the VBI elements (acted upon by vehicles) and non-VBI elements (free of vehicles) for a bridge. The main reason for placing this material in the appendix rather in the main text is to not bring unnecessary intrusion to the main flow of presentation.

We are indebted to a number of friends in preparation of this book. Our work would not be complete without an acknowledgment of this debt and a particular offering of thanks by the senior author to the following:

To the late Professor William McGuire, Cornell University, for introducing him to the interesting field of structural stability and dynamics and for inspiring him to conduct researches that have eventually led to the outcome of this book.

To Professor J.D. Yau, Tamkang University, for the collaboration of research on VBI problems that partially lays the foundation of this book.

To Dr. C.W. Lin, CECI Engineering Consultants, Inc., Taiwan for his first and successful attempt on the VSM for bridges, which has paved the way for future research along these lines.

To Dr. C.S. Chan, National Yunlin University of Science and Technology, Taiwan, for his skills in designing the hand-drawn cart, which forms an essential part of the experimental study presented in Chapter 8.

To the following former graduate students whose work has contributed to the development of material in the book: Dr. K.C. Chang (now in Kyoto University), Mr. Y.C. Li, Dr. W.F. Chen, and Mr. H.W. Yu from National Taiwan University, and Mr. Y. Qian from Chongqing University.

As for the Wiley side, we would like to express our appreciation to Ms. Anne Hunt for her handling of the book proposal in the initial stage, to Ms. Jemima Kingsly for serving as the contact point of Wiley, and to Mr. Steven Fassioms, project editor, and Mr. Hari Sridharan, production editor, for their timely and efficient editorial assistance in making the book come true.

Yeong-Bin Yang
Judy P. Yang
Bin Zhang
Yuntian Wu
Chongqing, China, 2019

# Acknowledgments

Parts of the materials presented in this book have been revised from the papers published by the authors and their co-workers in a number of technical journals. Efforts have been undertaken to update, digest, and rewrite the materials acquired from their sources, such that a unified and progressive style of presentation can be maintained throughout the book. In particular, the authors would like to thank the copyright holders for permission to use the materials contained in the following papers in their order of appearance in the book:

Yang, Y.B., and Yang, J.P. (2018). State-of-the-art review on modal identification and damage detection of bridges by moving test vehicles. International Journal of Structural Stability and Dynamics 18(2): 1850025. © 2018 World Scientific, reproduced with permission.

Yang, Y.B., Lin, C.W., and Yau, J.D. (2004). Extracting bridge frequencies from the dynamic response of a passing vehicle. Journal of Sound and Vibration 272: 471–493. Reproduced with permission from Elsevier.

Yang, Y.B., and Lin C.W. (2005). Vehicle-bridge interaction dynamics and potential applications. Journal of Sound and Vibration 284(1–2): 205–226. Reproduced with permission from Elsevier.

Lin, C.W., and Yang, Y.B. (2005). Use of a passing vehicle to scan the bridge frequencies – An experimental verification. Engineering Structures 27(13): 1865–1878. Reproduced with permission from Elsevier.

Yang, Y.B., and Chang, K.C. (2009). Extraction of bridge frequencies from the dynamic response of a passing vehicle enhanced by the EMD technique. Journal of Sound and Vibration 322(4–5): 718–739. Reproduced with permission from Elsevier.

Yang, Y.B., Li, Y.C., and Chang, K.C. (2012a). Effect of road surface roughness on the response of a moving vehicle for identification of bridge frequencies. Interaction and Multiscale Mechanics 5(4): 347–368. Reproduced with permission from Techno Press.

Yang, Y.B., Li, Y.C., and Chang, K.C. (2012b). Using two connected vehicles to measure the frequencies of bridges with rough surface – a theoretical study. Acta Mechanica 223(8): 1851–1861. Reproduced with permission from Springer.

Yang, Y.B., Chang, K.C., and Li Y.C. (2013a). Filtering techniques for extracting bridge frequencies from a test vehicle moving over the bridge. Engineering Structures 48: 353–362. Reproduced with permission from Elsevier.

Yang, Y.B., Chen, W.F., Yu, H. W., and Chan, C.S. (2013b). Experimental study of a hand-drawn cart for measuring the bridge frequencies. Engineering Structures 57: 222–231. Reproduced with permission from Elsevier.

Yang, Y.B., Li, Y.C., and Chang, K.C. (2014). Constructing the mode shapes of a bridge from a passing vehicle: A theoretical study. Smart Structures and Systems 13(5): 797–819. Reproduced with permission from Techno Press.

Yang, Y.B., Zhang, B., Qian, Y., and Wu, Y.T. (2018). Contact-point response for modal identification of bridges by a moving vehicle. International Journal of Structural Stability and Dynamics 18(5), 1850073 (24 pages). © 2018 World Scientific, reproduced with permission.

Zhang, B., Qian, Y., Wu, Y.T., and Yang, Y.B. (2018). An effective means for damage detection of bridges using the contact-point response of a moving test vehicle. Journal of Sound and Vibration 419: 158–172. Reproduced with permission from Elsevier.

# 1
# Introduction

The idea of extracting bridge frequencies from the dynamic response of a moving test vehicle was theoretically proposed by Yang et al. (2004a) and verified in field test by Lin and Yang (2005). It was then extended to the construction of mode shapes (Yang et al. 2014) and damage identification of bridges. Previously, it was referred to as the indirect method for bridge measurement, in that it requires no vibration sensors to be mounted on the bridge, instead relying on one or few vibration sensors on the test vehicle. When compared with the conventional direct method that relies fully on the response of the bridge fitted with vibration sensors and data acquisition systems, the advantage of the indirect method is obvious: mobility, economy, and efficiency.

Over the years, there have been numerous studies on the indirect method for bridge measurement. Significant advances have been made on various aspects of application. Recently, a state-of-the-art review of the related research works conducted worldwide, including those by Yang and co-workers, was compiled (Yang and Yang 2018), which forms the basis of this chapter. In this book, the term *vehicle scanning method*, instead of the indirect method, will be used to grossly refer to the techniques concerning the modal identification and damage detection of bridges using moving test vehicles fitted with vibration sensors, as it better conveys the ideas involved.

## 1.1 Modal Properties of Bridges

The modal parameters of a bridge (i.e., frequency, mode shape, and damping), are key parameters for various engineering applications. For a newly completed bridge, one may be interested in measuring the few frequencies of the bridge for the purpose of model updating (i.e., for removing the gap between the real model and finite element model used in design). Accurately modeling the real structure is necessary for structural control aimed at counterbalancing the external excitations due to natural disasters including earthquakes and wind gusts. It is also useful to future design of bridges with similar structures. On the other hand, for a bridge in use, a regular monitoring of the modal properties provides the most useful information for evaluating the degradation in various parts of the structure, including the drop in stiffness, breakage in connections, settlement in supports, or deterioration in materials of the structure, due to long-term overloading in traffic or environmental effects such as flooding, earthquakes,

*Vehicle Scanning Method for Bridges*, First Edition. Yeong-Bin Yang, Judy P. Yang, Bin Zhang and Yuntian Wu.

strong wind gusts, or weathering. Undoubtedly, a proper understanding of the structural condition of a bridge is essential to its maintenance, rehabilitation, and proper function during the service life.

This chapter reviews previous works for tackling the following two aspects of the vehicle scanning method for bridges, i.e., using instrumented test vehicles: (i) *modal identification*, which identifies frequencies, mode shapes, and damping ratios of bridges; and (ii) *damage detection*, which detects local damage and severities of bridges by various means.

Conventionally, on-site measurement methods such as the ambient vibration test, forced vibration test, and impact vibration test have been employed to identify the modal parameters of a bridge. These methods are referred to as the *direct method* for bridge measurement, since they rely directly on the response measured from the bridge, which requires quite a large number of vibration sensors to be installed on the bridge.

Starting from the 1970s, a tremendous amount of research has been carried out on the direct method. Among others, ongoing traffic or controlled vehicular movement measurements were adopted as the source of excitation in early works (McLamore et al. 1971; Abdel-Ghaffar and Housner 1978; Ward 1984; Mazurek and DeWolf 1990; Casas 1995; Ventura et al. 1996; Huang et al. 1999a). Normally, the direct approach is designed on a *one-system-per-bridge* basis, which requires quite a large number of vibration sensors to be mounted on the bridge of concern if the mode shapes of the bridge are desired. However, on-site instrumentation on bridges is generally costly, risky, laborious, and not maintenance free. The other drawback is that the monitoring system tailored for one bridge can hardly be transferred to another bridge. As such, only a small percentage of the bridges in operation are fully covered by some monitoring systems, while for bridges equipped with fixed monitoring systems, the problem of *sea-like data* generated by the sensing devices continuously is an issue that cannot be easily overcome. Ironically, it is hard to guarantee that the lifespan of the monitoring system installed on a bridge, including the sensors and central data logger, is longer than that of the bridge, such that each bridge can be monitored from its birth to death.

To resolve these problems, Yang et al. (2004a), explored the idea of extracting bridge frequencies from a passing test vehicle in which only the first frequency of the bridge is considered. Lin and Yang (2005) verified the feasibility of the idea the following year using a field test. Later, the procedure for constructing the mode shape of the bridge using the instantaneous amplitude generated by the Hilbert transform of the component response of the bridge's frequency of concern was developed (Yang et al. 2014). The technique of using the vehicle-collected data was broadly referred to as the *indirect method* for bridge measurement, since *no* sensors are required to be deployed on the bridge, while only a small amount of sensors are needed on the vehicle. This method has the advantage of mobility, economy, and efficiency, compared with the conventional direct approach.

This chapter reviews basic concept of the technique for using the moving test vehicle to extract the bridge dynamic properties, particularly for highway bridges. Then the works conducted by Yang and co-workers along these lines dated back to the 1990s will be briefed. Research available up to about 2018 will be divided into three parts concerning the *theoretical analysis and simulation*, *laboratory tests*, and *field tests*. Finally, concluding remarks along with future works will be outlined. In each part of the review, all the related works will be cited in a chronological order according to their year of publication.

## 1.2 Basic Concept of the Vehicle Scanning Method

In this section, the basic idea behind the procedures for extracting the *frequencies* of vibration and for constructing the mode shapes of the bridge will be briefly described (Yang et al. 2004a). Although only a simply supported beam has been used in the formulation, as well as in most previous studies, the key parameters involved provide insight for unveiling the underlying mechanism and for applying moving test vehicles to scan bridge dynamic properties.

### 1.2.1 Bridge Frequency Extraction

Consider a vehicle moving over a bridge, illustrated in Figure 1.1. The bridge is modeled as a simple beam with a smooth surface in the theoretical formulation, for the purpose of deriving closed-form solution. The vehicle moving at speed $v$ is modeled as a *single-degree-of-freedom (*SDOF*) system*, containing a sprung mass $m_v$ and stiffness $k_v$. One essential assumption here is that the mass of the test vehicle is very small compared with that of the bridge, implying that the inertia effect of the vehicle can be ignored in determining the deflection of the beam. The function of the test vehicle is twofold in the initial study. First, it serves as an *exciter* to the bridge during its movement over the bridge, because it behaves as a horizontally moving load. Second, it plays the role of *message receiver* of the bridge, as it is set into (vertical) motion exclusively by the vertical vibration of the bridge.

As the vehicle is set into (vertical) motion by the bridge, the frequencies of vibration of the bridge will be transmitted into the vehicle through the contact point. Naturally, the frequency of the vehicle itself will appear in its response as well. The other frequency that may appear in the vehicle's response is the driving frequency representative of the speed of the vehicle. Whether the frequencies of the bridge can really be observed in the vehicle's response depends on factors such as the ratio of vehicle's frequency to each frequency of the bridge, the frequencies implied by the frequencies of the roughness profile of the bridge, and whether that might pollute the vehicle response, among others.

Damping of both the vehicle and bridge was neglected for the sake of obtaining elegant closed-form solutions in the initial stage of developing the theory, and also due to the fact that damping is not important for transient problems.

The adoption of an SDOF system for the vehicle via the single sprung mass model is crucial to derivation of the vehicle response during its passage over the bridge. It is with such a simple model that the response of the moving test vehicle over the bridge can be

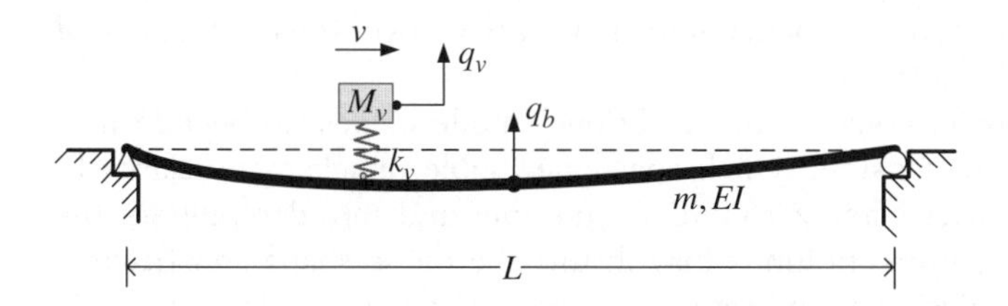


**Figure 1.1** Sprung mass moving over a beam.

presented in closed-form and that the feasibility of the vehicle scanning method for bridges can be mathematically demonstrated. However, it also implies that only vehicles complying with the SDOF system can be practically utilized to extract bridge frequencies, as the theory has revealed. In other words, the SDOF system adopted for the vehicle in the theoretical formulation should be interpreted as a *full-car model with a single axle*. It should not be interpreted as a *half-car model* of a two-axle vehicle, since the interaction between the two axles of the vehicle will make it difficult in extract bridge frequencies from the vehicle response.

### 1.2.2 Bridge Mode Shapes Construction

It is well known that merely using the frequencies of a bridge is not sufficient for identification damages to the bridge. To identify the location and severity of damage, information such as mode shapes or their derivatives should be supplemented. To this end, it is essential to develop a technique for recovering the mode shapes using the data collected by the moving test vehicle. The fact that the entire bridge axis is sensed by the moving vehicle (i.e., the full spatial coverage in the field test) is an advantage over the conventional method, for which only the locations of the bridge installed with sensors can be measured.

Theoretically speaking, in extracting the bridge frequencies, we deal mainly with the frequency components of the measured vehicle response. However in constructing the bridge mode shapes, we shall deal with the amplitude components of the vehicle response. The following is a brief description of the procedure for constructing bridge mode shapes based on Yang et al. (2014).

The first step here is to retrieve the *modal component response* from the vehicle response for the specific mode shape of concern of the bridge. If a closed-form solution is available, this can be directly obtained from the vehicle response with no difficulty. However, if the vehicle response is only available in discrete form as measured from the field test, one can first perform a fast Fourier transformation (FFT) to the vehicle response, and then single out from the spectrum the modal component response of concern using the proper filters, such as the band-pass filter (BPF) (Yang et al. 2013a).

For a specific time series (i.e., the component response), we can strengthen its most local characteristic by performing the Hilbert-Huang transform (HHT). The original time series and its HHT can then be combined to yield an analytical function in complex form, of which the time-dependent amplitude is referred to as the *instantaneous amplitude* or envelope of the original time series and the associated phase is the *instantaneous phase*. These definitions are physically meaningful only when the time series is mono-component or narrow-band (Huang et al. 1998, 1999b), which works well for the current case of a component response. By transforming the instantaneous amplitude history from the time domain to the spatial domain, we can recover the mode shape of the bridge of concern in absolute sense.

The preceding procedure for constructing the bridge's mode shapes has been formulated under the conditions of constant vehicle speed, negligible vehicle mass relative to the bridge, smooth road surface, and zero initial conditions and zero damping for the bridge. Any variations in these conditions may distort the mode shape constructed. However, a parametric analysis has indicated that the result is generally acceptable (Yang et al. 2014).

## 1.3 Brief on the Works Conducted by Yang and Co-Workers

By introducing the term *vehicle-bridge interaction* (VBI) in the 1990s, Yang and co-workers have directed research toward the response of moving vehicles (Yang and Lin 1995; Yang et al. 1995; Yang and Yau 1997; Yau et al. 1999), in addition to the bridge response, as was conventionally the case with previous researchers. The vehicle-bridge system is like a child–mother system. The behavior of the child (i.e., the test vehicle), being much smaller in mass, is primarily dominated by and reflective of the frequency contents of the mother (i.e., the bridge). Such a point was supported by the fact that the use of a moving mass or moving load model for vehicles has little effect on the bridge responses (Yau et al. 1999), and that the vehicle response is sensitive to the existence of roughness on the bridge surface. It is with such an understanding that the concept of the vehicle scanning method for extracting bridge frequencies from a moving test vehicle over the bridge was proposed by Yang et al. (2004a).

In the initial theoretical study (Yang et al. 2004a), only the first mode of vibration of the bridge is included, while both the moving vehicle and bridge are assumed to be damping free, as the first attempt was focused on the feasibility of the technique (i.e., to examine if the bridge frequency is truly contained in the vehicle response). The feasibility of the idea for extracting the bridge frequencies was soon verified in a field test in northern Taiwan (Lin and Yang 2005), in which a *one-axle cart* (i.e., the trailer towed by a tractor) is adopted, resembling closely the *SDOF* model used for the vehicle in the theoretical study. The vertical vibration of the cart is measured by a seismometer mounted right above the axle. The linkage between the cart and tractor is designed as a universal joint that can basically avoid transmission of vibrations from the tractor. In this study, it was verified that the first bridge frequency can be clearly extracted from the spectral response of the vehicle, while *ongoing traffic* is beneficial for the realization, as it tends to amplify the bridge vibration. However, the bridge frequency may be blurred by some high-frequency components resulting from *pavement roughness* and other environmental factors.

The previous closed-form formulation was extended to include the effect of *multi-modes of vibration* of the bridge. Yang and Lin (2005) offer a complete theoretical basis for extracting the bridge frequencies from the acceleration response of a moving test vehicle. Particularly, the second frequency of the bridge was shown to be visible, aside from the first frequency, along with other applications of the results identified.

In order to extract higher frequencies of the bridge, the vehicle response was first processed by the *empirical mode decomposition* (EMD) to generate the intrinsic mode functions (IMFs), and then by the fast Fourier transform (Yang and Chang 2009a). One feature with the EMD technique is that frequencies of higher modes can be made more visible by sifting. The procedure was adopted to extract bridge frequencies from the recorded response, as well as the simulation result, of a passing vehicle. It was demonstrated that higher frequencies of the bridge can be successfully extracted via preprocessing of the vehicle response by the EMD. Further, a parametric study was conducted to examine the vehicle's key parameters, aimed at increasing the success rate for identifying the frequencies of the supporting bridge (Yang and Chang 2009b). Various amplitude ratios were defined in terms of the key parameters and used as indicators for evaluating the success rate in identification of bridge frequencies from the passing vehicle. For the cases studied, it was shown that the initial amplitude ratio plays a role more important than the others for successful identification of bridge frequencies.

Road roughness has been annoying to extraction of bridge frequencies from the vehicle's response in numerical simulations, largely due to use of the *point model* for the wheels of vehicles. This is unrealistic as the wheels are of finite size in practice and cannot touch the bottom of valleys in the roughness profile expressed in terms of the *power spectral density* (PSD) functions (ISO 1995). To this end, the wheel was modeled as a massless *disk model* of finite size, neglecting the tire deformations (Chang et al. 2011). It was found that the point model tends to introduce some unrealistic high-frequency vibrations on the system responses, while underestimating the amplitude of vehicle's frequency in the vehicle response.

To address the effect of *road surface roughness*, an approximate theory in closed form was presented for physically interpreting the role and influence of surface roughness on identification of bridge frequencies (Yang et al. 2012a), with extension to include the action of an accompanying vehicle. The road surface profiles of various levels adopted are based on the PSD functions of ISO 8608 (ISO 1995). However, the amplitudes of roughness so obtained are too large to be reflective of the road surface for use in the VBI simulation. Thus, the geometric mean of the functional value provided by ISO 8608 was simply scaled down by self-judgment. Through the numerical simulation, it was observed that better resolution can be achieved for the bridge frequencies when the frequency ratio of the vehicle to the first bridge mode is greater than one.

The existence of road roughness may excite the vehicle frequency to a level of amplitude higher than the bridge frequencies, making it difficult to identify the latter. To overcome this drawback, *two connected vehicles* identical in shape were used in the numerical study for removing the blurring effect of road roughness (Yang et al. 2012b). The idea is that by subtracting the spectral response of the following vehicle from the leading vehicle, one can obtain a *residual spectrum* that is generally roughness free. Based on this, the bridge frequencies can be identified with generally higher resolution.

One problem encountered in the vehicle scanning method is that the vehicle frequency may appear as a dominant peak in the spectrum, thereby rendering the bridge frequencies invisible. Attempts have been made to filter out the vehicle frequency using the BPF, *singular spectrum analysis* (SSA), and the *singular spectrum analysis with band-pass filter* (SSA-BPF) (Yang et al. 2013a). The SSA-BPF technique was demonstrated to be most effective for extracting the bridge frequencies, as there is no need to select the number of singular values, unlike the SSA, while the unexpected peaks encountered by the BPF are circumvented.

A closed-form expression is presented for the variation of the instantaneous frequencies of the bridge under a moving vehicle considering their coupling effect (Yang et al. 2013c). It is suggested that the *frequency variation* caused by vehicles moving on the bridge be taken into account, when using the moving vehicle for detecting bridge frequencies or damages. Such an effect is crucial for non-negligible vehicle/bridge mass ratios or when the vehicle/bridge resonance condition is approached.

A *hand-drawn test cart* allows us to measure the structural frequencies in a human-controlled, efficient, and mobile way (Yang et al. 2013b). In this study, focus is placed on the elastic properties of the cart tires and the reliability of the bridge frequencies is extracted. It was demonstrated that the hand-drawn cart can be reliably used in the field for measuring the bridge frequencies. Moreover, qualitative guidelines were drawn for the design of test vehicles.

The *stochastic subspace identification* (SSI) was modified to deal with the time-variant coupled noisy VBI system (Yang and Chen 2016). The governing equations for the vehicle and bridge were expressed first in state space, and then discretized and transformed to a form suitable for SSI by separating the known from the unknown parameters. Using the Hankel matrix along with the orthogonal projection theorem and singular value decomposition, the observability matrix derived with the vehicle effect suppressed was used to identify the bridge frequencies. It was demonstrated that (i) the proposed SSI approach is more effective for identifying bridge frequencies below 20 Hz, compared with the conventional one; (ii) adding a small amount of *damping* to the test vehicle can help suppress the vehicle frequency; and (iii) *ongoing traffic* is beneficial for amplifying bridge frequencies with rough surface.

The *wavelength characteristic* is a useful clue for locating and assessing the severity of slope discontinuity in beams (Yau et al. 2017). By modeling the slope discontinuity of a beam by an internal hinge restrained by a rotational spring, the wavelength of the beam is calculated from the vehicle's response and used to identify the location and severity of the *discontinuity* in the beam. It was demonstrated that the wavelength-based technique offers a promising, alternative approach for detection of damages in girder type bridges.

The effect of *vehicle damping* on the identification of the first bridge frequency was investigated for three different levels of road roughness by the EMD technique (Yang and Lee 2017). Through the numerical simulation, it was shown that higher vehicle damping tends to suppress the vehicle frequency, and the inclusion of vehicle damping helps suppress the roughness effect, thereby making the first bridge frequency more visible.

Yang et al. (2018a) proposed using the *contact-point response* of the test vehicle with the bridge, rather than the vehicle itself, for modal identification of bridges. The contact-point response is free of the vehicle frequency, an annoying effect for bridge frequencies extraction, and it allows more bridge frequencies to be identified. For all the scenarios studied, i.e., for varying vehicle speeds or frequencies, for smooth or rough road surfaces, with or without existing traffic, the contact-point response was verified to outperform the vehicle response in extracting either the frequencies or mode shapes of the bridge. They proceeded further to calculate the *instantaneous amplitude squared* (IAS) of the driving component of the contact-point response by the Hilbert transform (Zhang et al. 2018; Yang et al. 2018b). The feasibility of using the IAS peaks to detect the location and severity of damages of the bridge was verified. In the presence of ongoing traffic, the damages of the bridge are identified from the repeated or invariant IAS peaks generated for different traffic flows on the bridge.

Yang et al. (2019) first attempted a conceptual, *virtual vehicle* moving eccentrically along a suspended footbridge to detect the vibration modes of the bridge inherent with clusters of frequencies along the direction of the *virtual sensor* mounted on the vehicle. Such a virtual device allows us to interpret the complicated coupling flexural-torsional vibration of the suspension footbridge.

## 1.4 Works Done by Researchers Worldwide

Only papers that utilize the response or action of the moving test vehicle to evaluate the modal property, damages, or other properties of the beam will be reviewed. All the papers to be reviewed will be divided into the following three groups: theoretical analysis and simulation, laboratory test, and field investigation. If a paper contains a laboratory test part, it will be included in the group of laboratory test, even though the paper contains some theoretical analysis. Similarly, if a paper consists of both theoretical analysis and field investigation, it will be included in the group of field investigation. Papers falling in each group will be reviewed according to their year of publication. Also, only papers that are readily available to the authors mainly through accessible technical journals will be reviewed. The authors apologize for any relevant papers that were missed.

### 1.4.1 Theoretical Analysis and Simulation

Majumder and Manohar (2004) developed a scheme for detecting the local and/or distributed *stiffness loss* in beams using the vibration data generated by a moving oscillator. This scheme properly considers the time-varying structural matrices, structural nonlinearities, and the spatial incompleteness of measurement data. In the numerical simulation, the moving oscillator was modeled as a SDOF system and the two ends of the beams were assumed to have partially immobile bearings.

Bu et al. (2006) used the measured dynamic response of a vehicle moving over a bridge in the *damage assessment* of a bridge deck modeled as an Euler–Bernoulli beam, inspired by Yang et al. (2004a). The vehicle was modeled as a SDOF system. The damage was represented by a reduction in flexural stiffness of the beam element. Based on the dynamic response sensitivity analysis, the identification algorithm is realized with a regularization technique for the measured vehicle acceleration. Measurement noise, road surface roughness, and model errors were included in the simulations. The results indicated that the proposed algorithm is computationally stable and efficient in the presence of environmental noise and road surface profile.

Kim and Kawatani (2008) investigated the damage identification of a bridge, considering the coupling between the bridge and moving vehicle, including the effect of roadway-surface roughness. The idea is to identify damage directly from the changes in element stiffness using a pseudo-static formulation derived from the equations of motion for the coupled system. The *element stiffness index* (ESI) is adopted as the damage indicator, which is defined as the ratio of the flexural rigidity of a damaged element to the intact one. The proposed procedure provides an alternative choice for damage identification of bridges.

Gonzalez et al. (2008) proposed the collection of data from accelerometers fixed in a specific vehicle type, which can be used to estimate the *road roughness* for car manufacturers. This approach is based on the relationship between the power spectral densities of road surface and vehicle accelerations via a transfer function. This paper showed how road profiles can be accurately classified using the axle and vehicle-body accelerations from a range of simulated vehicle-road dynamic scenarios.

McGetrick et al. (2009) investigated the use of an instrumented vehicle fitted with accelerometers on its axles to monitor the dynamics of bridges. A simplified quarter-car–bridge interaction model is used in theoretical simulations, and the natural frequency of the bridge is extracted from the spectra of the vehicle accelerations. The accuracy is better at lower speeds and for smooth road surface. The *structural damping* of the bridge was also monitored for smooth and rough road profiles. The magnitude of peaks in the PSD of the vehicle accelerations decreased with increasing bridge damping, and this decrease was easier to detect for smoother road profiles.

Nasrellah and Manohar (2010) proposed a strategy based on the dynamic state estimation method, which employs the particle filtering algorithms, to tackle the parameter identification of a beam-moving vehicle system based on the time histories of strains and displacements of the beam. The vehicle was modeled as a SDOF oscillator, and the beam was restrained by rotational springs to account for partial fixity at the ends. Factors considered include measurement noise, guideway unevenness, spatially incomplete measurements, finite element models for the supporting structure and moving vehicle, and imperfections in the formulation of the mathematical models.

Nguyen and Tran (2010) applied the *wavelet transform* to the displacement response of a moving vehicle for detecting multi-cracks on the bridge as a beam structure with smooth surface. Numerical simulation was conducted using a cracked beam element and a 4-DOF *half-car vehicle* model. The locations of the cracks were pinpointed by positions of peaks of the wavelet transform and the velocity of the moving vehicle. Low vehicle speed was recommended.

In briefing the needs and strategies for bridge monitoring in Japan, Fujino and Siringoringo (2011) described the concept of *structural health monitoring* (SHM) as an essential part of risk reduction. The strategies implemented for bridge monitoring in Japan were categorized into three main groups, according to the purpose of monitoring: natural hazard and environment condition, effective stock management, and failure prevention. Examples of bridge monitoring systems that implement these strategies and the lessons learned in this regard were also presented.

Lu and Liu (2011) presented an iterative technique for identifying both the bridge damages and vehicle parameters based on the bridge dynamic response using the *penalty function method* with regularization. The effects of measurement noise, different vehicle models, measurement time duration and modeling error on the identification results were investigated in the numerical simulation. It was shown that the proposed method has the potential for real application to damage detection and parameter identification.

Yin and Tang (2011) used a moving vehicle to identify the tension loss in cables and deck damage of a cable-stayed bridge. In this study, the vertical displacement response of the vehicle moving over the damaged bridge relative to the intact one was used. The vehicle was modeled as a sprung mass unit. No road roughness was considered. Discussions were given concerning the feasibility and limitation of the proposed detection technique.

Roveri and Carcaterra (2012) utilized the HHT to identify the presence and location of damages for bridges under a traveling load, based on a *single point measurement* of the bridge. The effect of ambient noise was also taken into account. Damage location is revealed by direct inspection of the *first instantaneous frequency* generated by the HHT, which shows a sharp crest at the damaged section. The analysis results are not very sensitive to the crack depth and ambient noise, but are sensibly affected by the damage location and speed of the moving load.

Khorram et al. (2012) used the displacement response obtained either at the midspan of the beam or on the moving load, referred to as the fixed and moving sensor approaches, respectively, to detect the location and size of a crack in a beam under a moving load. Each crack was modeled by an internal hinge with rotational spring. The responses obtained were processed by the *continuous wavelet transform* (CWT). No VBI effect was considered as the vehicle was modeled as a moving force. The effect of road surface profile was also ignored. The moving-sensor approach was found to be more effective than the fixed-sensor approach.

Meredith et al. (2012) applied the *moving average filter* and EMD to the acceleration response of a beam under a moving load to locate damage of the beam. With the EMD technique, a sudden loss of stiffness in a structural member will cause a discontinuity in the measured response that can be detected through a distinctive spike in the filtered IMF. The technique is further tested using the response of a discretized beam with multi damaged sections modeled as localized loss of stiffness. They found that the use of a moving average filter on the acceleration response, prior to applying the EMD, can improve the sensitivity to damage.

In their numerical study, Gonzalez et al. (2012) proposed a six-step algorithm for identifying the *damping* of a bridge from the acceleration responses of the two axles of a vehicle moving over the bridge, with the vehicle simulated by the *half-car* model to account for the VBI. The proposed method was tested for its accuracy for a range of parameters, including bridge span, vehicle velocity, road roughness, vehicle's initial vibration condition, signal noise, modeling error, and frequency matching between the vehicle and bridge.

Guebailia et al. (2013) solved the free vibration equation of a multi-span thin plate by separation of variables to obtain an equation of motion with varying coefficients for the corresponding beam, from which the frequencies and mode shapes are estimated. This method is very efficient compared with the previous methods. Because of its analytical nature, it avoids carrying out several integrations as with the Rayleigh-Ritz method. The results obtained agree well with previously published results mainly for the bending modes, whereas the difference becomes important for the mode shapes of the second order.

Gonzalez and Hester (2013) divided the acceleration response of a beam into the static and dynamic components, and where the beam has a local loss in stiffness, an additional damage component. The combination of these components establishes how the damage singularity appears in the total response. For a given damage severity, the amplitude of the damage component will depend on how close the damage is to the sensor, and its frequency content will increase with higher velocities of the moving force. It was noted that a thorough understanding of the relationship between the static and damage components contributes to estimation of the occurrence, location, and severity of a damage.

Keenahan et al. (2014) used the instrumented trailer of a truck-trailer system for detecting the *change in damping* of a bridge as an indication of the occurrence of damage. The influence of road profile roughness on the vehicle vibration was overcome by analyzing the spectra of the difference in the accelerations recorded of the two axles of the trailer. The effectiveness of the approach in detecting damage simulated as a loss in stiffness was investigated, along with its sensitivity to the following factors: vehicle speed, road roughness class, bridge span length, changes in the axle properties, and noise.

Based on the generalized pattern search algorithm, Li et al. (2014) presented an optimization method for identifying the bridge parameters indirectly by a passing vehicle. Using the proposed method, the first frequency and stiffness of the bridge can be efficiently identified with reasonable accuracy for various noise levels.

Li and Au (2014) presented an approach for identifying the damage location of a continuous bridge from the acceleration response of a vehicle moving on a bridge with rough surface for both the intact and damaged cases. The damage detection method is implemented by the modal strain energy based method and genetic algorithm (GA). First, the modal strain energy-based method is used to estimate the damage location using a damage indicator based on the frequencies extracted from the vehicle responses for both the intact and damaged bridges. Second, the identification problem is transformed into a global optimization problem and solved by GA. For each pass of the vehicle, the location of damage can be identified until sufficient accuracy is met.

Malekjafarian and OBrien (2014) processed the signals from the accelerometers mounted on a vehicle traveling over a bridge using two instrumented axles. The short-time frequency domain decomposition (FDD) is used to estimate bridge *mode shapes* from the dynamic response of the vehicle. The FDD is employed in a multistage manner sequentially to each of the segments of the bridge. A rescaling process is used to construct the global mode shape vector. The effect of road roughness is alleviated by applying external excitation to the bridge or by subtracting signals in the axles of successive trailers towed by the vehicle.

Oshima et al. (2014) presented an approach for assessing the state of a bridge using the *mode shapes* extracted from the responses of multiple single-axle monitoring carts. Heavy trucks were used to amplify the bridge vibrations. Two types of damages, immobilization of a support and decrease in beam stiffness at the center, were evaluated for varying levels of road roughness and measurement noise. The feasibility of the proposed approach was verified through simulations of interactive vibration between a 2D beam and the passing vehicles, each modeled as a sprung mass. It was found that the damage state can be recognized by the estimated mode shapes for beams with severe damage, such as immobilization of rotational support. However, the proposed approach has low robustness against noise.

OBrien et al. (2014) presented a method for monitoring the transport infrastructure such as pavements and bridges through the analysis of vehicle accelerations. An algorithm was developed to identify the interaction forces from the response of the half-car model moving over the bridge. The potential of the method to identify the global bending stiffness of the bridge, while predicting the pavement roughness, was presented. A range of bridge spans is considered in the simulations with the effects of road roughness and signal noise investigated.

Chen et al. (2014) presented the concept of classification, a signal processing technique, with semi-supervised learning on graphs for the indirect bridge health monitoring. The purpose is to design a map that relates each input to a predefined class label. The proposed method has been validated using the dataset of 6240 acceleration signals generated by the laboratory scale test, which includes 30 acceleration signals for each of 13 different bridge conditions, 8 different speeds, and 2 vehicles of different weights (Cerda et al. 2014).

Keenahan and OBrien (2014) studied a drive-by bridge inspection approach using a beam in free vibration. They pointed out that the travel time for the test vehicle to move over the bridge at highway speeds may be too short for the bridge to go through a full cycle of vibration, which is not good for spectral analysis. An optimization approach was proposed in the simulation as an alternative to standard signal processing techniques to overcome the challenges of short signals collected from the moving vehicle and the nonlinear nature of the drive-by system.

Kong et al. (2015) performed a study on damage detection in the frequency domain using methods based on the modal data, transfer function or frequency response function, and transmissibility. The *transmissibility* of a vehicle-bridge coupled system was first derived for a system consisting of a simply supported beam and a SDOF vehicle. Then, a numerical study was conducted to investigate the feasibility of detecting bridge damages using the vehicle transmissibility. Several damage indicators were constructed based on the transmissibility from both the bridge and vehicle responses, and compared. Two methods were proposed to measure the transmissibility of the vehicle responses for two vehicles. Method I used one reference vehicle and one moving vehicle; Method II used two moving vehicles at a constant distance along the bridge axis. With these, modal data such as natural frequencies and modal shape squares were successfully extracted for damage detection.

Li and Au (2015) presented a GA-based method to identify the damage of girder bridges from the response of a moving vehicle. Starting with the formulation of an optimization problem, a guided GA was used to search for the global optimal value. Frequencies of the bridge at the intact and damaged states were extracted from the vehicle responses, from which the possible locations of damages were roughly estimated. These locations are not unique as frequencies alone are insufficient to identify them. However, the initial results were used to narrow down the search region for the GA. The proposed strategy was used to identify the damage location for simply supported and continuous girder bridges in the presence of road roughness and measurement noise.

Using a 3D VBI model, Hester and Gonzalez (2015) showed that the area under the filtered acceleration response of the bridge increases with increasing damage, even at highway load speeds. Once a datum reading is established, the area under subsequent readings can be monitored and compared with the baseline reading, as an increase in the area may indicate the presence of damage. The sensitivity to road roughness and noise was tested for several damage scenarios, with damage successfully identified in favorable scenarios.

OBrien and Keenahan (2015) used a vehicle equipped with *traffic speed deflectometers* (TSDs), a device developed for pavement deflection measurement, in their damage detection for bridges. A 4-DOF vehicle model containing two TSDs was simulated crossing a simply supported beam with a damage simulated as a stiffness loss in one of its elements. Using the displacements recorded by the sensors, the apparent profile was determined for the bridge by an optimization algorithm. The results showed that the time-shifted difference in the apparent profile can be reliably used as a damage indicator of the bridge in the presence of noise.

Malekjafarian et al. (2015) conducted a review of indirect bridge monitoring using the passing vehicles. They noted that most previous works have been directed toward identifying the dynamic properties of the bridge from the responses measured on the vehicle, such as natural frequencies, mode shapes, and damping. Meanwhile, some works have been extended to detection of bridge damages utilizing the interaction effect between the vehicle and bridge. A total of 73 articles up to 2015 has been cited. This paper provides a critical review of the indirect methods used for bridge monitoring, along with discussions and recommendations on the challenges to be overcome for successful implementation in practice.

Another review paper was presented by Zhu and Law (2015) for SHM using the VBI data. As noted, the VBI-based approach allows the target bridges to be monitored or assessed under operating conditions, which is an advantage compared with the conventional approach. The bridge damage identification techniques based on VBI were divided into three categories based on the bridge responses, the vehicle responses, and both the vehicle and bridge responses. Focus was placed on the challenges for the general implementation of SHM in practice.

Zhu and Law (2016) proceeded to give an overview on the inverse problems of VBI, where the dynamic interaction force exists between the vehicle and bridge deck. The *identification of moving load* is a typical inverse problem, for which the solution approaches can be categorized into those based on the analytical model or finite element model with a focus on the solution technique. Structural parameter identification with moving loads as the excitation is another kind of inverse problems for the VBI. The bridge and/or vehicle responses can be taken to identify the parameters of the structure as part of the structural condition assessment. These two topics were reviewed with examples and experimental results to illustrate their effectiveness and limitation, along with future directions of research highlighted.

Yin (2016) presented a semi-analytical solution to the problem of a simply supported beam subjected to a moving sprung mass with initial velocity and constant acceleration or deceleration. A finite element modeling procedure was adopted to tackle the VBI, and the responses of the vehicle and bridge were computed by various schemes. It was found that the dynamic responses of the beam and sprung mass were mainly dominated by the initial velocity of the sprung mass, but not by the acceleration or deceleration.

Kong and Cai (2016) studied the *scour effect* on the responses of the bridge, including the super- and sub-structure, and even the responses of vehicles traveling on the bridge, which in turn can be used to detect scouring. A field bridge with scour history was adopted, and the bridge-vehicle-wave interaction was considered by including the wave force as a time-variant force. A three-axle truck was simulated as a two-axle model by condensing the rear two axles into one axle. The free vibration and dynamic analyses under wave loads were conducted on a single pile or piles with different scour depths. This paper only numerically studied the idea of using vehicles for scour detection; the feasibility is yet to be verified.

In their numerical study, Kong et al. (2016) used a specialized test vehicle consisting of a tractor and two following trailers to eliminate the effects related to driving frequencies and road roughness. The responses of one trailer were subtracted by the other with a time shift to obtain the residual responses, which were processed with the FFT and *short-time Fourier transform* to extract the bridge modal properties. The proposed method was verified to be effective and robust in extracting bridge modal properties, especially under ongoing traffic flows and different road surfaces. Besides, vehicle parameters, such as the trailers' mass and stiffness, spacing, and traveling speed, were studied for the design of a proper test vehicle for field applications.

Elhattab et al. (2016) used the acceleration histories of an instrumented truck to calculate changes in the bridge displacement profile, which was shown to be sensitive to structural damage. Three different bridges modeled as 1D beam elements and 2D plate elements were considered. The truck was represented as a simple quarter-car model crossing over a rough profile. Also, the sensitivity to signal noise, corrupted truck properties, initial conditions, and transverse vehicle position was studied.

Li et al. (2016a) presented an approach for identification of distributed damage due to cracks, based on the dynamic response sensitivity of a moving vehicle with nonlinear springs. Both simply supported beam and continuous two-span bridge were considered. The results showed that the distributed damage(s) can be identified accurately by the measured acceleration data of four virtual moving acceleration sensors, which are insensitive to measurement noise.

Feng and Feng (2016) proposed a bridge damage detection procedure that utilizes vehicle-induced displacement response of the bridge, particularly, the curvature of the first mode shape, without requiring prior knowledge about the traffic excitation and road roughness. The first mode shape was extracted by directly analyzing the PSD functions of measured bridge displacement responses under vehicle excitations. Three damage scenarios were considered, i.e. damage at single, double, and multiple locations, each involving several extents of damage defined by the reduction in element stiffness. The results revealed that all the damage cases can be successfully identified. Besides, the damage detection performance was evaluated for cases involving different levels of road roughness and less measurement points.

OBrien and Malekjafarian (2016) used the mode shapes estimated from a passing vehicle for damage detection of bridges. The bridge response at the moving coordinate was measured from an instrumented vehicle (simulated as a half-car model) with laser vibrometers and accelerometers. A modified version of the short-time FDD method was applied to the measured responses to estimate the mode shape. A damage index based on mode shape squares (MOSS) was used to detect the presence and location of the damage. It was shown that the presence and location of the damage can be detected with acceptable accuracy when the vehicle moves very slowly, not over 8 m/s.

He and Zhu (2016) regarded the dynamic response of a beam under a moving load as the superposition of two components, i.e. the moving-frequency component and the natural-frequency component of the beam. They derived the closed-form solution of the dynamic response of a damaged simply supported beam subjected to a moving load and found that the moving-frequency component is preferred in damage localization. A multi-scale discrete wavelet transform was employed to separate the moving-frequency component from the total dynamic response and to subsequently locate the damage. Examples with single or multi damages were used to validate the efficacy of the response calculation algorithm and the damage localization method. The effects of the vehicle's velocities and vehicular dynamics on damage localization were investigated.

Hester and Gonzalez (2017) examined the theoretical feasibility and practical limitations of a drive-by system in identifying damage due to localized stiffness loss. They divided the total vehicle response into the static, dynamic, and damage components for bridges with a damage. By comparing the effect of the damage component with other effects such as vehicle speed, road profile, and noise on a wavelet contour plot, the workable frequency range was identified. The algorithm used the specific frequency bands to improve the sensitivity to damage. Recommendations on selection of the mother wavelet and frequency band were provided, together with discussions on the impact of noise, road profile and periodic measurements.

OBrien et al. (2017a) used the IMFs processed by the EMD for the moving speed component of the vehicle' response to detect the damage in a bridge. This technique can identify the damage location in the absence of road roughness. However, it is sensitive to changes in the road profile as might be expected from changes in the lateral position of the vehicle. It was concluded that the damage may be located if the differences in the acceleration signals between the damaged and intact bridges are small for the case with rough road profiles.

Wang et al. (2017a) identified the moving vehicle parameters using the bridge acceleration response, based on the Bayesian theory application of a particle filter. The bridge pavement roughness used in the parameter identification is estimated in advance using the response of a instrumented car considering the VBI. Numerical results show that the vehicle parameters, including the weight, are estimated with high accuracy and robustness against noise and modeling error. Also, the method is validated through field measurement.

To estimate the bridge mode shapes, Malekjafarian and OBrien (2017) used a truck-trailer system equipped with an external excitation to excite one of the bridge frequencies. They showed that the amplitude of the signals of the two identical axles processed by the Hilbert transform contains can be used to estimate the bridge mode shapes. The concept of subtraction for the responses measured from the two axles is used to remove the effect of road roughness.

Wang et al. (2017c) showed that the axle information is contained in the wavelet coefficient curve of the scale that corresponds to driving frequency. They computed the specific wavelet scale through iterating with no subjective selection for automatic identification of axle distribution information. The results showed that the proposed method acquires precise axle information from the responses of an axle-insensitive structure (e.g., girder) and reduces the requirement of equipment for bridge weight-in-motion measurement.

Bao and Liu (2017) presented a review of techniques for *scour detection* derived from vibration-based damage detection by investigating the natural frequency spectrum of a bridge or a bridge component. In this review, the technique for extracting bridge frequencies using a passing vehicle was also commented.

Wang et al. (2017b) used the frequency shift to identify the damaged supports of railway tracks. The rail-sleeper-ballast system is modeled as an Euler beam evenly supported by springs, the stiffness of which is reduced when the fastener is loose or missing and the ballast under the sleepers is damaged. An auxiliary mass is utilized, which when mounted on the beam, will result in change of the frequencies of the whole system with respect to its location. It was found that the frequencies induced by the auxiliary mass change periodically when the supports are undamaged, whereas the periodicity will be broken due to damaged supports, particularly, a clear drop in frequencies is observed when the auxiliary mass moves over the damaged support. Both numerical and experimental examples were carried out to validate the proposed method.

OBrien et al. (2017b) compared the instantaneous curvature (IC) with the moving reference curvature (MRC) to determine a local loss of stiffness in a bridge through the data collected from a passing instrumented vehicle. It is assumed that the absolute displacements on the bridge can be measured by the vehicle. A half-car model is used to represent the passing vehicle, and the damage is represented as a local loss of stiffness in different parts of the bridge.

Qi and Au (2017) started with simulation of the dynamic responses of a moving vehicle under impact excitation. With the component response of each frequency of the bridge extracted by a filter, they construct the associated mode shape by the Hilbert transform pair. While only the information of the moving vehicle is used in the computation, the additional impact on the vehicle helps to alleviate the effects of measurement noise and road surface roughness.

Nagayama et al. (2017) proposed a new frequency estimation strategy utilizing two ordinary vehicles through signal processing of cross-spectral density functions. The feasibility of such an approach was confirmed in numerical analyses using a VBI model for various conditions. Also, an experimental study featuring synchronized sensing of two different vehicles was performed, indicating that the first natural frequency of the bridge can be identified under various driving speed combinations.

Tan et al. (2017) presented a new frequency identification technique based on wavelet analysis. The wavelet transform is characterized by its high-frequency resolution and can be used to visualize the bridge damage indicated by a change of the fundamental frequency of the bridge. The acceleration signals generated from the VBI model were processed by wavelet analysis to extract the bridge frequency. Also, the use of a subtracted signal from two consecutive axles was examined, which can substantially remove the effect of the road roughness from the recorded acceleration history.

Quirke et al. (2017) detected the bridge damage via comparison of apparent profiles sensed by the passing vehicle. The cross entropy optimization method is used to determine the apparent profiles that generate a vehicle dynamic response most similar to that of a measured input. The "measured" bogie vertical accelerations are generated using a 3D train–bridge interaction model built in Abaqus and used as input to a 2D algorithm built in Matlab. The damage of the bridge is represented by a local reduced width of the bottom flange of a bridge beam. Apparent profiles for various damage scenarios are inferred and compared over time to detect damage, considering sensor noise and track irregularity.

Zhu et al. (2018) used the accelerations at both the axle and vehicle body to determine the local damages. The numerical results with the vehicle moving over simple or continuous beams show that the acceleration responses from the vehicle or the bridge structure are less sensitive to the local damages than the interaction force between the wheel and the structure. Other effects considered include the movement patterns and moving speed of the vehicle, and measurement noise. A heavier or slower vehicle was shown to be less sensitive to measurement noise, giving more accurate results.

OBrien et al. (2017b) presented a procedure for bridge damage identification through drive-by monitoring. The IC is used to determine a local loss of stiffness in a bridge through measurements collected from a passing instrumented vehicle, compared with the MRC. It is assumed that the absolute displacements on the bridge can be measured by the vehicle. The bridge is represented by a finite element model, and the vehicle by a half-car model. A generic road surface is assumed, as well as 1% random noise and no noise environments. The numerical simulations show that the local damage can be detected using the IC if the deflection responses can be measured with sufficient accuracy. The damage quantification can be obtained from MRC.

### 1.4.2 Laboratory Test

The fundamental ideas for identifying the frequencies and mode shapes of a bridge using the vertical acceleration response of a moving test vehicle have been described in Sections 1.2.1 and 1.2.2. The key consideration in the theoretical formulation is adoption of the SDOF system for the test vehicle, with which the vehicle touches the bridge during its journey through a *contact point*. Thus, the vibration of the vehicle is dominated fully by the bridge. Beside this, the vehicle receives no disturbance or coupling from other sources, when undergoing the vertical oscillation. Such a property should be strictly followed in preparing the vehicle model for laboratory test, if the purpose of the test is to verify the physical existence of the theory.

To prepare a scaled vehicle model that can remain strictly in the vertical position during horizontal movement, as shown in Figure 1.1, is generally difficult. The authors have recently observed a very skillful vehicle model design by Dr. Shota Urushadze (2017) of the Institute of Theoretical and Applied Mechanics, Czech Academy of Sciences, which can fulfill the property of a SDOF system, while maintaining the vertical stability during the movement. The other concern is the power to pull the vehicle, which should be designed in a way not to introduce any additional perturbation to the test vehicle.

It might at first look easy to construct a scaled bridge model that is simply supported. But in fact, care should be exerted in the design of a scaled bridge for it to have a reasonable *frequency ratio* relative to the scaled vehicle, which should not be far from that adopted in the theory. The other concern is to keep the vehicle model move along the bridge central line, so as to avoid the occurrence of twisting deformation not considered in the theory.

The following is a brief review of the relevant researches conducted along the lines of vehicle scanning method in the laboratory.

Law et al. (2006) presented the parameter identification of a vehicle moving on a multispan continuous bridge by the dynamic response sensitivity analysis. The road friendliness of the vehicles was studied for the highway pavement. The moving vehicle was modeled as a system with single, two, or four DOFs. The modified beam functions were used to calculate the response of the bridge. Starting with an initial guess on the unknown parameters, the identification was realized by the least-squares method and regularization technique using the measured strain, velocity, or acceleration from as few as a single sensor. Simulation and experimental studies using *a two-axle vehicle* indicated that the identified results are acceptable, and the responses reconstructed from the identified parameters agree well with the measured ones.

A moving vehicle with a *self-tapping device* was used by Zhang et al. (2012) for the damage detection of beam and plate structures. The amplitude of the spectrum obtained from the impedance of the point where the tapping force was applied was taken to be approximately proportional to the MOSS, and the damage index was defined as a function of the difference between the intact and damaged MOSS curves. This approach was reported to be good for vehicle speeds up to 2 m/s. However, it was not based on the acceleration response of the vehicle, as the concerned response was controlled by the tapping device and measured at the same time for computing the point impedance.

Cerda et al. (2012) compared the results of the indirect bridge health monitoring with the direct approach in a *scale model experiment*, focused on bridge frequency changes. Acceleration signals were collected from a vehicle and bridge system in a laboratory-scale experiment for four different bridge scenarios and five speeds. These signals were classified using a *short-time Fourier transform* meant to detect shifts in the fundamental frequency of the bridge due to changes in the bridge condition. The results showed near-perfect detection of changes when this technique was applied to signals collected from the bridge (direct monitoring), and promising levels of detection when signals from sensors on the vehicle (indirect monitoring) were used.

Casciati and Wu (2013) investigated the possibility of increasing the accuracy of local positioning by multiple readings and their differential elaboration. In the case study, noncontact laser displacement sensors were adopted (being installed in the plane of the motion). Two *laser displacement sensors* were mounted: one of them addressed its beam to an object at rest and the second was focused on the moving object. Elaborations from the readings of both sensors were shown to increase the measured accuracy.

McGetrick and Kim (2013) conducted a parametric study for a drive-by bridge inspection system using a vehicle fitted with accelerometers on its axles. The effectiveness of the proposed approach in detecting damage was investigated using a simplified VBI model in theoretical simulations and a scaled VBI model in a laboratory experiment. To identify the existence and location of damage, the vehicle accelerations were recorded and processed using a *continuous Morlet wavelet transform* and a damage index was established. Factors considered in the study include bridge span length, vehicle speed and mass, damage level, and surface roughness.

Zhang et al. (2013) extracted the operating deflection shape (ODS) of a bridge from the passing vehicle for damage detection of the bridge. Using the ODS curvature and assuming the intact structure to be smooth and homogenous, a damage detection algorithm called the global filtering method (GFM) was proposed and verified both numerically and experimentally. Compared with traditional methods using the ODS extracted from FRFs, the proposed method is time-saving, easier to implement, and produces greater accuracy, as it does not require many preinstalled sensors and solving eigenvector or singular value problems. It was found that the increase in vehicle speed will decrease the quality of the identification result.

Cerda et al. (2014) extended their previous study (Cerda et al. 2012) to include three different types of damage scenarios: changes in the support conditions (rotational restraint), additional damping, and an added mass at the midspan. The vehicle was instrumented with two accelerometers on the suspension shafts to record the acceleration at the wheel level and another two accelerometers on the suspension to acquire data filtered by the suspension system. A set of frequency features was used in conjunction with a support vector machine classifier on the data measured from the passing vehicle and those from the bridge for comparison. For each type of damage, four levels of severity were explored. The results showed that for each damage type, the classification accuracy based on the vehicle response is, on average, as good as or better than that based on the bridge response. Classification accuracy showed a steady trend for low (1–1.75 m/s) and high vehicle speeds (2–2.75 m/s), with a decrease of about 7% for the latter.

Chang et al. (2014b) proposed the pseudo-static damage identification method for a VBI system using a moving vehicle in the laboratory test. The stiffness index, defined as the ratio of flexural rigidity of the damaged to intact element, was adopted as the damage indicator. Three vehicle models of two axles and two vehicle speeds were considered. It was demonstrated that the locations and severities of damages are detectable using the proposed method in spite of the probable changes of roadway roughness and environmental conditions. In addition, adopting higher vehicle speed as well as a vehicle with frequency close to that of the bridge increased the probability of detecting damages.

Lederman et al. (2014) presented algorithms for diagnosing the severity and location of damages in a bridge model using the moving vehicle. On the vehicle, two sensors were mounted on the sprung portion of the vehicle (chassis sensors), and the other two sensors on the unsprung portion of the vehicle (wheel sensors). Signal processing and machine learning were used to analyze the vibration data collected both from the bridge and passing vehicle. Features were selected using principal component analysis (PCA), and a regression was performed using the kernel regression method. Various damage severities and positions were simulated by added mass on the bridge. Two experiments were conducted; one to detect the damage severity (i.e., size of mass) and the other the damage location (i.e., position of mass). In the first test, the magnitude of the mass was varied, while the location kept constant, and vice versa for the second test. In both cases, part of the data was used to train the algorithm, and the remaining to test its validity. It correctly quantified the nature of the mass from the test data as a mean square error.

McGetrick and Kim (2014a) conducted a wavelet-based approach for identifying the existence and location of damage from vehicle accelerations. First, in the theoretical simulation, a VBI model was used to investigate the effectiveness of the approach, by which a number of damage indicators were evaluated and compared. Second, a scaled laboratory experiment was carried out to validate the theoretical results and assess the ability of the selected damage indicators to detect changes on the bridge. The experimental setup included a scaled *two-axle vehicle* with axle spacing of 0.4 m, fitted with two accelerometers at axle centers to monitor the bounce motion.

Kim et al. (2014) presented a vibration-based health monitoring strategy for short span bridges utilizing an inspection vehicle. The feasibility of the strategy was investigated through a scaled laboratory experiment using a *two-axle vehicle*. Both the vehicle and bridge responses were measured either separately or together. The results indicated that the natural frequency and changes in damping of the bridge can be detected. The possibility of diagnosis of the damage location and severity of bridges by comparing patterns of identified bridge dynamic parameters through periodical monitoring was also observed.

McGetrick et al. (2015) studied the use of an instrumented vehicle to detect and monitor bridge dynamic parameters. In the laboratory experiment, the scaled *two-axle vehicle model*, fitted with accelerometers at the centers of its axles, was allowed to cross a simply supported steel beam with scaled road surface roughness. The study showed that beam stiffness can be identified with a reasonable degree of accuracy using the measured vehicle accelerations.

Li and Hao (2015) investigated the damage of shear connectors in a composite bridge model under moving loads by the relative displacements processed by the *CWT* and *HHT*. Both the numerical and experimental results indicated that both relative displacements and accelerations can be used to identify the location and occurrence damage in shear connectors when the bridge is under moving loads, but the relative displacement is a better quantity for structural health monitoring of composite bridges.

Kim et al. (2017) conducted laboratory experiments using a test vehicle system equipped with accelerometers. The vehicle system used comprised one tractor and two trailers. The tractor was a two-axle vehicle, heavier than the trailers, serving to excite the bridge into motion. The trailers were *two-axle vehicles*, which were excited by the bridge already set in motion, serving as receivers of the bridge vibration. The purpose is to verify the practical feasibility of three drive-by methods: (i) bridge-frequency extraction using the Fourier spectrum of a vehicle's dynamic response; (ii) damage detection using the change in a vehicle's spectral distribution pattern; and (iii) roadway surface profile identification.

Wang et al. (2017d) presented a method for extracting the bridge influence line (IL) from the dynamic response induced by a passing vehicle. Based on the ILs for different boundary conditions and the features of bridge vibration, a piecewise polynomial and a series of harmonic sinusoids were used to describe the IL of the measured points and the fluctuation part of the bridge response, respectively. A mixed function obtained by linear superposition of these factors was adopted to fit the bridge dynamic response, with the coefficients of the piecewise polynomial calculated by a least-square fitting. Numerical examples of vehicle–bridge models were adopted to validate the fitting method, while illustrating the efficiency via comparison with the direct inverse calculation method. The feasibility of this method was demonstrated by both a laboratory experiment and a field test.

### 1.4.3 Field Investigation

In the field test by Lin and Yang (2005) for the approach to extract the bridge frequencies from the moving vehicle response, the *trailer* of a tractor-trailer system was used as the test vehicle, which possesses the feature of a *single degree-of-freedom* system, in close resemblance of the theoretical model used (Yang et al. 2004a). The same concept of SDOF for the test vehicle was exactly followed in development of the hand-drawn cart (Yang et al. 2013b). The following is a brief review of the relevant field tests conducted along similar lines.

Fujino et al. (2005) described the vehicle intelligent monitoring system (VIMS) developed for detecting the conditions of road pavement and expansion joints of highways in Tokyo, using an *ordinary patrol car* installed with an accelerator and GPS together with a PC. The dynamic response of the vehicle was used to measure the road pavement condition and expansion joints, and the GPS to identify the location of the vehicle. A prototype of the VIMS was installed to a motor car and measurement was taken at real roads, for which the accuracy of measurement and effectiveness of the system were demonstrated.

Gomez et al. (2011) presented an approach for tracking a structure's long-term performance, comparing the current measured responses with the prior ones in a successive manner. Accelerometers were installed on a curved concrete box girder bridge to yield a huge amount of datasets. These data were supplemented by those recorded by accelerometers mounted on a Caltrans water truck. The findings from analyses of these datasets served to inform owners and managers as to the potential feedback from their instrumentation. The measured responses from the vehicle sensors were discussed for the potential of using an instrumented vehicle for bridge frequency measurement.

Miyamoto and Yabe (2011, 2012) assessed the condition of reinforced/prestressed concrete bridges using the acceleration response of a *public bus*. A prototype monitoring system that uses information technology and sensors was described in detail, along with examples of bridge condition assessment based on the vibrations recorded by an in-service public bus equipped with measurement devices. The results indicated that the proposed method can be used to prioritize the repair/strengthening works of existing bridges based on the collected data, which are helpful to bridge administrators for establishing their maintenance strategies.

Kim and Lynch (2012) proposed a wireless monitoring system with sensors installed on both the bridge and a *three-axle truck* to record synchronized data for the bridge and vehicle. The dynamic properties of the bridge were first identified from its free vibration response. Then, the vehicle–bridge response data was used to identify the time varying load imposed on the bridge by the vehicle. The truck was driven across the

Yeondae Bridge, Incheon, Korea, at constant velocity with both the bridge and vehicle responses measured. Excellent agreement was found to exist between the measured bridge response and that predicted by the VBI model.

Siringoringo and Fujino (2012) estimated the fundamental frequency of bridges from the response of an instrumented *two-axle, light commercial vehicle.* The theoretical formulation was first presented to demonstrate the feasibility of the method. Then an experimental verification was conducted on a full-scale simply-supported short span bridge using the test vehicle fitted with accelerometer. The first frequency of the bridge was estimated with reasonable accuracy from the spectrum of the vehicle moving at constant velocity. It was noted that the vehicle's dynamic response is dominated by the bouncing and pitching motions at the entrance/exit (i.e., expansion joint of the bridge), which is an issue to overcome.

Chang et al. (2014a) verified both theoretically and experimentally the variability in bridge frequency induced by a parked vehicle, and suggested such an effect to be considered in bridge-related engineering, especially for those cases with near vehicle-bridge resonance conditions or with large vehicle-to-bridge mass ratios. Moreover, an analytical formula was presented for estimating the variability range.

McGetrick and Kim (2014b) presented a bridge inspection method incorporating a *wavelet-based* damage indicator and pattern recognition. First, a simplified VBI model was used to investigate the effectiveness of the approach in detecting bridge damages from the vehicle acceleration. Using the coefficients of the CWT, different damage levels were distinguished by the damage indicator. In the field test, damage was applied artificially to a steel truss bridge traveled by a two-axle van at constant speed. It was found that the damage indicators obtained from the bridge and vehicle responses showed similar patterns, but it was difficult to distinguish between different artificial damage scenarios.

Tsai et al. (2015) used the vibration response of an *inspection train car,* along with the HHT, to inspect the railway track structure on site. The test results showed that the response measured by the moving vehicle cannot be easily used to identify the bridge characteristics and frequencies unless a sophisticated analysis method is used. In addition, the short duration of a vehicle passage over a single bridge span is a factor that should be considered for the on-board measurement system. It was shown through on-site testing of the tracks that the proposed data analysis method can be used in conjunction with the routine operation of trains to create a method for the monitoring of track defects.

Li et al. (2016b) investigated the dynamic property of a hybrid girder bridge by experimental and numerical means, especially under moving vehicles. Before the inauguration of the bridge, a dynamic field test was conducted. The vehicle was represented as a 16-DOF model. A 3D finite element model was built to represent the complex geometry of the bridge, which was verified by the measured data. The impact effect, ride and pedestrian comfort, and related parameters analysis for the bridge under moving vehicles were studied numerically and experimentally. The results indicated that the impact factor formula from design codes significantly underestimates the impact effect, which may result in unfavorable influence on the bridge safety.

Kafle et al. (2017) and Maizuar et al. (2017) presented a framework to assess the dynamic behavior of bridges by integrating computational modeling and noncontact radar sensor techniques (IBIS-S). The former adopted the weight-in-motion (WIM) technology to consider the real-time traffic information in monitoring the dynamic behavior of a bridge; the latter used the IBIS-S to assess the degradation of a bridge. 3D finite element models were developed for bridges in Australia. The result showed that the WIM along with IBIS-S can efficiently and accurately capture the realtime dynamic behavior and characteristics of the bridge. Also, the detection of structural degradation can be done efficiently by vibration monitoring using the IBIS-S, while this is not the case for frequency measurement of a bridge girder.

McGetrick et al. (2017) investigated the use of low-cost sensors incorporating global navigation satellite systems (GNSS) for implementation of the drive-by system in practice, via field trials with an instrumented vehicle. The potential of smartphone technology to be harnessed for drive-by monitoring is established, while smartphone GNSS tracking applications are found to compare favorably in terms of accuracy, cost, and ease of use to professional GNSS devices.

Kong et al. (2017) used the field test data of an existing bridge to verify the methodology proposed by Kong et al. (2016) for extracting the bridge modal properties. In their simulation, the bridge finite-element model was updated using the measured accelerations and strains of the bridge; two types of test vehicle models were used for the tractor-trailer system; and the measured surface-roughness profile was also included. Parametric studies were conducted to determine the trailer mass and stiffness. Vehicle Model I shows a good capability in extracting the bridge frequencies and the first two modal shapes with dominant vertical components. However, Vehicle Model II performed better than Vehicle Model I on the extraction of bridge modal shapes that are dominant in lateral bending.

Cantero et al. (2017) reported an experimental campaign aimed at measuring the evolution of bridge modal properties during the passage of a vehicle. It investigated not only frequency shifts due to various vehicle positions, but also changes in the mode shapes of vibration. Two different bridges were instrumented and loaded by traversing trucks or trucks momentarily stationed on the bridge. The measurements were analyzed by means of an output-only technique and a novel use of the CWT first presented here. The analysis revealed the presence of additional frequencies, significant shifts in frequencies and changes in the mode shapes, which were theoretically verified using a simplified numerical model. This paper offers an interpretation of VBI of two case studies, showing that the modal properties of the vehicle and bridge do change with varying vehicle position.

## 1.5 Concluding Remarks

Since its inauguration in 2004, the vehicle scanning method for bridges has quickly progressed. It started with extraction of bridge frequencies and was expanded to construction of bridge mode shapes, as well as to damage detection of the bridges. Based on the review of the related previous works, and especially the experience of the authors and co-workers, we can draw the following conclusions:

1) The key parameters of the vehicle scanning method have been identified for extracting the bridge dynamic properties, including frequencies and mode shapes, from a moving test vehicle by the analytical formulation, which forms the foundation for further research.

2) The effect of road roughness is always annoying to the measurement by moving vehicles, which can be alleviated by means such as EMD technique, filtering techniques, dual connected vehicles, inclusion of vehicle damping, consideration of ongoing traffic, etc.
3) The vehicle model is simplified as a SDOF system in the theoretical framework, and it is for this that the VSM was proved to be feasible. If other models with multiple DOFs are used for the vehicle, either in numerical simulation, laboratory test, or field test, the coupling effect of the multi-DOFs of the vehicle itself has to be considered. Besides, the use of point model for the vehicle-bridge contact in numerical simulation may result in exaggeration of high frequency oscillations for road roughness defined by the PSD function.
4) The vertical acceleration of the vehicle has been adopted as the response for extracting the bridge properties, simply because it can be easily measured. In fact, the vehicle acceleration is the result of transmission from the bridge, via the wheel tires, suspension system, and then to the vehicle body, plus pollution from roughness and environment. The path of transmission is a factor to be considered in the design of test vehicles. It was shown that the contact-point response can perform better than the vehicle response, as the vehicle frequency was excluded.
5) With regard to damage detection of bridges, the limited amount of previous researches indicates that the vehicle scanning method is more sensitive to damages of the bridge, partly due to the fact the vehicle can touch each point on the bridge span, compared with the conventional approach that requires numerous sensors to be deployed on the bridge.
6) As for laboratory test, to prepare high-quality scaled test vehicle is generally difficult, concerning its stability in movement and the disturbance brought by the driving force on the moving vehicle. Also, it should be borne in mind that the vehicle mass is negligibly small relative to the bridge mass in reality.
7) There is a lack of field tests in the literature, partly due to the fact that few test vehicles can really resemble the theoretical model shown in Figure 1.1. Here, we are not saying that the model shown in Figure 1.1 should be strictly adopted in the field test. However, if other multi-DOF models are to be used, then the mechanical coupling between the multi-DOFs of the vehicle itself, in addition to that by the bridge frequencies, should be thoroughly studied.
8) It is realized that the vehicle scanning method for modal identification and damage detection of bridges, when maturely developed, possesses the advantages of mobility, efficiency, and efficiency, compared with the conventional approach of bridge measurement.

# 2

# Vehicle Scanning of Bridge Frequencies

Simple Theory

The frequencies of bridge vibration represent a kind of information most useful to modal identification and health assessment of bridges. Traditional vibration tests aimed at measuring the bridge frequencies, known as the direct approach, require installation of certain vibration sensors and data logger on the bridge, which is not only costly but also inconvenient. The idea of using a moving test vehicle to extract the dynamic properties, particularly, the frequencies of a bridge was first theoretically attempted by Yang et al. (2004a). This approach was previously known as the indirect approach and renamed as the vehicle scanning method for bridge measurement. In this chapter, the simplest and fundamental theory for extracting bridge frequencies from a moving test vehicle is presented. In order to identify the key parameters dominating the vehicle-bridge interaction (VBI) response, while unveiling the key phenomena involved, assumptions that lead to closed-form solutions are adopted. Particularly, a vehicle is modeled as a sprung mass, and a bridge as a simply supported beam. Only the first mode of bridge vibration is included in the formulation. The concept of extracting bridge frequencies from a passing vehicle, however, is not restricted by these assumptions, as will be demonstrated in the finite element analysis with virtually no assumptions. The materials presented in this chapter are based primarily on the paper by Yang et al. (2004a).

## 2.1 Introduction

The measurement of frequencies of vibration of a bridge, especially the one of the fundamental mode, is an issue of meaning in engineering. For a newly completed bridge, one is interested in its first few frequencies of vibration, as they serve as useful parameters for comparison with those predicted by the numerical (finite element) model used in design. How well the measured frequencies agree with the predicted ones is an indication of the adequacy of the numerical model used in design. Such information provides useful clues for calibrating the models to be used in future designs of similar bridges, to possibly account for the uncertainties in material properties, structural shapes and boundary conditions, etc. For the purpose of maintenance or rehabilitation, it is often required that the frequencies of vibration, plus other dynamic properties, of the bridge be measured. Theoretically, the frequencies of vibration of the bridge, especially of the fundamental one, when monitored regularly over a long period, serve as a

*Vehicle Scanning Method for Bridges*, First Edition. Yeong-Bin Yang, Judy P. Yang, Bin Zhang and Yuntian Wu.

useful reference for evaluating the degradation in stiffness or strength of the bridge, and even for identifying possible damage to the bridge, say, due to support settlement or dislocations on the bridge caused by earthquakes, impacts, or overloading by daily running traffic.

Conventionally, measuring vibration frequencies of a bridge requires some on-site instrumentation, which may be costly, time-consuming, and even dangerous in practice, depending on the size, shape, and location of the bridge. One common approach is to mount all the vibration sensors, including the seismometers, at selected cross sections of the bridge and have them connected to a computer-driven data acquisition system. If ongoing traffic cannot be partially terminated for the purpose of field testing, the task of equipment mounting and data acquisition can be rather risky and inconvenient. Nevertheless, on-going traffic can be a source of excitation to the bridge in data-taking, usually large enough to excite some of the higher modes of interest. Researches conducted along this line, utilizing on-going traffic or controlled vehicular excitations as the source of excitation, include McLamore et al. (1971), Abdel-Ghaffar and Housner (1978), Mazurek and DeWolf (1990), Casas (1995), Paultre et al. (1995), Ventura et al. (1996), Conner et al. (1997), and Farrar and James III (1997), among others.

For the case where the traffic can be temporarily terminated for experiment or before a bridge is opened to traffic, either an ambient vibration test (in the sense of no on-going traffic in this chapter), an impact test or a forced vibration test can be conducted on the bridge. In the ambient vibration test, the frequencies of vibration of the bridge are measured with the bridge remaining basically nonexcited by any artificial force (Farrar and James III 1997; Fujino et al. 2000; Chang et al. 2001). As a result, such tests may not yield bridge frequencies with high amplitudes, and those of the higher modes tend to become invisible due to the pollution of environmental noises.

For structures that are susceptible to wind loads, such as cable-stayed and suspension bridges, researchers conducted field measurements showing rather high peaks for the first several frequencies under wind-induced vibrations (Brownjohn et al. 1994; Xu et al. 2000). In the impact test, the bridge is excited by an impulse or impact force generated by devices such as heavy hammers or by letting the rear wheels of a truck drop from a wooden block. The level of impact forces is usually large enough to excite the first several modes (Douglas and Reid 1982; Huang et al. 1999). An alternative is to conduct a forced vibration test on the bridge, using devices such as the vibration shaker (Okauchi et al. 1992; Farrar and James III 1997; Maragakis et al. 1998; Fujino et al. 2000). This method allows us to identify the first few frequencies of vibration of the bridge, as the input energy is usually large enough to excite the modes of concern.

Although the techniques for conducting the aforementioned tests are generally mature, the effort (i.e., time, cost, and manpower) required to conduct each of such tests is tremendous. It is with this in mind that the feasibility of extracting the frequencies of vibration of a bridge, especially of the fundamental one, from the dynamic response of a moving vehicle over the bridge was attempted by Yang et al. (2004a). The basic idea is that a vehicle moving over a bridge at a specific speed can excite the bridge to a certain level, thereby playing the role of a *vibration shaker*. The vehicle in the meantime serves as a *message receiver*, in the form of a moving sprung mass, whose dynamic response is totally dominated by the frequencies of the supporting bridge. Thus, if we can record the vertical dynamic (acceleration) response of the vehicle during its passage of the bridge, say, using seismometers installed inside the vehicle, we can analyze the frequency content of the recorded vehicle response, eliminating those related to the vehicle itself, and obtain the frequencies associated with the supporting bridge.

There exists great academic interest to exploit the feasibility of extracting the bridge frequencies from a passing test vehicle, as it was not attempted before 2004, and in this regard it is essential to start working on the first frequency of the bridge in a preliminary, theoretical manner, which forms the objective of this chapter. In order to identify the key parameters dominating the vehicle-bridge interaction (VBI) response, while unveiling the key phenomena involved, assumptions that lead to closed-form solutions are adopted in the present analytical study. To this end, the vehicle is modeled as sprung mass and only the first mode of vibration is adopted for the bridge as a simply supported beam. In the same way, in an independent finite element analysis, virtually all the modes of vibration, rather than only the first mode, of the bridge are considered in the simulation. It is confirmed that the key phenomena observed in the simplified analytical study have their counterpart in the numerical, finite element study.

## 2.2 Formulation of the Analytical Theory

For simplicity, the vehicle is modeled as a *single degree of freedom* (DOF), as shown in Figure 2.1. This should not be regarded as a half of the two-axle vehicle, but as an independent single-axle vehicle (a trailer) by itself, in the sense that no coupling exists with other axles or other parts of the vehicle. As shown in Figure 2.1, a vehicle of speed $v$ is modeled as a lumped mass $m_v$ supported on a spring of stiffness $k_v$, moving across a simply supported bridge of length $L$ with smooth pavement. If the damping of the bridge is ignored, the equations of motion for the sprung mass and the bridge can be written as:

$$m_v\ddot{q}_v + k_v\left(q_v - u\big|_{x=vt}\right) = 0 \tag{2.1}$$

$$m\ddot{u} + EIu'''' = f_c(t)\delta(x - vt) \tag{2.2}$$

where $q_v$ denotes the vertical deflection of the sprung mass, $\delta$ is the delta function, a dot and a prime respectively denote differentiation with respect to time $t$ and coordinate $x$ of the beam, and $E$ denotes the elastic modulus, $I$ the moment of inertia, $u(x, t)$ the displacement, and $m$ the mass per unit length of the beam. The contact force $f_c$ existing between the sprung mass and the beam can be expressed as

$$f_c(t) = k_v(q_v - u\big|_{x=vt}) + m_v g \tag{2.3}$$

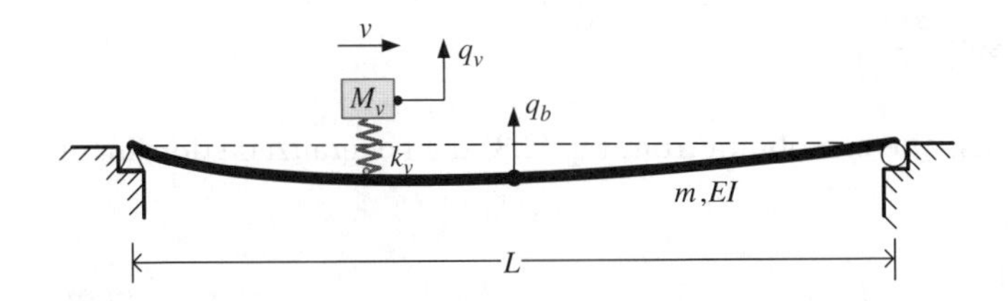


**Figure 2.1** Sprung mass moving over a beam.

where $g$ is the acceleration of gravity. It should be noted that the vehicle displacement $q_v$ is measured from the static equilibrium position of the vehicle. Because of this, no gravity term appears in Eq. (2.1). In contrast, the motion of the beam may be affected by the gravitational load of the vehicle, as implied by the contact force $f_c$ in Eq. (2.3).

For the moving load problem, which is transient in nature, the response of the beam can be well simulated by considering only the first mode of vibration (Biggs 1964). Also, for the purpose of investigating the feasibility of using a test vehicle to scan the bridge frequencies, it is appropriate to concentrate only on the first mode of vibration of the bridge. In accordance, the displacement $u(x, t)$ of the beam can be approximated as

$$u(x,t)=q_b(t)\sin\left(\frac{\pi x}{L}\right) \tag{2.4}$$

where $q_b(t)$denotes the generalized coordinate (or the midspan displacement) of the first mode for the beam. Substituting the preceding expression for $u$ into Eqs. (2.1) and (2.2), multiplying both sides of Eq. (2.2) by $\sin(\pi x/L)$ and integrating with respect to length $L$ of the beam, one obtains

$$m_v\ddot{q}_v+\left(\omega_v^2 m_v\right)q_v-\left[\omega_v^2 m_v \sin\left(\frac{\pi vt}{L}\right)\right]q_b=0 \tag{2.5}$$

$$\begin{aligned}&\frac{mL}{2}\ddot{q}_b+\left[\frac{mL}{2}\omega_b^2+\omega_v^2 m_v \sin^2\left(\frac{\pi vt}{L}\right)\right]q_b-\left[\omega_v^2 m_v \sin\left(\frac{\pi vt}{L}\right)\right]q_v\\&=-m_v g\sin\left(\frac{\pi vt}{L}\right)\end{aligned} \tag{2.6}$$

where $\omega_v$ and $\omega_b$ denote the natural vibration frequency (in rad/s) of the vehicle and the bridge, respectively:

$$\omega_v=\sqrt{\frac{k_v}{m_v}},\quad \omega_b=\frac{\pi^2}{L^2}\sqrt{\frac{EI}{m}} \tag{2.7}$$

In the following, approximate closed-form solutions will be sought for the vehicle and bridge based on some practical assumptions.

## 2.3 Single-Mode Analytical Solution

Assuming the vehicle mass $m_v$ to be much less than the bridge mass $mL$, i.e., $m_v/mL << 1$, one can approximate Eq. (2.6) by the following:

$$\ddot{q}_b+\omega_b^2 q_b=\frac{-2m_v g}{mL}\sin\left(\frac{\pi vt}{L}\right) \tag{2.8}$$

Using zero initial conditions, one can obtain from Eq. (2.8) the generalized coordinate $q_b$ of the beam as

$$q_b=\frac{\Delta_{st}}{1-S^2}\left[\sin\left(\frac{\pi vt}{L}\right)-S\sin\left(\omega_b t\right)\right] \tag{2.9}$$

where $\Delta_{st}$ denotes approximately the static deflection of the midspan of the beam under the gravity action of the mass $m_v$ at the same point:

$$\Delta_{st} = -\frac{2m_v g L^3}{\pi^4 EI} \tag{2.10}$$

which is very close to the exact value of $-m_v gL^3/(48EI)$. The *speed parameter* $S$ is defined as the ratio of half the *driving frequency*, $\pi v/L$, to the bridge frequency $\omega_b$:

$$S = \frac{\pi v}{L\omega_b} \tag{2.11}$$

Substituting Eq. (2.9) for the bridge coordinate $q_b$ into Eq. (2.5), we can solve from Eq. (2.5) the vehicle response $q_v$ by Duhamel's integral as

$$q_v(t) = \frac{\omega_v \Delta_{st}}{2(1-S^2)} \left\{ \frac{1}{\omega_v}(1-\cos\omega_v t) - \frac{\omega_v\left(\cos\left(\frac{2\pi vt}{L}\right) - \cos(\omega_v t)\right)}{\omega_v^2 - \left(\frac{2\pi v}{L}\right)^2} - S\left[\frac{\omega_v\left(\cos\left(\left(\frac{\pi v}{L} - \omega_b\right)t\right) - \cos(\omega_v t)\right)}{\omega_v^2 - \left(\frac{\pi v}{L} - \omega_b\right)^2} - \frac{\omega_v\left(\cos\left(\left(\frac{\pi v}{l} + \omega_b\right)t\right) - \cos(\omega_v t)\right)}{\omega_v^2 - \left(\frac{\pi v}{L} + \omega_b\right)^2}\right]\right\} \tag{2.12}$$

Evidently, the vehicle response is dominated by four specific frequencies, that is, the vehicle frequency, $\omega_v$, driving frequency of the moving vehicle, $2\pi v/L$, and two shifted frequencies of the bridge, $\omega_b - \pi v/L$, $\omega_b + \pi v/L$. By letting $\mu$ denote the frequency ratio of the bridge to the vehicle, i.e., $\mu = \omega_b/\omega_v$, and by using the definition for the speed parameter $S$, i.e., $S = \pi v/(L\omega_b)$, one can rewrite the preceding equation as

$$q_v(t) = \frac{\Delta_{st}}{2(1-S^2)}\left[(1-\cos\omega_v t) - \frac{\cos{}^{2\pi vt}/_L - \cos\omega_v t}{1-(2\mu S)^2} - S\cdot\frac{\cos\left(\omega_b - {}^{\pi v}/_L\right)t - \cos\omega_v t}{1-\mu^2(1-S)^2} + S\cdot\frac{\cos\left(\omega_b + {}^{\pi v}/_L\right)t - \cos\omega_v t}{1-\mu^2(1+S)^2}\right] \tag{2.13}$$

which can be differentiated once to yield the velocity of the vehicle as

$$\dot{q}_v(t) = \frac{\Delta_{st}\omega_v}{2(1-S^2)}\left[\sin\omega_v t + \frac{2\mu S\cdot\sin{}^{2\pi vt}/_L - \sin\omega_v t}{1-(2\mu S)^2} + S\cdot\frac{\mu(1-S)\cdot\sin\left(\omega_b - {}^{\pi v}/_L\right)t - \sin\omega_v t}{1-\mu^2(1-S)^2} - S\cdot\frac{\mu(1+S)\cdot\sin\left(\omega_b + {}^{\pi v}/_L\right)t - \sin\omega_v t}{1-\mu^2(1+S)^2}\right] \tag{2.14}$$

and twice to yield the acceleration of the vehicle as

$$\begin{aligned}\ddot{q}_v(t) = \frac{\Delta_{st}\omega_v^{\,2}}{2\left(1-S^2\right)}\Bigg[&\cos\omega_v t + \frac{(2\mu S)^2\cdot\cos 2\pi vt/L - \cos\omega_v t}{1-(2\mu S)^2}\\ &+S\cdot\frac{\mu^2(1-S)^2\cdot\cos\left(\omega_b - \pi v/L\right)t - \cos\omega_v t}{1-\mu^2(1-S)^2}\\ &-S\cdot\frac{\mu^2(1+S)^2\cdot\cos\left(\omega_b + \pi v/L\right)t - \cos\omega_v t}{1-\mu^2(1+S)^2}\Bigg]\end{aligned} \tag{2.15}$$

Apparently, the vertical displacement, velocity, and acceleration of the vehicle depend on both the parameters $S$ and $\mu$. In contrast, the central displacement of the bridge, as given in Eq. (2.9), depends only on the speed parameter $S$. Here, only the acceleration of the vehicle is of concern, since it is the parameter that can be easily measured by seismometers in practice.

The maximum displacement response of the vehicle has been drawn with relation to the speed parameter $S$ and frequency ratio $\mu$ using a tri-phase plot in Figure 2.2a, of which the contour projection on the horizontal plane was given in Figure 2.2b. As can be seen, the vehicle achieves its maximum in the region with $S = 0.3\sim0.6$ and $\mu = 0.5\sim1.5$, especially in the vicinity of $\mu = 1$, when the bridge frequency equals the vehicle frequency. This can be easily conceived, since under such a condition resonance will occur on the vehicle-bridge system, and a significant amount of kinetic energy will be transmitted to the vehicle, a smaller subsystem compared with the bridge. For the purposes of extracting the bridge frequency from the vehicle response, it is preferable that the vehicle response be enlarged as much as possible or that the peak regions noted above can be reached.

Similarly, Figures 2.3a and b show the tri-phase plot and contour lines, respectively, for the vertical velocity of the vehicle. Clearly, the maximum response of the vehicle velocity occurs in the region with $S = 0.3\sim0.6$ and $\mu = 0.5\sim1.0$ and in the vicinity of the point with $S = 0.3$ and $\mu = 1.0$. In addition, Figure 2.4a and b show the tri-phase plot and contour lines, respectively, for the vehicle acceleration, which indicates that the maximum response occurs in two regions indicated by ($S = 0.3\sim0.4$, $\mu = 0.55\sim1.5$) and ($S = 0.3\sim0.5$, $\mu = 0\sim0.5$); the latter, however, can hardly be reached in practice.

As already mentioned, in extracting the bridge frequency from the vehicle response, higher visibility can be achieved if the vehicle response can be magnified as much as possible. The above analyses offer some clues for selecting the related physical parameters of the sprung mass in practice. In the following, we shall investigate the relative influence of each of four specific frequencies on the response amplitude of the vehicle. To this end, one may rewrite the vehicle acceleration $\ddot{q}_v(t)$ as follows:

$$\begin{aligned}\ddot{q}_v(t) = \frac{\Delta_{st}\omega_v^2}{2\left(1-S^2\right)}\Bigg[&A_1\cos\omega_v t + A_2\cos\frac{2\pi v}{L} + A_3\cos\left(\omega_b - \frac{\pi v}{L}\right)\\ &+A_4\cos\left(\omega_b + \frac{\pi v}{L}\right)\Bigg]\end{aligned} \tag{2.16}$$

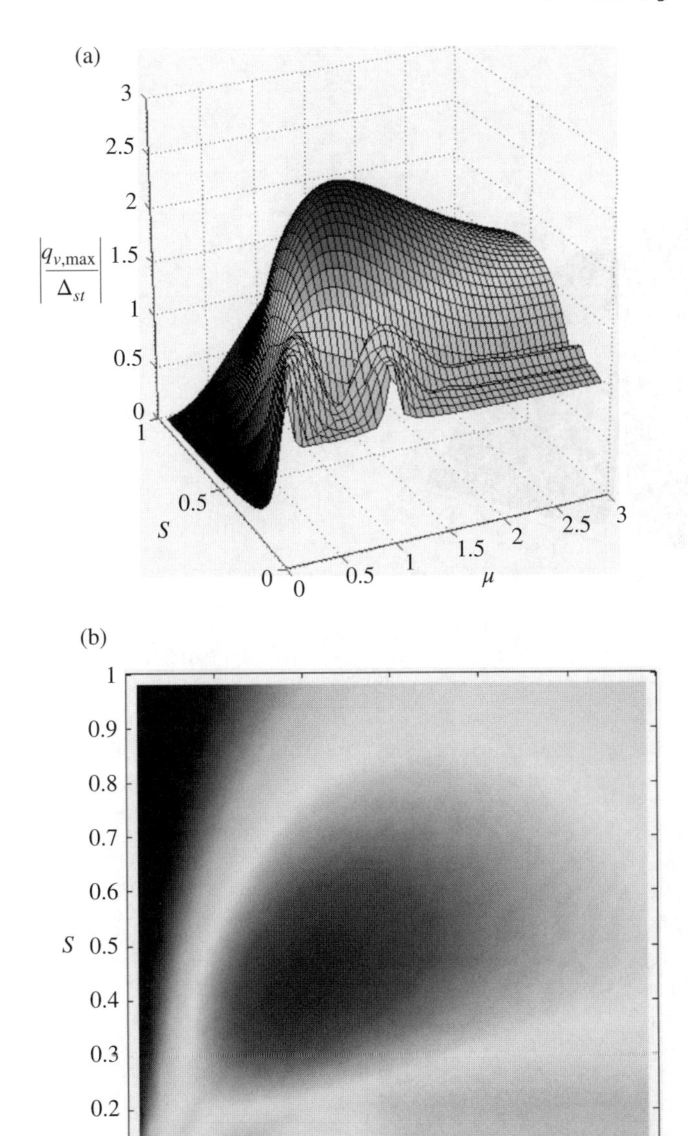


**Figure 2.2** Maximum displacement response of vehicle: (a) tri-phase plot; (b) plane projection.

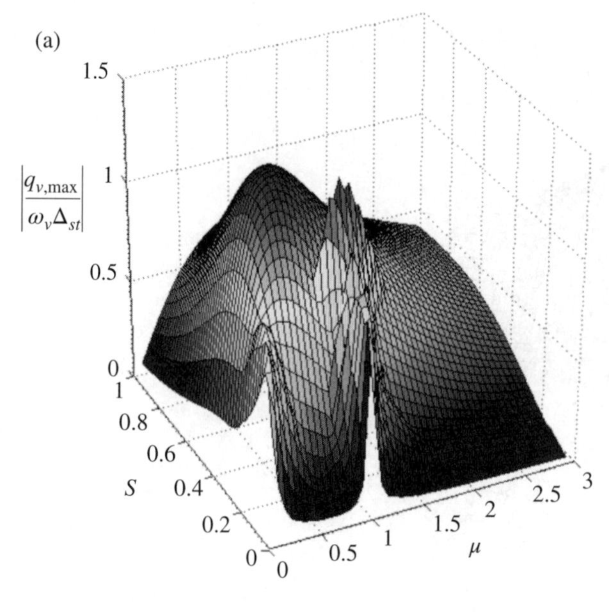


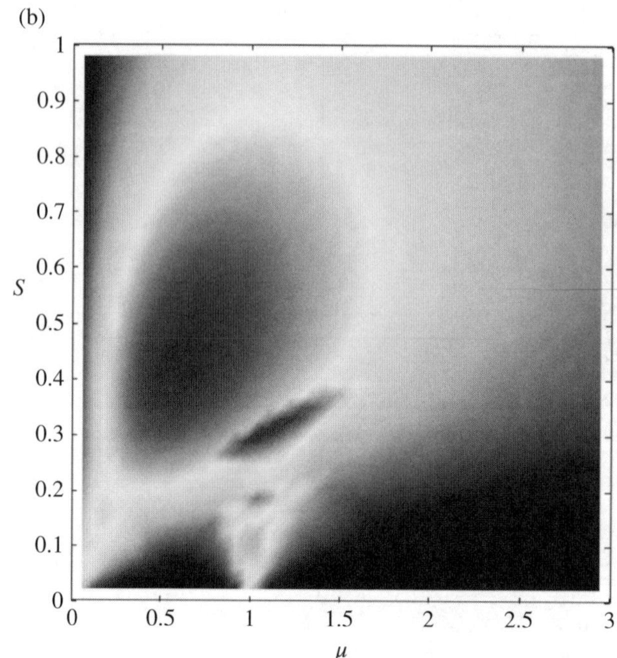


**Figure 2.3** Maximum velocity response of vehicle: (a) tri-phase plot; (b) plane projection.

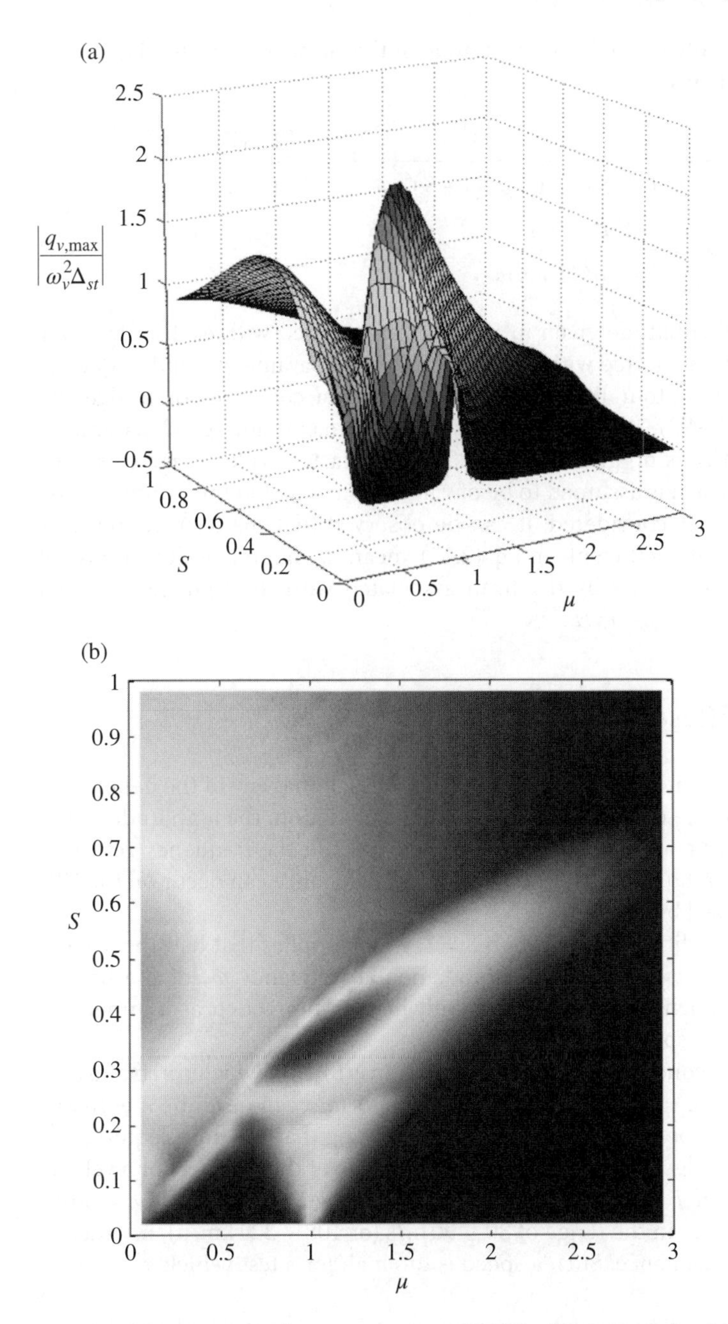


**Figure 2.4** Maximum acceleration response of vehicle: (a) tri-phase plot; (b) plane projection.

where $A_1$, $A_2$, $A_3$, $A_4$ denote the relative magnitude of the contribution associated with each of the four frequencies:

$$A_1 = 1 - \frac{1}{1-(2\mu S)^2} - \frac{S}{1-\mu^2(1-S)^2} + \frac{S}{1-\mu^2(1+S)^2}, \quad A_2 = \frac{(2\mu S)^2}{1-(2\mu S)^2}$$
$$A_3 = \frac{S\mu^2(1-S)^2}{1-\mu^2(1-S)^2}, \quad A_4 = -\frac{S\mu^2(1+S)^2}{1-\mu^2(1+S)^2} \tag{2.17}$$

In Figures 2.5–2.8, the amplitude given in Eq. (2.17) associated with each of the four major frequencies has been plotted with respect to the speed parameter $S$ and frequency ratio $\mu$, along with the contour lines. Using the field data collected for bridges by Dusseau and Dusaisi (1993), it can be shown that the maximum speed parameter $S$ encountered in practice is no greater than 0.3. In contrast, a wider range exists for the frequency ratio $\mu$, which is assumed to be of values from 0 to 3 in this study. For the range of parametric values considered, it can be observed that the term $A_4$ remains the largest among the four coefficients in Eq. (2.17), meaning that the peak response of the vehicle is dominated mainly by the term associated with the rightward shifted frequency of the bridge, i.e., $\omega_b + \pi v/L$.

## 2.4 Condition of Resonance

For our purposes, the condition of resonance is defined such that any of the denominators in the vehicle response equals zero. Under such a condition, the amplitude of the vehicle response reaches a local maximum or a local peak in the frequency response plot. As can be seen from Eq. (2.12), four conditions of resonance can occur on the VBI system, as will be analyzed below.

The first resonance condition occurs when $S = 1$, which implies that half the driving frequency of the vehicle, $\pi v/L$, equals the fundamental frequency $\omega_b$ of the bridge, according to Eq. (2.11). Such a condition can hardly be met in practice, as it implies an unreasonably high vehicle speed.

The second resonance condition occurs when $\omega_v^2 - (2\pi v/L)^2 = 0$, which can be further split into two as $\omega_v + 2\pi v/L = 0$ and $\omega_v - 2\pi v/L = 0$. The former can never be met mathematically. The latter can be rewritten as $T_v = L/v$, meaning that the time required for the vehicle to pass the bridge, $L/v$, must equal the period $T_v$ of the vehicle. For a vehicle with a vibration period of $T_v = 1 \sim 0.3$ second passing a bridge, say, of length $L = 30$ m, the vehicle speed $v$ must be in the range of 30 ~ 90 m/s (or 108 ~ 324 km/h), in order to excite the condition of resonance. Such a speed is too high for a test vehicle to meet in practice.

The third resonance condition occurs when $\omega_v^2 - (\pi v/L - \omega_b)^2 = 0$, which can be factored as $\omega_v + (\omega_b - \pi v/L) = 0$ and $\omega_v - (\omega_b - \pi v/L) = 0$. The former cannot be met in practice, since half the driving frequency, $\pi v/L$, is generally smaller than the bridge frequency $\omega_b$. The latter condition can possibly be met if the bridge frequency $\omega_b$ happens to be slightly larger than the vehicle frequency $\omega_v$.

The fourth resonance condition occurs when $\omega_v^2 - (\pi v/L + \omega_b)^2 = 0$, which can be factored as $\omega_v + (\omega_b + \pi v/L) = 0$ and $\omega_v - (\omega_b + \pi v/L) = 0$. The former can never be met

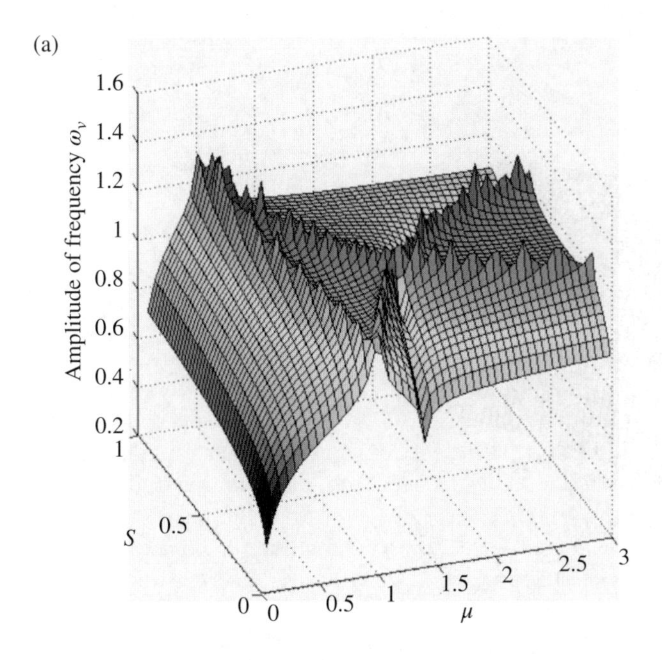


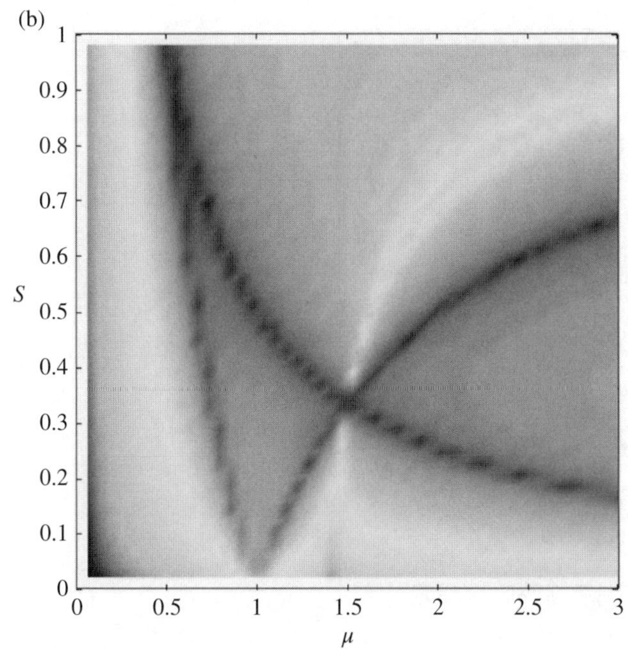


**Figure 2.5** Amplitude of frequency $\omega_v$: (a) tri-phase plot; (b) plane projection.

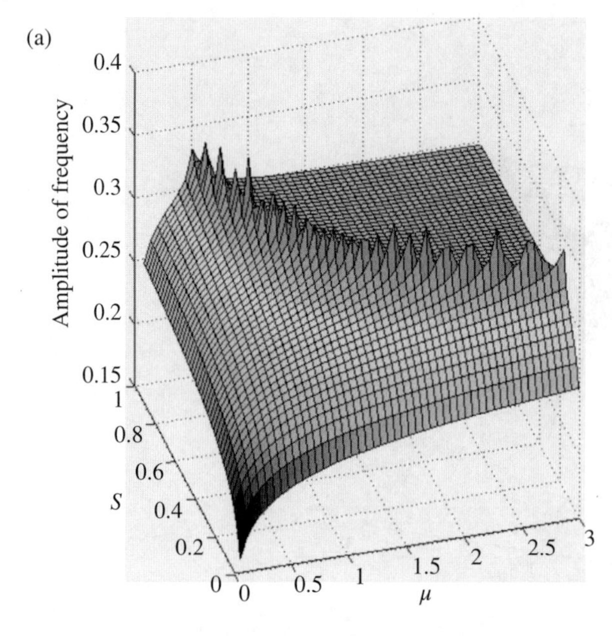


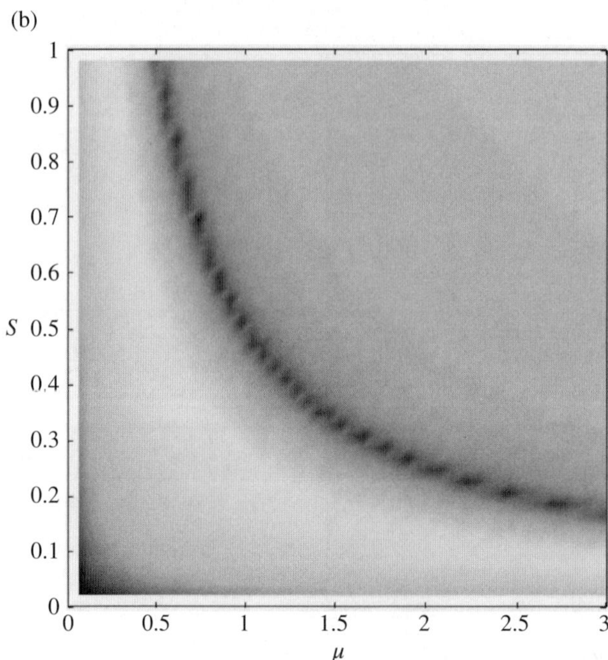


**Figure 2.6** Amplitude of frequency $2\pi v/L$: (a) tri-phase plot; (b) plane projection.

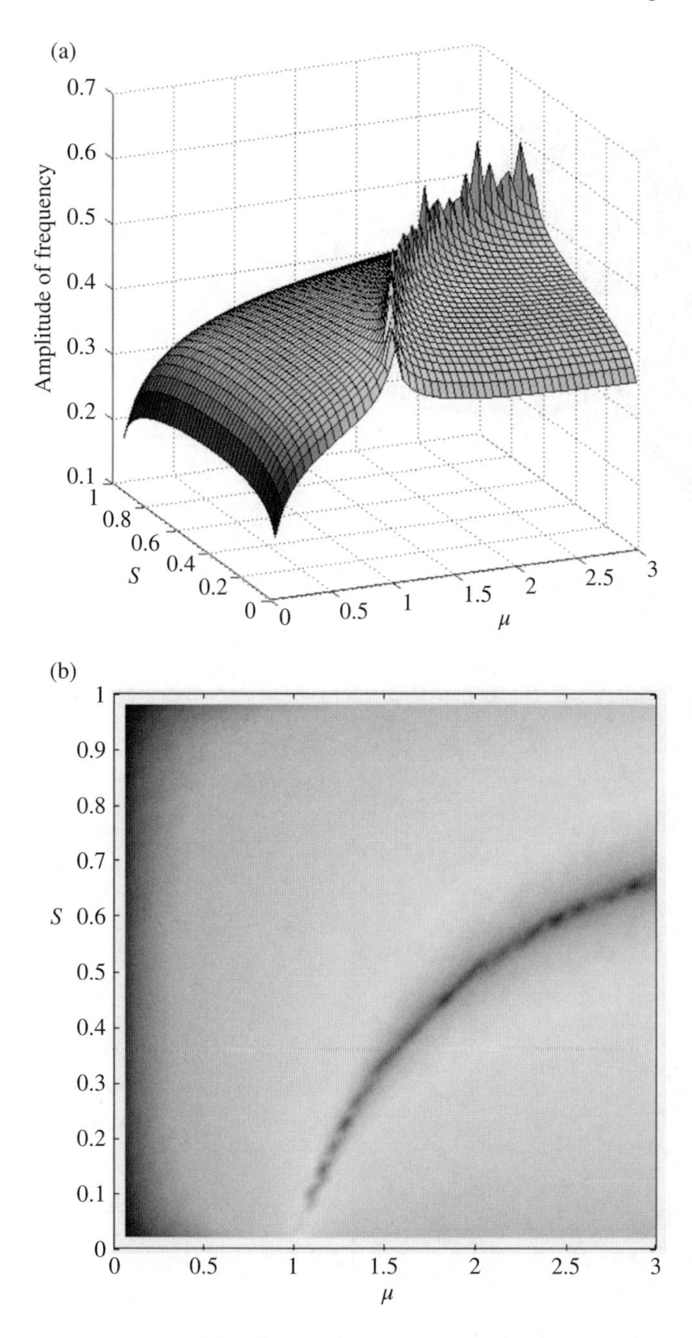


**Figure 2.7** Amplitude of frequency $\omega_b - \pi v/L$: (a) tri-phase plot; (b) plane projection.

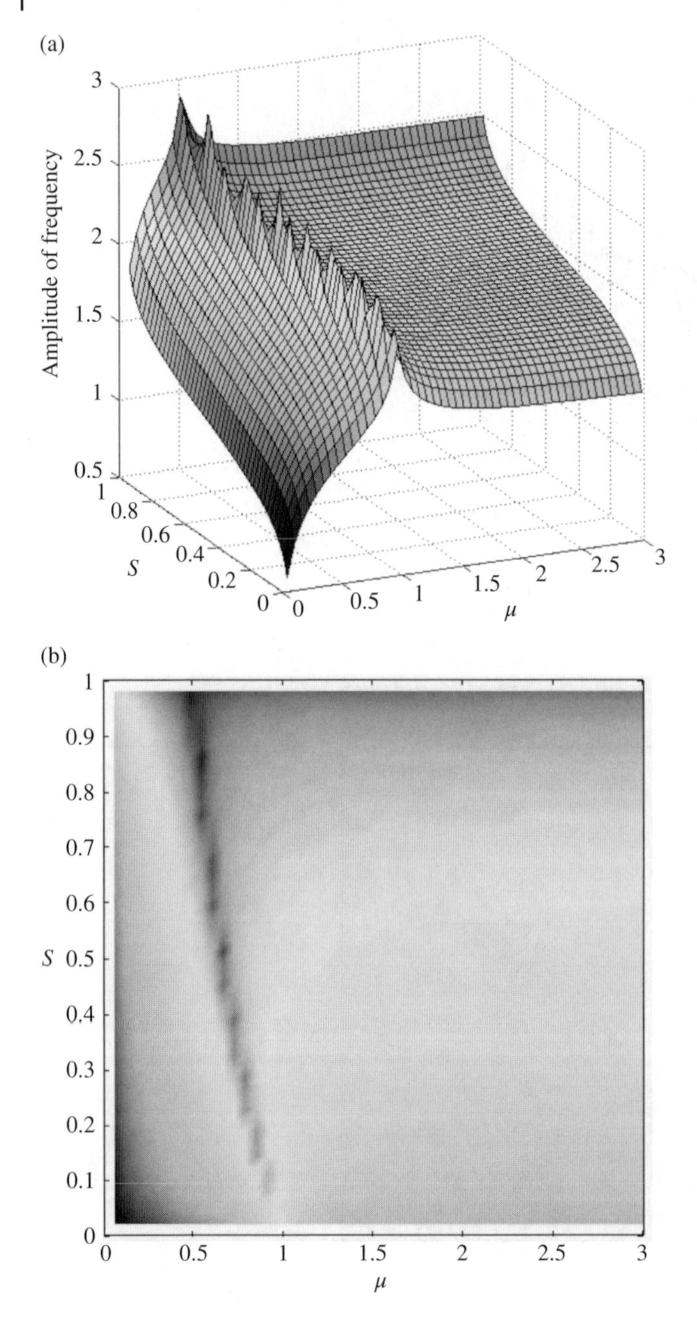


**Figure 2.8** Amplitude of frequency $\omega_b + \pi v/L$: (a) tri-phase plot; (b) plane projection.

mathematically. The latter condition can possibly be met, if the bridge frequency $\omega_b$ happens to be slightly smaller than the vehicle frequency $\omega_v$.

From the above analysis, it is concluded that if the bridge frequency $\omega_b$ is close to the vehicle frequency $\omega_v$, then the vehicle can be excited to resonance through adjustment of the vehicle speed $v$, such that either of the following two conditions are met: $\omega_v - (\omega_b - \pi v/L) = 0$ and $\omega_v - (\omega_b + \pi v/L) = 0$, as will be demonstrated in the numerical study. Moreover, the vehicle response is dominated by the four specific frequencies: $\omega_v$, $2\pi v/L$, $\omega_b - \pi v/L$, and $\omega_b + \pi v/L$, using the present single-mode approximation.

## 2.5 Simulation by the Finite Element Method (FEM)

The above analytical study, though unavoidably based on some assumptions, serves to identify the key parameters dominating the VBI response. It provides the guidelines for simulating the same problem using more sophisticated or realistic approaches, say, the FEM or experimental means. In fact, the idea of extracting bridge frequencies from the vehicle response is not restricted by any of the assumptions adopted in the analytical study, since the bridge frequencies, by nature, should always be present in the response of vehicles traveling over the bridge, as will be demonstrated in the finite element simulation to follow, for which most assumptions are removed. The main question in developing the technology is how to extract or interpret the bridge frequencies from the response "measured" (as in the field test) or "calculated" (as in the analytical or numerical study) for the passing vehicle.

In the finite element analysis, one may discretize the beam into a number of elements and assume the sprung mass to be acting at only one of the elements. Such an element consisting of the beam element and the acting sprung mass $m_v$ is referred to as the VBI element, as shown in Figure 2.9. For our purposes, the effect of the wheel mass $m_w$ is simply ignored. In this case, the sprung mass is regarded as a single-DOF system. Following basically the procedure of Yang and Yau (1997), but with the modifications by Chang et al. (2010), the equations of motion for the VBI element can be written as (see Appendix, Section A.1 for the derivation):

$$
\begin{aligned}
&\begin{bmatrix} m_v & 0 \\ 0 & [m_b] \end{bmatrix} \begin{Bmatrix} \ddot{q}_v \\ \{\ddot{u}_b\} \end{Bmatrix} + \begin{bmatrix} c_v & -c_v\{N\}^T \\ -c_v\{N\} & [c_b]+c_v\{N\}\{N\}^T \end{bmatrix} \begin{Bmatrix} \dot{q}_v \\ \{\dot{u}_b\} \end{Bmatrix} \\
&+ \begin{bmatrix} k_v & -vc_v\{N'\}^T - k_v\{N\}^T \\ -k_v\{N\} & [k_b]+vc_v\{N\}\{N'\}^T + k_v\{N\}\{N\}^T \end{bmatrix} \begin{Bmatrix} q_v \\ \{u_b\} \end{Bmatrix} = \begin{Bmatrix} 0 \\ -m_v g\{N\} \end{Bmatrix}
\end{aligned}
\tag{2.18}
$$

where $\{u_b\}$ denotes the nodal displacement vector, and $[m_b]$, $[c_b]$ and $[k_b]$, respectively, the mass, damping, and stiffness matrices of the bridge element, and $\{N\}$ is a vector containing cubic Hermitian interpolation functions associated with the transverse

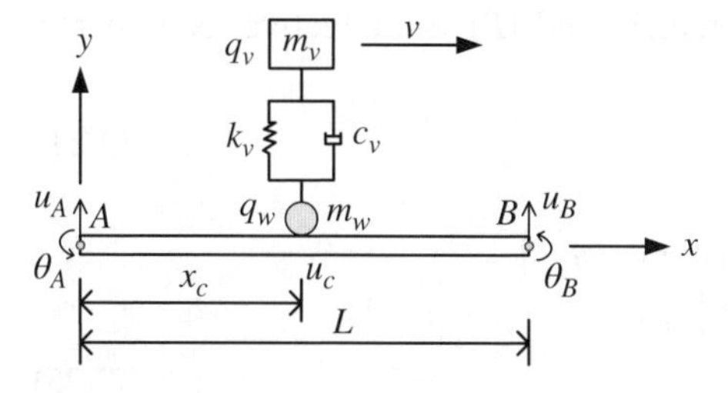


**Figure 2.9** Vehicle-bridge interaction (VBI) element.

displacement of the element, but evaluated at the contact point position $x_c$ of the sprung mass.

Let us assume that the beam has $N$ DOFs. The entire VBI system is said to have $N+1$ DOFs, due to addition of the sprung mass. Following the procedure presented in Appendix, Section A.2, the VBI element derived above can be assembled with all the other ordinary beam elements of the structure not directly in contact with the sprung mass to yield the $(N+1)$-dimensional structural equations as:

$$[M]\{\ddot{q}\}+[C]\{\dot{q}\}+[K]\{q\}=\{F\} \tag{2.19}$$

where $\{q\}$ denotes the system displacement vector, which contains all the DOFs of the beam plus the sprung mass DOF, $\{F\}$ the corresponding force vector, and $[M]$, $[C]$, $[K]$, respectively, the $(N+1)$-dimensional mass, damping, and stiffness matrices of the system assembled from those of the VBI element and noncontact beam elements. For convenience of computer coding, the vehicle DOF has been arranged as the last DOF of the VBI system in Appendix, Section A.2. Apparently, all the system matrices and vectors are functions of the acting position $x_c$ of the sprung mass, which therefore are time-dependent and should be updated at each time step of the time-history analysis.

The system equations as given in Eq. (2.19) can be solved by Newmark's $\beta$ method of direct integration (Newmark 1959). Consider a typical time step from $t$ to $t+\Delta t$ in the nonlinear time-history analysis. The acceleration and velocity vectors of the system at $t+\Delta t$ can be discretized as follows:

$$\{\ddot{q}\}_{t+\Delta t}=a_0\left(\{q\}_{t+\Delta t}-\{q\}_t\right)-a_2\{\dot{q}\}_t-a_3\{\ddot{q}\}_t \tag{2.20}$$

$$\{\dot{q}\}_{t+\Delta t}=\{\dot{q}\}_t+a_6\{\ddot{q}\}_t+a_7\{\ddot{q}\}_{t+\Delta t} \tag{2.21}$$

where the coefficients are

$$\begin{aligned}
&a_0=\frac{1}{\beta\Delta t^2}; \quad a_1=\frac{\gamma}{\beta\Delta t}; \quad a_2=\frac{1}{\beta\Delta t}; \quad a_3=\frac{1}{2\beta}-1\\
&a_4=\frac{\gamma}{\beta}-1; \quad a_5=\frac{\Delta t}{2}\left(\frac{\gamma}{\beta}-2\right); \quad a_6=\Delta t(1-\gamma); \quad a_7=\gamma\Delta t
\end{aligned} \tag{2.22}$$

In this study, the two parameters are selected as $\beta=0.25$ and $\gamma=0.5$, which implies a constant average acceleration with unconditional numerical stability.

Substituting Eqs. (2.20) and (2.21) into the system equations, Eq. (2.19), one obtains, after some manipulations, the following equivalent linear equations:

$$\left[\bar{K}\right]_t\{q\}_{t+\Delta t}=\{\bar{F}\}_{t+\Delta t} \tag{2.23}$$

where $[\bar{K}]$ represents the effective stiffness matrix and $\{\bar{F}\}$ the effective load vector, defined as follows:

$$\left[\bar{K}\right]_t=[K]_t+a_0[M]_t+a_1[C]_t \tag{2.24}$$

$$\begin{aligned}
\{\bar{F}\}_{t+\Delta t}=\{F\}_{t+\Delta t}&+[M]_t\left(a_0\{q\}_t+a_2\{\dot{q}\}_t+a_3\{\ddot{q}\}_t\right)\\
&+[C]_t\left(a_1\{q\}_t+a_4\{\dot{q}\}_t+a_5\{\ddot{q}\}_t\right)
\end{aligned} \tag{2.25}$$

where $[K]_t$, $[M]_t$ and $[C]_t$ are, respectively, the stiffness, mass, and damping matrices of the system evaluated at time $t$. The force vector $\{F\}_{t+\Delta t}$ denotes the external loads of the system at time $t+\Delta t$.

For a vehicle with an assumed speed $v$, the procedure for solving the system equations, Eq. (2.19), at each time step, i.e. at time $t+\Delta t$, is as follows: (i) Use the system matrices $[M]_t$, $[C]_t$, $[K]_t$ at time step $t$ to compute the effective stiffness matrix $[\bar{K}]_t$; (ii) Calculate the acting position $x_c$ of the sprung mass; (iii) Calculate the external force vector $\{F\}_{t+\Delta t}$ and the effective load vector $\{\bar{F}\}_{t+\Delta t}$; (iv) Solve the equivalent system equations, Eq. (2.23), for the displacements $\{q\}_{t+\Delta t}$; (v) Compute the system accelerations and velocities from Eqs. (2.20) and (2.21); (vi) Update the system matrices $[M]_t$, $[C]_t$, $[K]_t$ for the next time step; (vii) Repeat steps (i)–(vi).

## 2.6 Verification of Accuracy of Analytical Solutions

In this section, the accuracy of the single-mode closed-form solution obtained for the VBI system, in particular, the vehicle response, will be verified by the finite element solution for a typical example. Consider a simply supported beam of length $L = 25\,\text{m}$, with the following properties: cross-sectional area $A = 2.0\,\text{m}^2$, moment of inertia $I = 0.12\,\text{m}^4$, mass per unit length $m = 4800\,\text{kg/m}$, and elastic modulus $E = 27.5\ \text{GN/m}^2$. The following data are adopted for the vehicle: mass $m_v = 1200\,\text{kg}$, spring stiffness $k_v = 500\ \text{kN/m}$, and zero damping. For this example, the vehicle to bridge mass ratio is 1/100. In the finite element analysis, 10 beam elements are used for the bridge. The fundamental frequency of vibration of the bridge computed is $\omega_b = 2.08$ Hz and the vehicle frequency is $\omega_v = 3.25$ Hz.

For the case where the vehicle passes through the beam at a speed of $v = 10\,\text{m/s}$, the vertical displacements of the vehicle and the bridge midpoint obtained by the two approaches have been plotted in Figures 2.10a and b, respectively. As can be seen from Figure 2.10b, the solutions obtained by the two approaches show a high degree of coincidence for the bridge response. Although slight deviations exist between the two solutions obtained for the vehicle response in Figure 2.10a, the analytical results are considered acceptable for the purpose of identifying the key parameters involved, especially the bridge frequency.

The vertical velocity responses of the sprung mass and the midpoint of the beam are shown in Figures 2.11a and b, respectively, and the acceleration responses in Figures 2.12a and b. As can be seen, generally accurate solutions have been obtained for all cases by the single-mode analytical approach, except for the midpoint bridge acceleration, where drastic oscillations due to higher modes were missing, compared with the finite element solution. Aside from the higher modes, both approaches reveal similar trends for the fundamental mode concerning the midpoint bridge acceleration.

A general conclusion from the results shown in Figures 2.10–2.12 is that the single-mode analytical solutions can be reliably used to simulate both the vehicle and bridge responses, except the midpoint acceleration of the bridge. Since the primary goal of this chapter is to conceptually develop a technique for extracting *the fundamental* frequency of the bridge from the response of the passing vehicle, rather than from the bridge response, the inherent lack of capability of the single-mode approach to deal with the higher modes of the bridge is not considered a handicap.

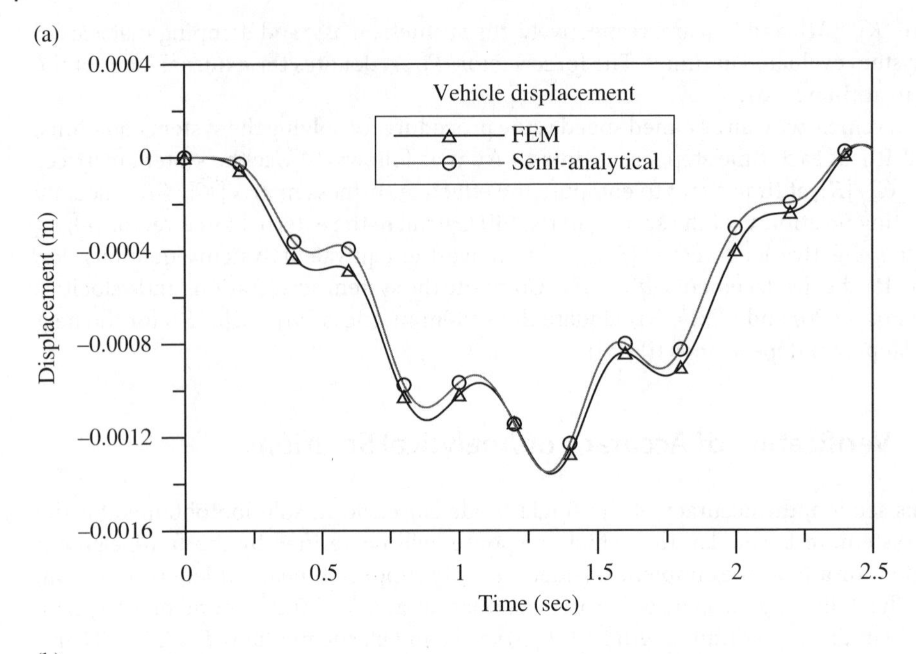


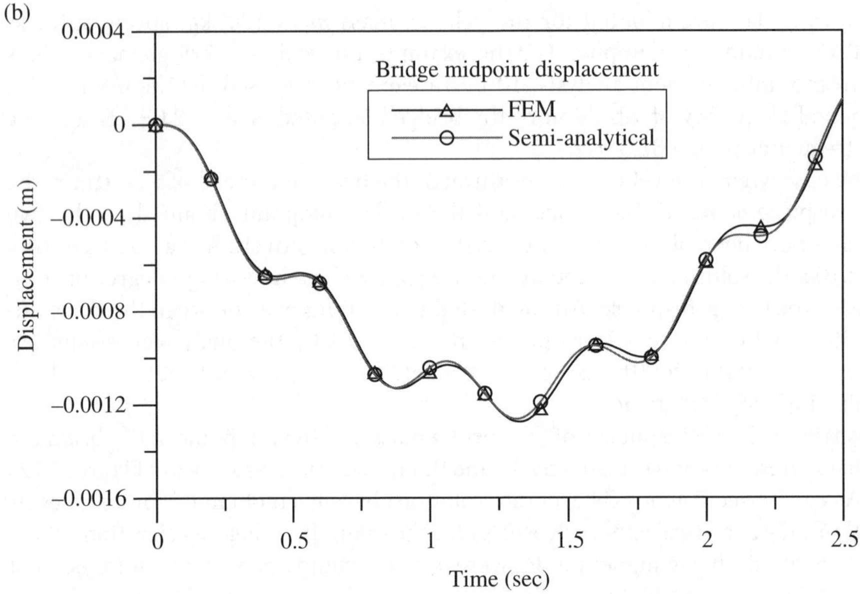


**Figure 2.10** Vertical displacement response of: (a) vehicle; (b) bridge midpoint ($v = 10\,\text{m/s}$).

## 2.7 Extraction of Fundamental Frequency of Bridge

In this section, we shall try to extract the fundamental frequency of the bridge from the time-history vertical vibration response of the vehicle obtained by the FEM, which is more "realistic" than the analytical one due to inclusion of the high-mode effect. Again,

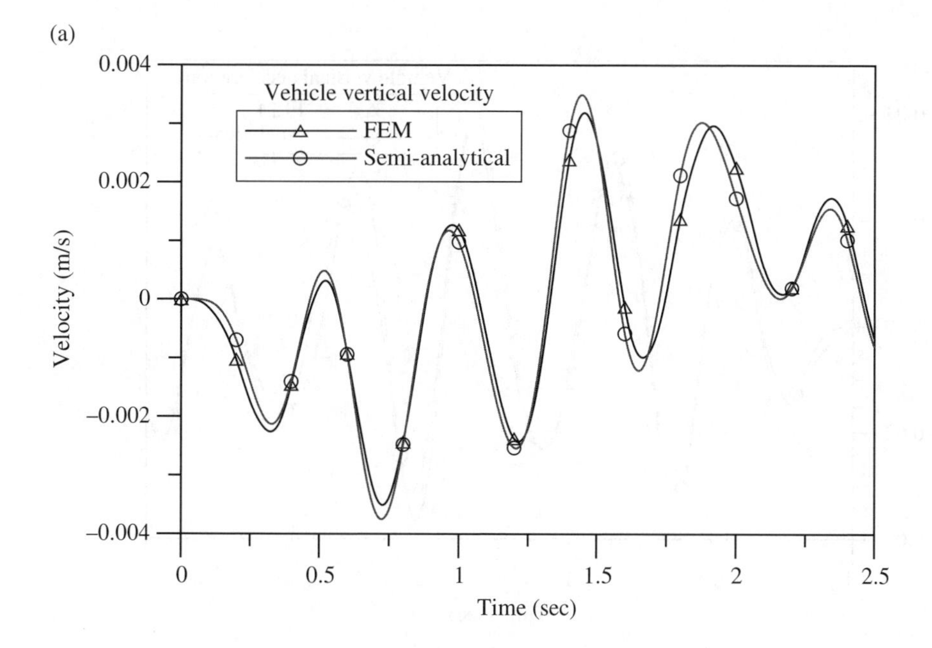


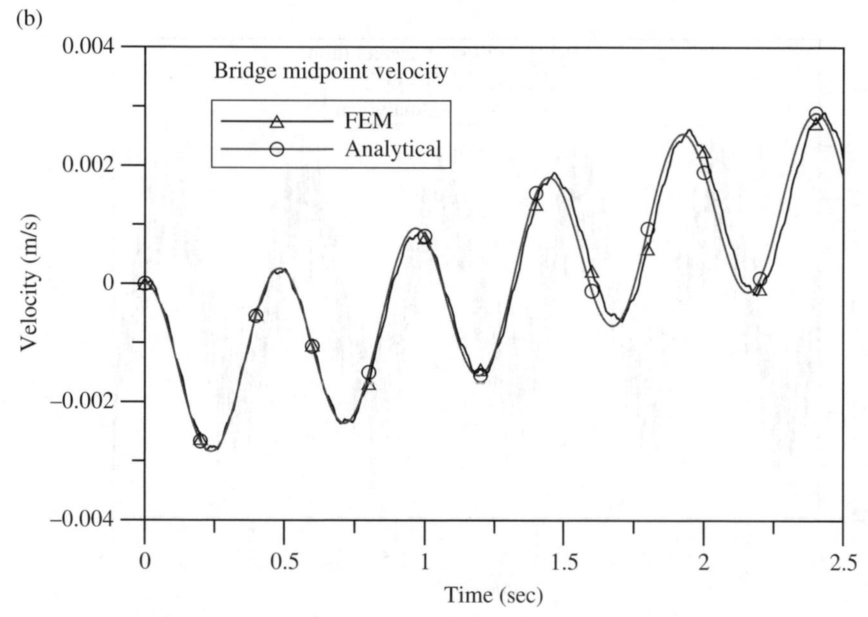


**Figure 2.11** Vertical velocity response of: (a) vehicle; (b) bridge midpoint ($v = 10\,\text{m/s}$).

the idea is to demonstrate that the bridge frequency can be successfully extracted from the "simulated" vehicle response, before we go for the field test to extract the bridge frequency from the "recorded" response of a moving vehicle (that can be modeled as a sprung mass) during its passage over a real bridge.

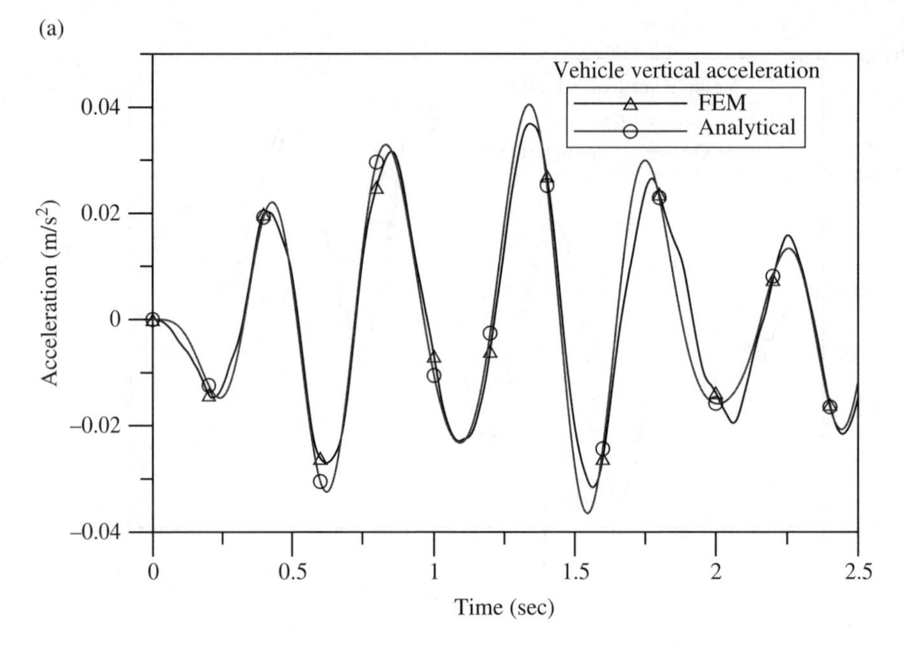


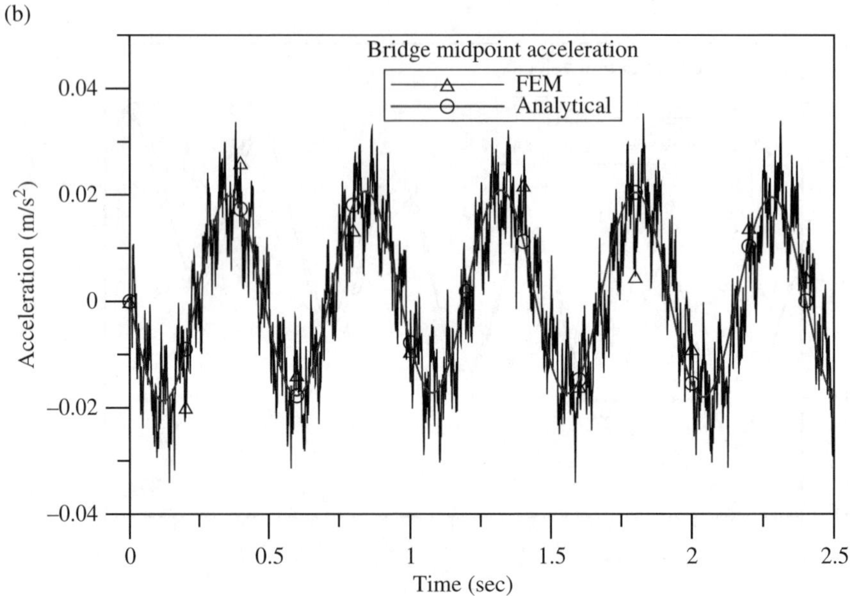


**Figure 2.12** Vertical acceleration response of: (a) vehicle; (b) bridge midpoint ($v = 10\,\text{m/s}$).

The frequency responses for the vertical acceleration of the vehicle and the bridge midpoint have been plotted in Figures 2.13a and b, respectively, in which the fundamental frequency of the bridge is indicated by a vertical dashed line. For convenience in engineering applications, the circular frequencies ($f$) in Hz (cycles/s) were used mostly in the spectral diagrams, which can be related to the natural frequency ($\omega$) used in the theoretical derivation by $\omega = 2\pi f$. Zero damping is assumed for the bridge. As can be

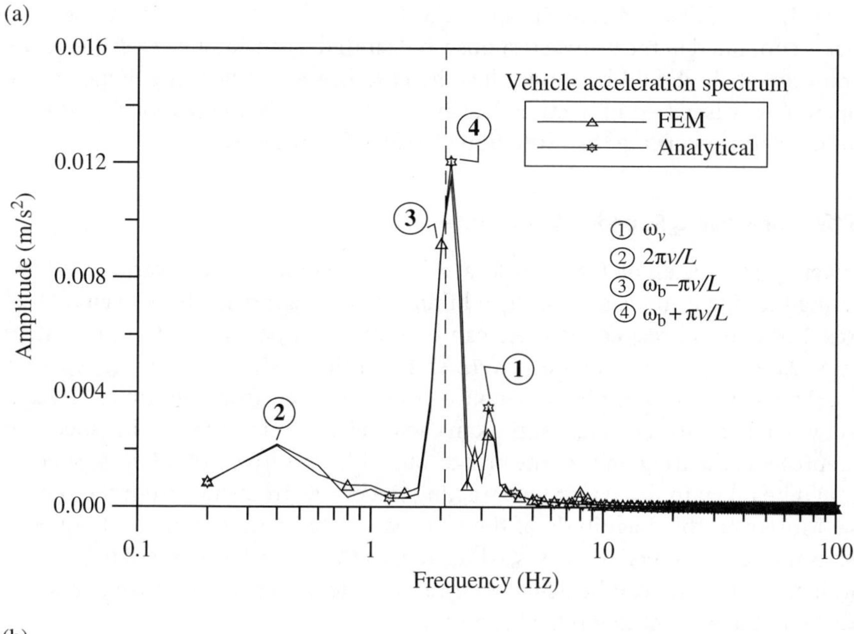


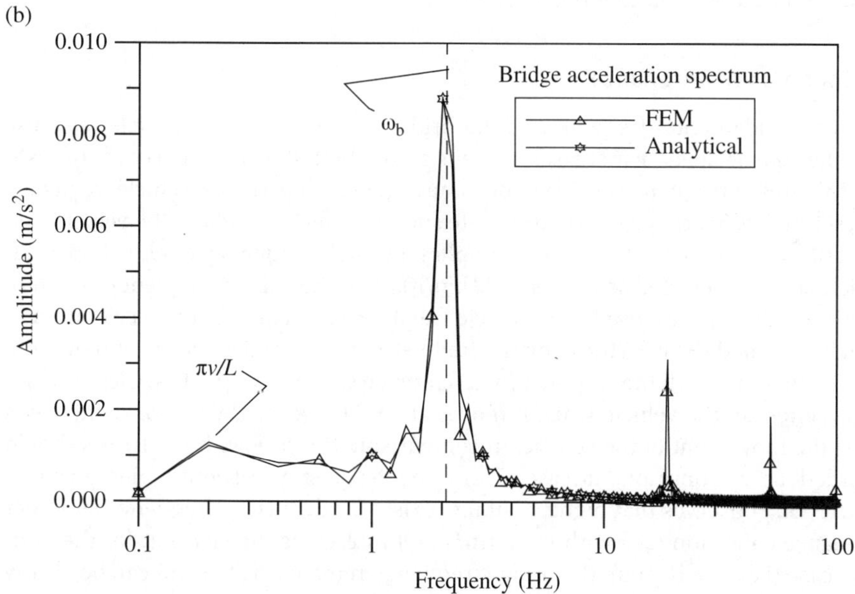


**Figure 2.13** Vertical acceleration spectrum of (a) vehicle; (b) bridge ($v = 10\,\text{m/s}$).

seen, both the single-mode and finite element approaches yield almost identical results, except that the high-mode frequency contents were missing from the bridge response. Of interest is the fact that half the driving frequency of the vehicle, $\pi v/L$ (= 0.2 Hz), can be observed as the lowest peak in the bridge spectrum shown in Figure 2.13b.

From the vehicle acceleration spectrum shown in Figure 2.13a, one observes that the four system frequencies dominating the vehicle response, i.e., $2\pi v/L$, $\omega_b - \pi v/L$, $\omega_b + \pi v/L$

and $\omega_v$, or 0.4, 1.88, 2.08, and 3.24 Hz, appear as local peaks. Evidently, the bridge frequency $\omega_b$ is contained in the vehicle response, but shifted by an amount $\pi v/L$ due to the vehicle movement. It should be noted that the case presented herein belongs to the nonresonant case discussed in Section 2.4. In the following, some parametric studies will be conducted concerning the extraction of bridge frequencies.

### 2.7.1 Effect of Moving Speed of the Vehicle

The frequency responses of the vehicle acceleration computed for various vehicle speeds using the single-mode approach and finite element approach have been plotted in Figures 2.14a and b, respectively. As can be seen from part (a), the three system frequencies $2\pi v/L$, $\omega_b - \pi v/L$, and $\omega_b + \pi v/L$ shift continuously as the vehicle speed $v$ increases. The same shifting phenomenon can be observed from the finite element results shown in part (b). The highest peak in each curve (i.e., for one specific speed) in part (b) represents the frequency of the bridge, but with a positive shift of $\pi v/L$. Such an effect should be taken into account in extracting the bridge frequency from the vehicle response. Moreover, the magnitude of the peak associated with the bridge frequency increases as the vehicle speed increases, which means that, from the point of field measurement, higher visibility can be achieved if the vehicle is allowed to move at a faster speed based on the present theoretical analysis.

### 2.7.2 Condition of Resonance

Because the fundamental frequency of the bridge contained in the vehicle response shifts as the vehicle speed $v$ increases, there is a possibility for the occurrence of resonance if the shifted frequency $\omega_b + \pi v/L$ becomes equal or close to the vehicle frequency $\omega_v$, as described in Section 2.4. To investigate such an effect, we adjust the vehicle suspension stiffness $k_v$ to 296 kN/m, which implies a vehicle frequency of $\omega_v$ = 2.50 Hz. If the vehicle speed is selected as 20 m/s (=72 km/h), then the shifted frequency $\omega_b + \pi v/L$ is 2.48 Hz, which is quite close to the vehicle frequency $\omega_v$. The vertical acceleration of the vehicle obtained for $v$ = 20 m/s by the finite element approach has been plotted in Figure 2.15, together with the response for a reference speed of $v$ = 10 m/s. Clearly, resonance is excited on the vehicle when it moves at $v$ = 20 m/s, as the response increases following the movement of the vehicle. In comparison, the case with $v$ = 10 m/s should be regarded as a nonresonant case. From the frequency response plot given in Figure 2.16, one observes that higher visibility exists for the bridge frequency $\omega_b$ under the resonance condition (i.e., with $v$ = 20 m/s). Of interest is the fact that for the nonresonant case (i.e., $v$ = 10 m/s), the bridge frequency remains visible and can be clearly identified. This is the case commonly encountered in practice.

The vertical response of the bridge midpoint under the same resonant speed $v$ = 20 m/s was plotted in Figure 2.17, together with that for the nonresonant speed $v$ = 10 m/s. Clearly, no resonance can be observed for the bridge at the speed $v$ = 20 m/s. Such a result is not surprising if one realizes that the vehicle (a smaller subsystem) serves to dissipate the kinetic energy of the bridge (a larger subsystem) under resonance, in a role similar to the tuned mass damper.

As a side note, if the first resonance condition mentioned in Section 2.4, i.e., $S$ = 1, is to be met for the present case, the vehicle has to move at a speed of $v$ = 104 m/s = 374 km/h,

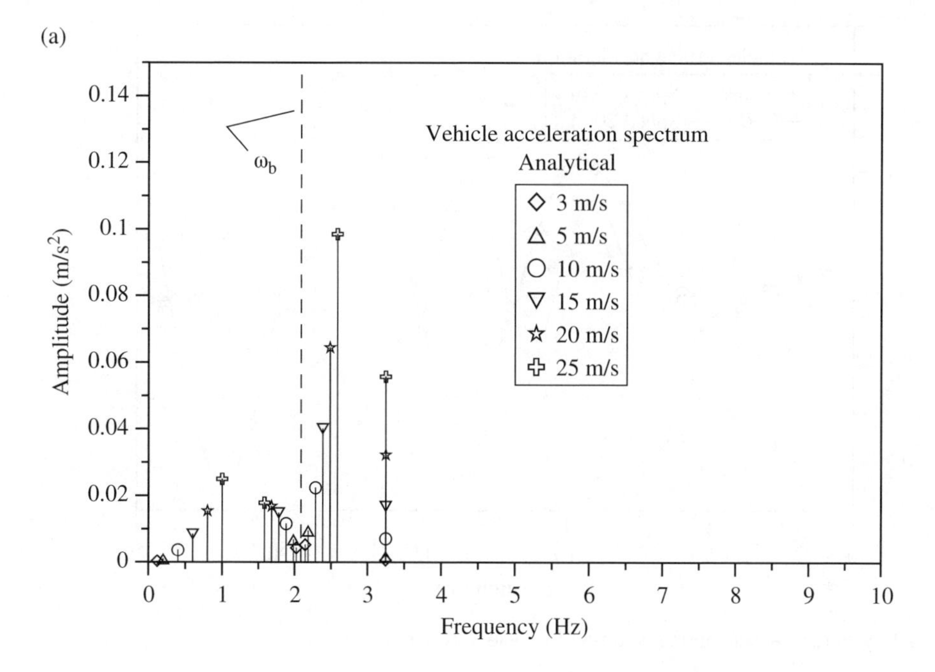


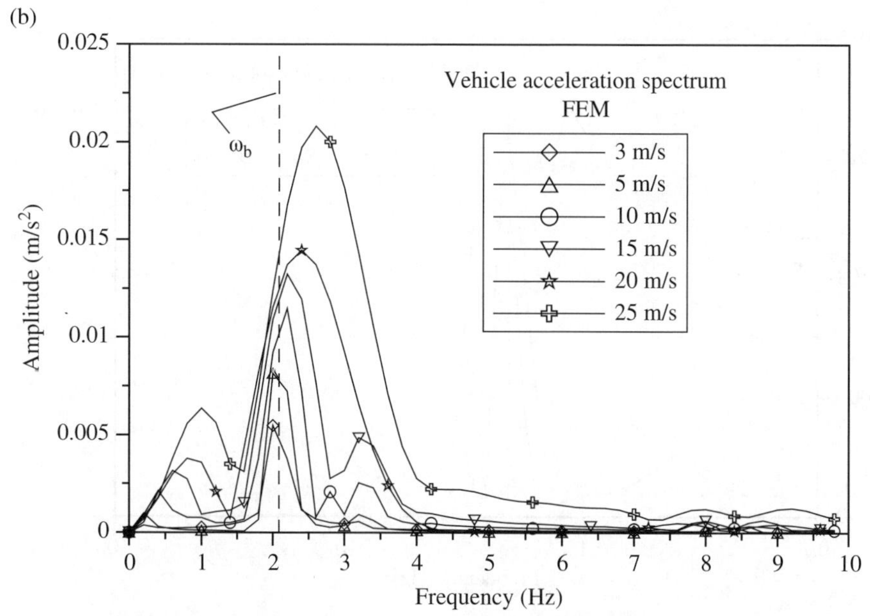


**Figure 2.14** Vertical acceleration spectrum of vehicle: (a) analytical, (b) finite element method (FEM).

which can hardly be reached in practice for a test vehicle in measurement. On the other hand, if use is to be made of the second resonance condition, i.e., $\omega_v^2 - (2\pi v/L)^2 = 0$ or $\omega_v - 2\pi v/L = 0$, then the vehicle has to move at a speed of $v = 62.5\,\text{m/s} = 225\,\text{km/h}$, which is still too high to be adopted in a field measurement.

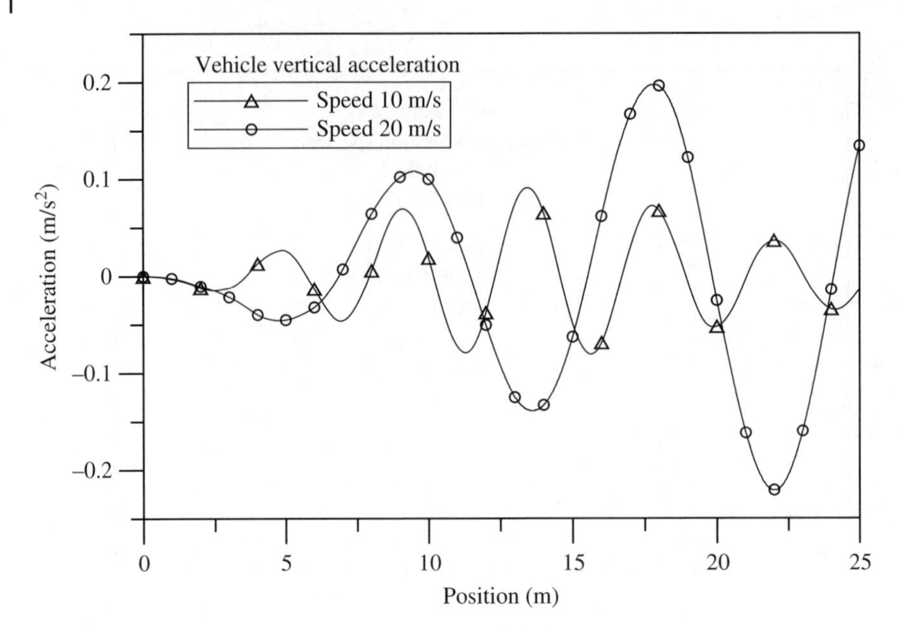


**Figure 2.15** Vertical acceleration response of vehicle (resonant case).

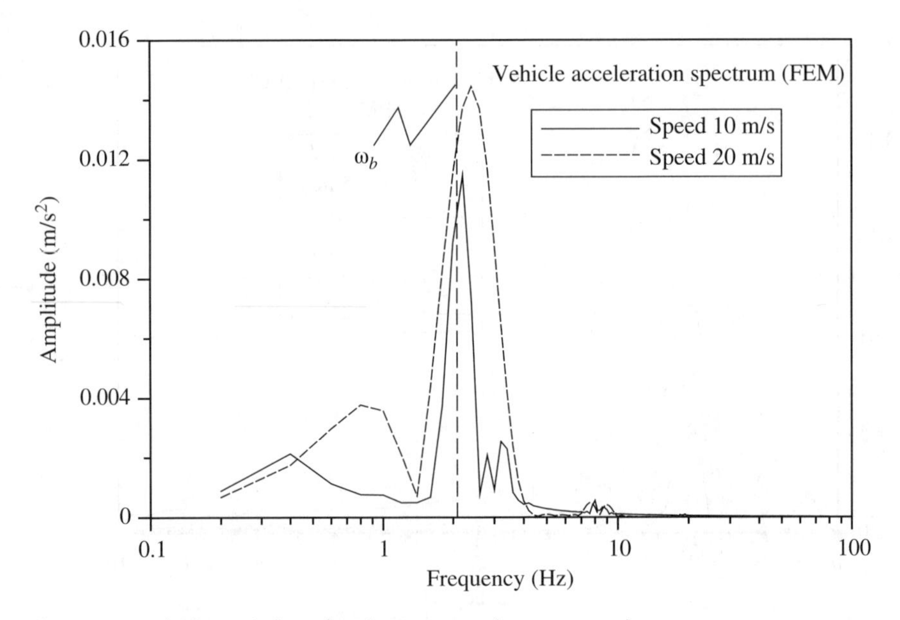


**Figure 2.16** Vehicle vertical acceleration spectrum (resonant case).

### 2.7.3 Effect of Damping of the Bridge

To investigate the effect of bridge damping on the vehicle response, three values of damping ratio are considered, 0%, 2%, and 5%. The other data are the same as those used in Section 2.6. From the result plotted in Figure 2.18, it is certain that the visibility of the bridge frequency decreases due to the presence of damping. Nevertheless, it can still be identified from the vehicle acceleration spectrum with no difficulty.

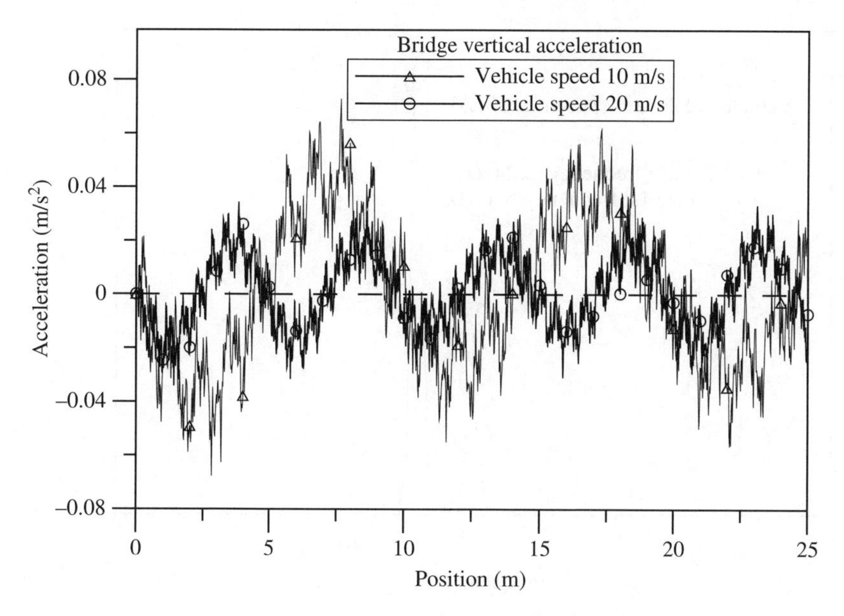


**Figure 2.17** Vertical acceleration response of bridge (resonant case).

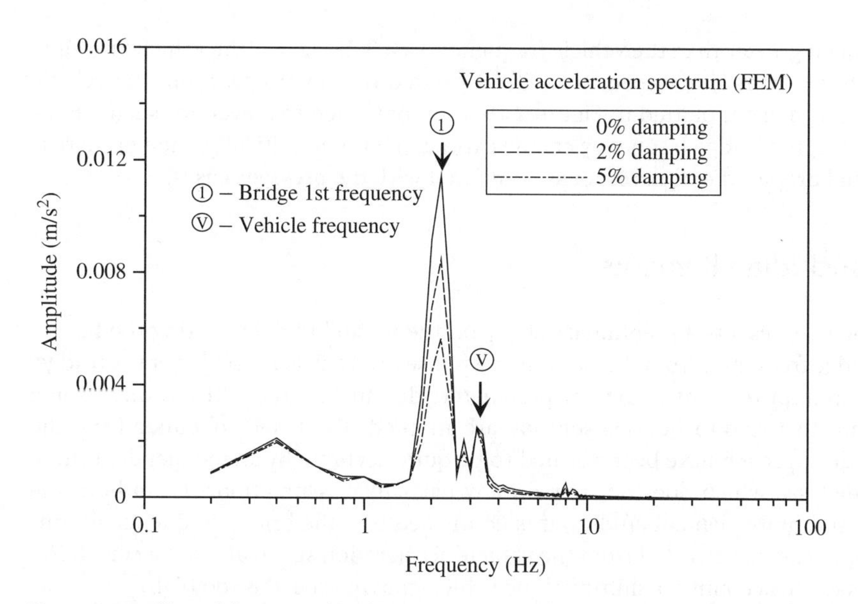


**Figure 2.18** Effect of damping of bridge.

### 2.7.4 Effect of a Vehicle Traveling over a Stiffer Bridge

In practice, a well-designed bridge is so arranged that it cannot be easily excited by vehicles traveling at "normal" speeds. This is why in some design codes, the fundamental frequency of the highway bridge is recommended to lie outside of the vehicle frequencies in the range of 2–5 Hz. To reflect such a situation, we have chosen to increase the stiffness of the bridge studied in Section 2.6 to a value of $EI$ = 11.25 GN.m$^2$, while keeping all the other data unchanged. The first frequency of the bridge is $\omega_b$ = 5.44 Hz,

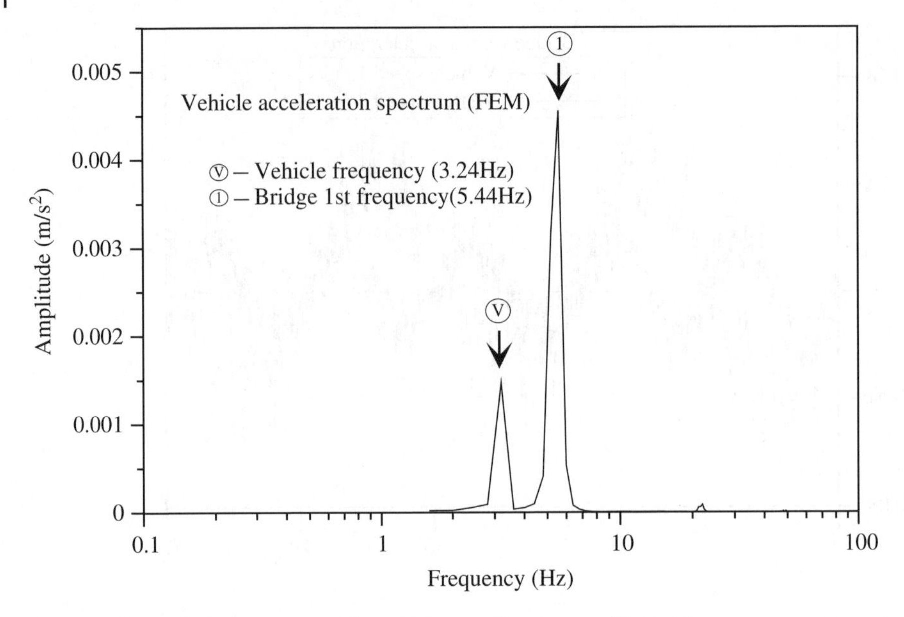


**Figure 2.19** Acceleration spectrum for vehicle traveling over a stiffer bridge.

which is much greater than the vehicle frequency $\omega_v = 3.25$ Hz. For the vehicle traveling with speed $v = 10$ m/s, no resonance will be excited on the bridge. From the vehicle acceleration spectrum plotted in Figure 2.19, it is confirmed that even for such a nonresonant case, the bridge frequency can be extracted with no difficulty, though there is a substantial drop in the peak response compared with the previous cases.

## 2.8 Concluding Remarks

This chapter represents a preliminary study on the feasibility of extracting the fundamental bridge frequency from the dynamic response of a vehicle passing over the bridge. As a first attempt to identify the key parameters dominating the VBI response, some assumptions that lead to a closed solution are adopted. The results obtained from the single-mode approach have been verified to be quite accurate by an independent finite element analysis, which does not rely on any particular assumptions. From both the analytical and finite element studies, it is confirmed that the bridge frequency is contained in and can be extracted from the vehicle acceleration spectrum, but a correction must be made to account for shifting if the vehicle moving speed is too high.

Further, higher vehicle speeds can result in higher amplitudes for the bridge frequency, which implied higher visibility and therefore is good for signal processing. The resonance condition for the moving vehicle to achieve the maximum response was discussed in detail, which, however, might not be easily met in practice. Noteworthy is the fact that the visibility of the bridge frequencies remains good even for the nonresonant case or in the presence of bridge damping. Future research should be conducted to address the factors not covered in this preliminary study, including, in particular, engine vibrations, pitching and rolling motions, damping and suspension mechanisms of the vehicle, pavement roughness, multi-span effect, multi-lane effect of the bridge, and existing traffic effect.

# 3

# Vehicle Scanning of Bridge Frequencies

## General Theory

This chapter differs from Chapter 2 in that all modes of vibration of the bridge are included in formulation of the general theory for the dynamic interaction between the moving vehicle and bridge. By assuming the vehicle/bridge mass ratio to be small, approximate but highly accurate closed-form solutions are obtained for the vertical responses of both the bridge and moving vehicle by the method of modal superposition. For both the bridge and vehicle responses, it is confirmed that rather accurate solutions can be obtained by considering only the first mode. The displacement, velocity, and acceleration of the bridge are governed at different extents by two frequency sets pertaining to the driving frequency of the vehicle and natural frequencies of the bridge. From the spectrum for the bridge displacement, the vehicle speeds can be shown to be associated with some low-frequency pikes. On the other hand, the vehicle responses are governed by five distinct frequencies that appear as driving frequencies, vehicle frequency, and bridge frequencies with shift. From the vehicle's acceleration spectrum, the first bridge frequency (with shift) is shown to have rather high visibility and can be easily identified. Potential applications of the present results are also indicated. The materials presented in this chapter are based primarily on the paper by Yang and Lin (2005) with necessary corrections on mathematical derivations and related results.

## 3.1 Introduction

The problem of a vehicle traveling over a bridge is commonly encountered in transportation facilities such as highway bridges, railroad bridges, aircraft/taxiway bridges in an airport, and so on. When a vehicle passes over a bridge, certain impact or dynamic amplification effect will be induced on the bridge, which needs to be taken into account in the design of the latter. Attention drawn to this subject dates back to the works of Willis (1849) and Stokes (1849) in the mid-nineteenth century. It has long been observed that, when a bridge is subjected to moving loads, the induced dynamic deflections and stresses can be significantly higher than those observed for the static case (Inglis 1934). In this aspect, the majority of the literature has been devoted to investigation of the bridge vibrations using the so-called *moving load* (Timoshenko 1922; Tan and Shore 1968; Fryba 1972; Olsson 1991; Yang et al. 1997), *moving mass* (Jeffcott 1929; Ting et al. 1974; Stanisic 1985; Sadiku and Leipholz 1987; Akin and

*Vehicle Scanning Method for Bridges*, First Edition. Yeong-Bin Yang, Judy P. Yang, Bin Zhang and Yuntian Wu.

Mofid 1989; Lee 1996; Foda and Abduljabbar 1998), and *moving sprung mass* models (Biggs 1964; Fryba 1972; Yang and Yau 1997; Pesterev et al. 2001; Yang and Wu 2001) for the vehicles. The following is a summary of the properties associated with each vehicle model (Yang et al. 2004b).

The *moving load* is the simplest model that can be conceived for a vehicle in studying the bridge vibrations. With this model, the essential dynamic characteristics of the bridge caused by the vehicle's moving action can be captured with a sufficient degree of accuracy. However, it suffers from the drawback that the interaction between the bridge and moving vehicle is ignored. For this reason, the moving load model is strictly valid only for the case where the mass of the vehicle is very small relative to that of the bridge, particularly when the interaction between the two subsystems can be neglected. The *moving mass* model represents an improvement over the moving load model in that effect of inertia of the vehicle is taken into account. Nevertheless, it does not allow consideration of the bouncing action of the moving vehicle relative to the bridge. Such an effect is expected to be significant in the presence of pavement roughness or for vehicles moving at rather high speeds.

To account for the bouncing or suspension action of the moving vehicle, a spring can be attached to the moving mass to result in the *sprung mass* model shown in Figure 3.1. This is the simplest model that can be used to study the *dynamic interaction* between the moving vehicle and the supporting bridge, the so-called vehicle-bridge interaction (VBI).

In the past two decades, researchers have developed vehicle models of various complexities to account for the dynamic properties of the vehicle; see, for instance, those listed in (Chu et al. 1986; Chatterjee et al. 1994; Fafard et al. 1998; Tan et al. 1998; Yang et al. 1999; Xia et al. 2000; Yang and Wu 2001). To the knowledge of the writers, however, a great majority of the previous related works focused on the dynamic response of bridges prior to the 1990s, with little attention paid to the dynamic behavior of the moving vehicle or to the frequency contents of the interaction between the two subsystems.

One exception is the series of works carried out by Yang and co-workers on the vertical response of moving vehicles, in which equal emphases have been placed on the bridge and vehicle responses, with the vehicle response used as the indicator for the riding comfort of passengers (Yang and Yau 1997; Yang et al. 1999; Yau et al. 1999; Yang and Wu 2001). Actually, it was based on more than 10 years of studies on the VBI problems that the idea of extracting the bridge frequencies from a moving passing vehicle was proposed by Yang et al. (2004a).

This chapter focuses exclusively on the *dynamic interaction*, especially the frequency aspects of such an interaction, between the vertical vibrations of the moving vehicle and

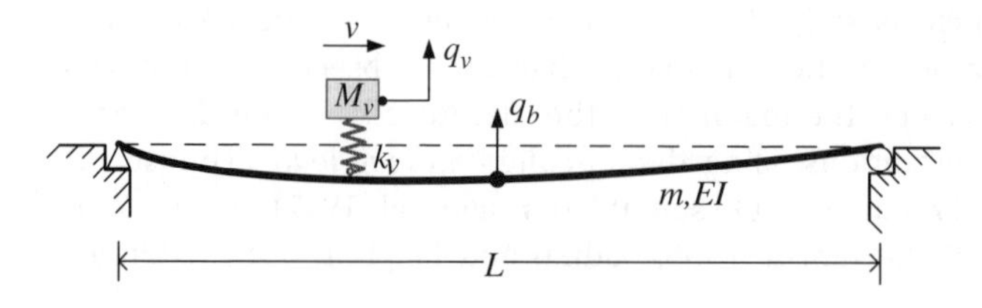


**Figure 3.1** Sprung mass moving over a simple beam.

the bridge, with equal emphases placed on both of them. To this end, two sets of second-order differential equations of motion are written, one for the moving vehicle and the other for the bridge. It is the interaction or contact force existing between the two subsystems that makes the two sets of equations nonlinear and coupled, as the contact force moves (and thereby varies) from time to time during their interaction, even though the two subsystems by themselves may be linear.

In order to highlight the effect of interaction, a simply supported beam subjected to a moving sprung mass is adopted. The mass of the vehicle is assumed to be small relative to that of the bridge. By the method of modal superposition, together with the use of convolution integrals, closed-form solutions will be obtained for both the bridge and the moving vehicle, which are approximate in the sense that iterations are not performed for updating the contact force existing between the two subsystems to consider the mutual interaction. The accuracy of the solutions obtained are compared with those obtained by independent finite element analyses. By transforming all the time-history responses into the frequency domain, some interesting phenomena can be observed. For example, from the displacement response of the bridge, one can identify the vehicle speeds as those appearing as low-frequency pikes. On the other hand, from the vehicle's acceleration spectrum, it is seen that the first bridge frequency can be identified with high visibility. Such implications are important and have their potential areas of applications. It is suggested that future research be carried out along the lines of interaction dynamics to investigate the practical aspects of the ideas presented herein.

## 3.2 Physical Modeling and Formulation

Consider the simply supported beam of length $L$ subjected to a vehicle moving at speed $v$ in Figure 3.1. The vehicle is represented as a concentrated mass $m_v$ supported by a spring of stiffness $k_v$ with the effect of damping of the suspension system neglected. The beam is assumed to be of the Bernoulli-Euler type with constant cross sections. No consideration will be made of the damping property or pavement irregularity of the bridge. Using a model as shown in Figure 3.1, one can derive the approximate, closed-form solutions for the VBI system considering all modes of vibration of the beam.

The equations of motion governing the transverse or vertical vibration of the beam and the moving vehicle are

$$\bar{m}\ddot{u} + EIu'''' = p(x,t) \tag{3.1}$$

$$m_v\ddot{q}_v + k_v q_v = k_v u\big|_{x=vt} \tag{3.2}$$

where $\bar{m}$ denotes the mass per unit length, E the elastic modulus, $I$ the moment of inertia of the cross section of the beam, $m_v$ and $k_v$ denote the mass of vehicle and stiffness of the suspension system, respectively, $q_v$ is the vertical displacement of the vehicle, $u(x, t)$ is the vertical displacement of the beam, and $p(x, t)$ is the applied force acting on the beam through the contact point with coordinate $vt$, which moves along with the vehicle. The applied force $p(x, t)$ on the beam can be expressed as follows:

$$p(x,t) = f_c(t)\cdot\delta(x - vt) \tag{3.3}$$

where $\delta(x - vt)$ is the Dirac delta function evaluated at the contact point, $x = vt$, and the *contact force* $f_c(t)$ is equal to the sum of the vehicle's weight and the elastic force of the suspension system, i.e.

$$f_c(t) = -m_v g + k_v (q_v - u|_{x=vt}) \tag{3.4}$$

with $g$ denoting the gravity of acceleration.

It should be noted that the vertical displacement of the vehicle is measured from its static equilibrium position. As such, there is no additional force acting on the vehicle mass except the elastic force resulting from the shortening or elongation of the supporting spring from the bottom. The solution to Eq. (3.1) will be expressed in terms of the modal shapes of the beam, $\phi_n(x)$, and associated modal coordinates, $q_{bn}(t)$, as

$$u(x,t) = \sum_n \phi_n(x) q_{bn}(t) \tag{3.5}$$

For the simply supported beam considered herein, the mode shapes of the beam are known to be of the sinusoidal type. Therefore, Eq. (3.5) becomes

$$u(x,t) = \sum_n \left[ \sin\frac{n\pi x}{L} \cdot q_{bn}(t) \right] \tag{3.6}$$

Substituting Eq. (3.6) for the displacement $u(x, t)$ into Eq. (3.1), multiplying both sides of the equation by $\phi_m(x)$, and integrating with respect to $x$ over the length $L$ of the beam, one obtains

$$\int_0^L \bar{m}\phi_m \sum_n (\phi_n \ddot{q}_{bn}) dx + \int_0^L EI\phi_m \sum_n (\phi_n q_{bn}) dx = \int_0^L f_c(t)\delta(x - vt)\phi_m dx \tag{3.7}$$

By use of the orthogonality conditions for the modal shapes and changing the subscript $m$ into $n$, Eq. (3.7) reduces to

$$\ddot{q}_{bn} + \omega_{bn}^2 q_{bn} = \frac{f_c(t) \cdot \int_0^L \delta(x - vt)\phi_n(x) dx}{\bar{m}\int_0^L \phi_n^2(x) dx} \tag{3.8}$$

where $\omega_{bn}$ is the frequency of vibration of the $n$th mode of the simple beam:

$$\omega_{bn} = \frac{n^2\pi^2}{L^2}\sqrt{\frac{EI}{\bar{m}}} \tag{3.9}$$

Further, through manipulation of the right-hand side of Eq. (3.8), the $n$th modal equation of the beam can be obtained as follows:

$$\begin{aligned} &\ddot{q}_{bn} + \omega_{bn}^2 q_{bn} + \frac{2\omega_v^2 m_v}{\bar{m}L} \cdot \sin\frac{n\pi vt}{L} \cdot \sum_j \left[ \sin\frac{j\pi vt}{L} \cdot q_{bj} \right] \\ &- \left[ \frac{2\omega_v^2 m_v}{\bar{m}L} \sin\frac{n\pi vt}{L} \right] q_v = \frac{-2m_v g}{\bar{m}L} \sin\frac{n\pi vt}{L} \end{aligned} \tag{3.10}$$

Similarly, Eq. (3.2) can be rewritten as

$$\ddot{q}_v + \omega_v^2 q_v = \omega_v^2 \sum_n \sin\frac{n\pi vt}{L} q_{bn} \tag{3.11}$$

where $\omega_v$ is the frequency of vibration of the vehicle,

$$\omega_v = \sqrt{\frac{k_v}{m_v}} \tag{3.12}$$

By the assumption that the mass $m_v$ of the vehicle is much less than the mass $\bar{m}L$ of the beam, i.e. $m_v/\bar{m}L \ll 1$, the governing equation in Eq. (3.10) can be approximated as follows:

$$\ddot{q}_{bn} + \omega_{bn}^2 q_{bn} = \frac{-2m_v g}{\bar{m}L} \cdot \sin\frac{n\pi vt}{L} \tag{3.13}$$

which is identical to the one for a beam subjected to a single moving load at constant speed $v$ (Yang and Lin 1995).

## 3.3 Dynamic Response of the Beam

For zero initial conditions, the solution to the second-order linear differential equation of the beam in Eq. (3.13) can be obtained as

$$q_{bn}(t) = \frac{\Delta_{stn}}{1 - S_n^2}\left[\sin\left(\frac{n\pi vt}{L}\right) - S_n \cdot \sin(\omega_{bn} t)\right] \tag{3.14}$$

where $\Delta_{stn}$ is the static deflection of the $n$th mode of the beam caused by the moving vehicle:

$$\Delta_{stn} = \frac{-2m_v g L^3}{n^4 \pi^4 EI} \tag{3.15}$$

and $S_n$ is a nondimensional speed parameter:

$$S_n = \frac{n\pi v}{L\omega_{bn}} \tag{3.16}$$

Hence, the total displacement response of the beam to the vehicle moving at speed $v$ is

$$u(x,t) = \sum_n \frac{\Delta_{stn}}{1 - S_n^2}\left\{\sin\frac{n\pi x}{L} \cdot \left[\sin\frac{n\pi vt}{L} - S_n \cdot \sin\omega_{bn} t\right]\right\} \tag{3.17}$$

Differentiating Eq. (3.17) with respect to time $t$, we obtain the velocity response of the beam as follows:

$$\dot{u}(x,t) = \sum_n \frac{\Delta_{stn}}{1 - S_n^2}\left\{\sin\frac{n\pi x}{L}\left[\left(\frac{n\pi v}{L}\right) \cdot \cos\frac{n\pi vt}{L} - (\omega_{bn} S_n)\cos\omega_{bn} t\right]\right\} \tag{3.18}$$

which can further be differentiated to yield the acceleration response of the beam as

$$\ddot{u}(x,t)=\sum_n \frac{\Delta_{stn}}{1-S_n^2}\left\{\sin\frac{n\pi x}{L}\left[\left(\omega_{bn}^2 S_n\right)\cdot\sin\omega_{bn}t-\left(\frac{n\pi v}{L}\right)^2\sin\frac{n\pi vt}{L}\right]\right\} \tag{3.19}$$

As can be seen, each of the responses of the beam given in Eqs. (3.17)–(3.19) can be divided into two components pertaining to the *driving frequencies* of the vehicle, $n\pi v/L$, and *natural frequencies* of the bridge, $\omega_{bn}$. By substitution of $x = L/2$ into Eqs. (3.17)–(3.19), one obtains the mid-point responses of the beam as

$$u\left(L/2,t\right)=\sum_n \frac{\Delta_{stn}}{\left(1-S_n^2\right)}\sin\frac{n\pi}{2}\cdot\left[\sin\frac{n\pi vt}{L}-S_n\cdot\sin\omega_{bn}t\right] \tag{3.20}$$

$$\dot{u}\left(L/2,t\right)=\sum_n \frac{\Delta_{stn}}{1-S_n^2}\sin\frac{n\pi}{2}\left[\frac{n\pi v}{L}\cdot\cos\frac{n\pi vt}{L}-\frac{n\pi v}{L}\cos\omega_{bn}t\right] \tag{3.21}$$

$$\ddot{u}\left(L/2,t\right)=\sum_n \frac{\Delta_{stn}}{1-S_n^2}\sin\frac{n\pi}{2}\left[\frac{n\pi v\omega_{bn}}{L}\sin\omega_{bn}t-\left(\frac{n\pi v}{L}\right)^2\cdot\sin\frac{n\pi vt}{L}\right] \tag{3.22}$$

Using the definition in Eq. (3.15) for $\Delta_{st1}$, the coefficients of the two components mentioned above for the responses of the beam pertaining to the driving frequencies and natural frequencies have been listed in Table 3.1. Because the midpoint of the beam happens to be a stationary point for the modal shapes of even order, the amplitudes corresponding to all the even-order modes are merely equal to zero. Thus, only the odd-order modes need to be considered in Table 3.1.

The contributions of the two frequency components to the midpoint displacement, velocity and acceleration of the beam were also plotted in Figures 3.2a–c for the case of $S_1 = 0.1$ and $\omega_{b1} = 6\pi$, which means that the time needed for the vehicle to pass through the beam is 10 times the fundamental period of the beam. As can be seen, the amplitudes of each individual mode pertaining to the two frequency components decrease

**Table 3.1** Coefficients of frequency terms in beam responses.

| | **Driving frequency** $\frac{n\pi v}{L}$ | **Natural frequency** $\omega_{bn}$ |
|---|---|---|
| Displacement | $\frac{\Delta_{st1}}{n^4\left(1-\frac{S_1^2}{n^2}\right)}$ | $\frac{\Delta_{st1}S_1}{n^5\left(1-\frac{S_1^2}{n^2}\right)}$ |
| Velocity | $\frac{\Delta_{st1}S_1}{n^3\left(1-\frac{S_1^2}{n^2}\right)}\cdot\omega_{b1}$ | $\frac{\Delta_{st1}S_1}{n^3\left(1-\frac{S_1^2}{n^2}\right)}\cdot\omega_{b1}$ |
| Acceleration | $\frac{\Delta_{st1}S_1^2}{n^2\left(1-\frac{S_1^2}{n^2}\right)}\cdot\omega_{b1}^2$ | $\frac{\Delta_{st1}S_1}{n\left(1-\frac{S_1^2}{n^2}\right)}\cdot\omega_{b1}^2$ |

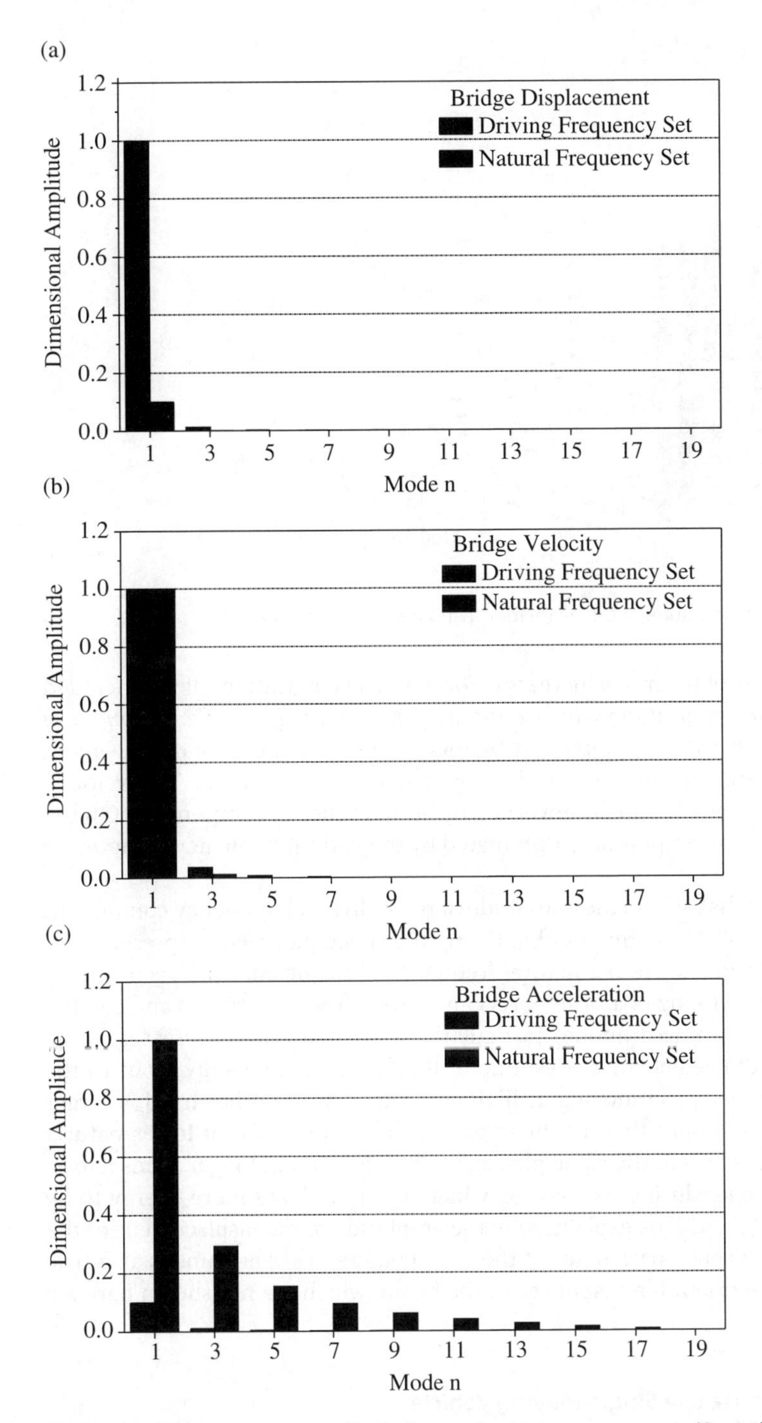


**Figure 3.2** Bridge response amplitude for two frequency components ($S = 0.1$) (a) displacement, (b) velocity, (c) acceleration.

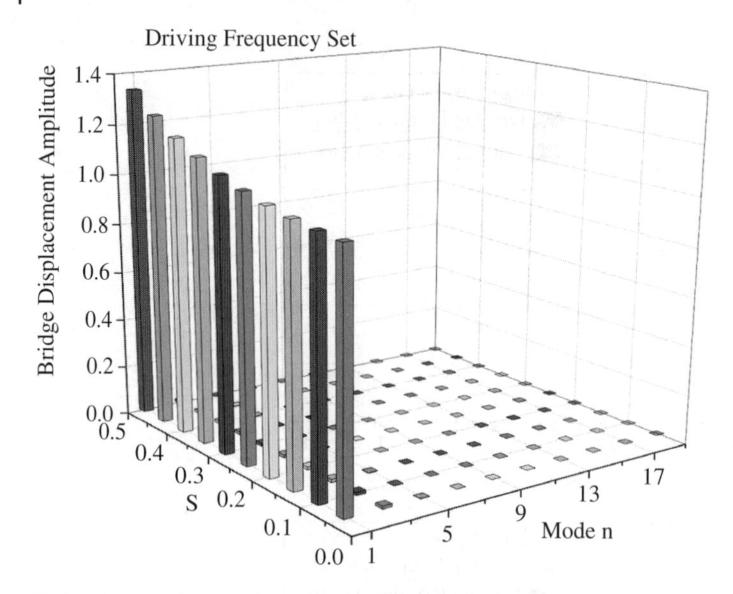


**Figure 3.3** Displacement amplitude of the beam for driving frequency component.

drastically as the order of the mode increases. For practical applications, it is concluded that *solutions of sufficient accuracy can be obtained for the beam if only the first mode of vibration is considered*, as was pointed out by Biggs (1964). Besides, the displacement amplitude is dominated mainly by the driving frequency component. The velocity amplitude, however, is nearly equally governed by both frequency components. On the contrary, the acceleration amplitude is dominated by the natural frequency component of the beam.

Another parameter that affects the amplitude of the individual frequency component is the speed parameter $S$. As defined in Eq. (3.16), the speed parameter represents the ratio of the driving frequency to the natural frequency of the beam, which is normally less than 0.5 in practice (Yang et al. 1995). As can be seen from Figures 3.3 and 3.4, the displacement amplitude of each mode, particularly of the first mode, increases as the speed parameter of the vehicle increases, due to the fact that the energy input to the beam is higher for the vehicle moving at higher speeds. On the other hand, a larger speed parameter also implies that a vehicle passes through a beam of lower natural frequencies or longer spans at the same physical speed. A beam of longer spans allows the vehicle to stay longer during its passage, which, in turn, allows more energy to be accumulated on the beam. This explains why the amplitude of the displacement of the beam increases as the speed parameter of the vehicle increases. The same is also true for the velocity and acceleration responses of the beam, which are not shown here for brevity.

### 3.3.1 Beam's Response to a Single Moving Vehicle

Figure 3.5 shows the midpoint acceleration response of the beam to a vehicle of $m_v = 1200\,\text{kg}$ and $k_v = 500\,000\,\text{N/m}$ moving at three different speeds, 5, 10 and 20 m/s. The following properties are adopted for the beam: $L = 25\,\text{m}$, $\bar{m} = 4800\,\text{kg/m}$, and

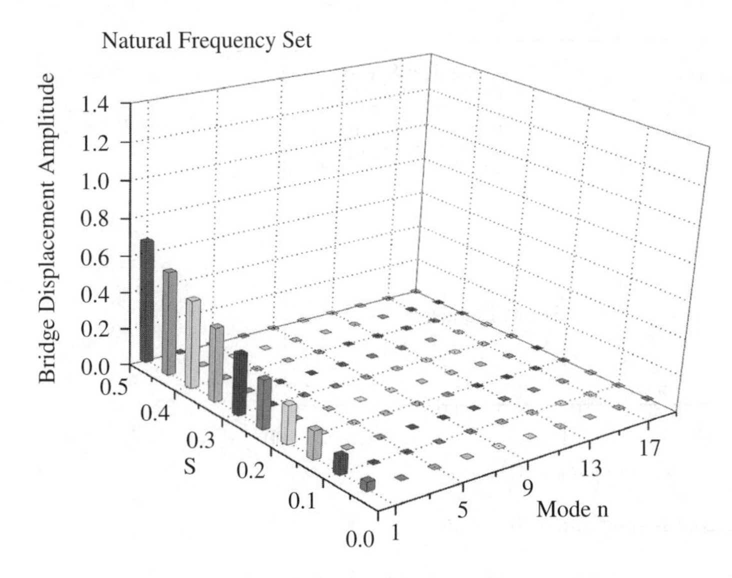


**Figure 3.4** Displacement amplitude of the beam for natural frequency component.

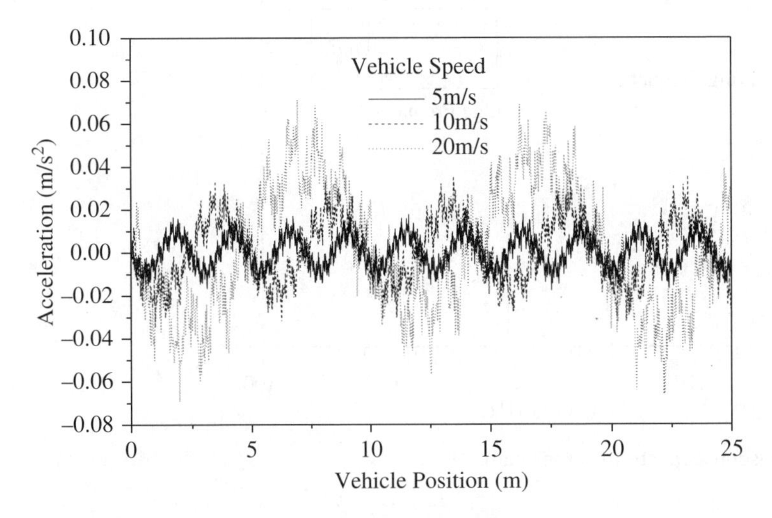


**Figure 3.5** Midspan acceleration response of beam.

$EI = 3.3 \times 10^9$ N-m$^2$. As was already mentioned, the displacement response of a beam is dominated mainly by the driving frequency component, and the velocity response is equally dominated by both the driving frequency and frequency components of the beam, but the acceleration response is dominated by the frequency component of the beam. Such a fact is confirmed by the spectra plotted in Figures 3.6–3.8, respectively, for the displacement, velocity, and acceleration of the midpoint of the beam. Clearly, the influence of the driving frequency component relative to the frequency component of the beam *decreases* as we go from displacement, to velocity, and then to acceleration of the beam.

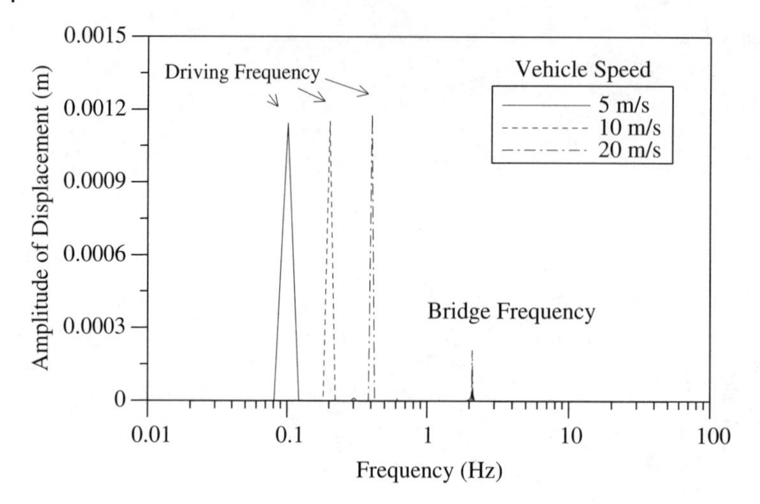


**Figure 3.6** Displacement response spectrum of the beam.

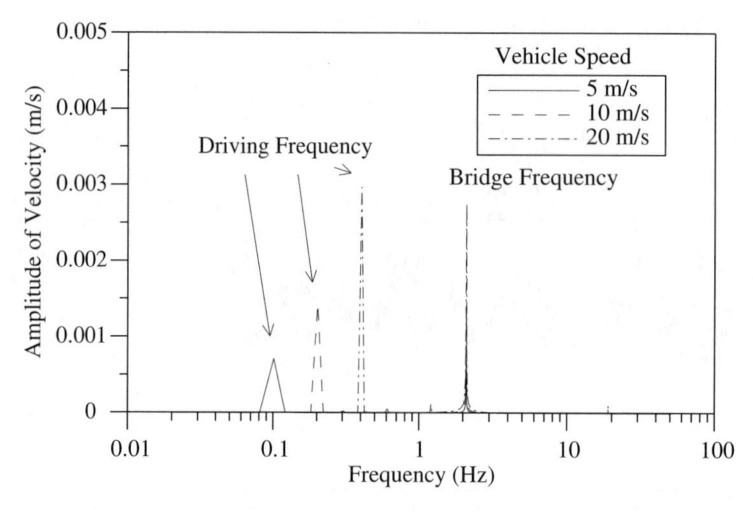


**Figure 3.7** Velocity response spectrum of the beam.

It is interesting to note that the three driving frequencies ($\pi v/L$) identified from Figures 3.6–3.8 are 0.1, 0.2, and 0.4 Hz, which correspond exactly to the three speeds 5, 10, 20 m/s of the moving vehicle. The three driving frequencies all fall below 1 Hz, much less than the frequency of the beam. Besides, higher response amplitudes are generated by vehicles moving at higher speeds, as indicated by the velocity and acceleration spectrum in Figures 3.7 and 3.8, respectively. One implication herein is that if the dynamic response of a beam under moving loads can be accurately monitored, the recorded data, when processed in real time, can be used for detecting the moving speeds of vehicles over the beam, at least from a theoretical point of view. The feasibility of such an idea may be of interest to the highway patrol officers involved in chasing

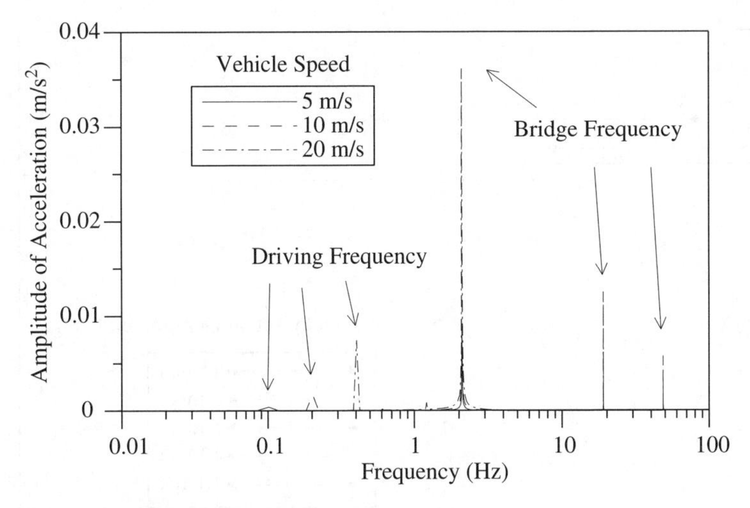


**Figure 3.8** Acceleration response spectrum of the beam.

speeding vehicles. It is suggested that further research be conducted in this regard concerning the technical aspects.

### 3.3.2 Beam's Response to Five Moving Vehicles

The same properties as those used in Section 3.3.1 are adopted for the beam and vehicle. In order to examine the effect of multivehicle loadings, however, we shall consider the case of five identical vehicles moving over the beam at five different speeds, 15, 8, 5, 12, and 10 m/s, with different entrance times. The solution to such a multivehicle case can be obtained by superposing the results obtained for each individual vehicle, with due account taken of the effect of time lag and different initial conditions (Yang et al. 1997). The displacement response obtained for the midpoint of the beam is plotted in Figure 3.9, along with the acting duration of each vehicle on the beam indicated.

Correspondingly, the spectrum of the displacement, velocity, and acceleration of the beam's midpoint is shown in Figures 3.10–3.12, respectively. From these figures, one observes that the five driving frequencies implied by the five vehicle speeds can be clearly identified from the displacement spectrum, not so clearly from the velocity spectrum, but almost invisible in the acceleration spectrum. Such an observation is consistent with the statement that the driving frequency component loses its influence as we go from displacement, to velocity, and then to acceleration of the beam. Therefore, if the speed of a moving vehicle is to be *detected* from the dynamic response of the supporting bridge, it is suggested that we work mainly on the displacement or velocity response. In this regard, further works should be conducted to develop techniques that are suitable for monitoring the displacement or velocity, rather than the acceleration response, of the bridge in the low-frequency range, as most vibration sensors available nowadays are of the acceleration or velocity type and are good for high-frequency vibrations.

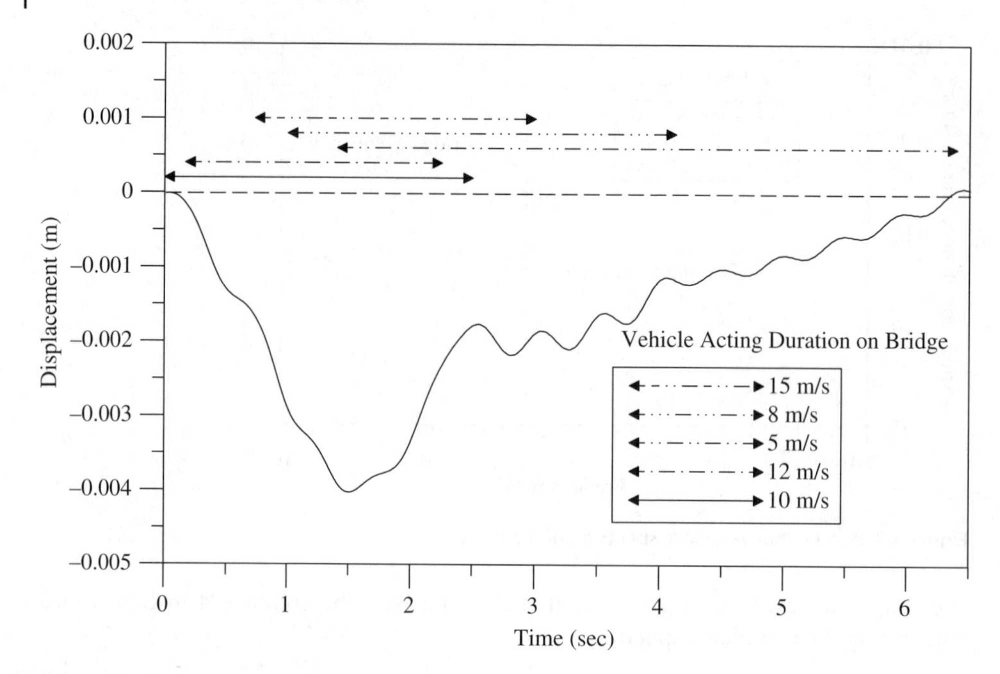


**Figure 3.9** Midspan displacement response of beam to five moving vehicles.

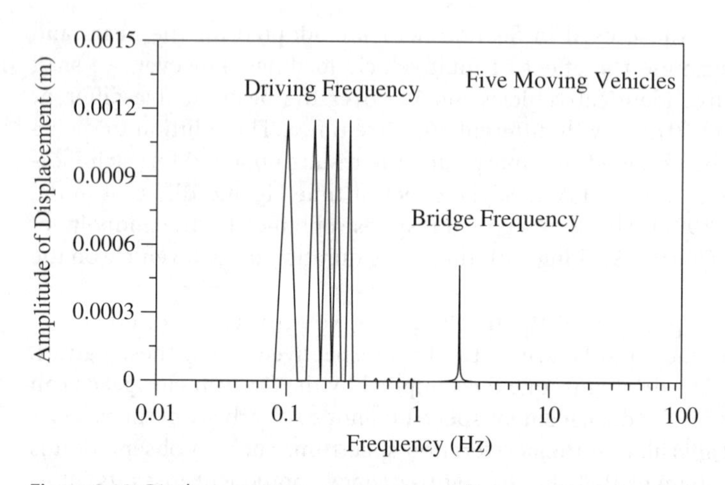


**Figure 3.10** Displacement response spectrum of beam subjected to five moving vehicles.

## 3.4 Dynamic Response of the Moving Vehicle

Substituting the expression in Eq. (3.14) for the displacement of the beam, $q_{bn}(t)$, into the right-hand side of Eq. (3.11), the equation of motion for the vehicle can be written as

$$\ddot{q}_v + \omega_v^2 q_v = \sum_{n=1}^{\infty} \frac{\Delta_{stn}\omega_v^2}{1 - S_n^2} \left\{ \sin\left(\frac{n\pi vt}{L}\right) \cdot \left[ \sin\left(\frac{n\pi vt}{L}\right) - S_n \cdot \sin\left(\omega_{bn} t\right) \right] \right\} \tag{3.23}$$

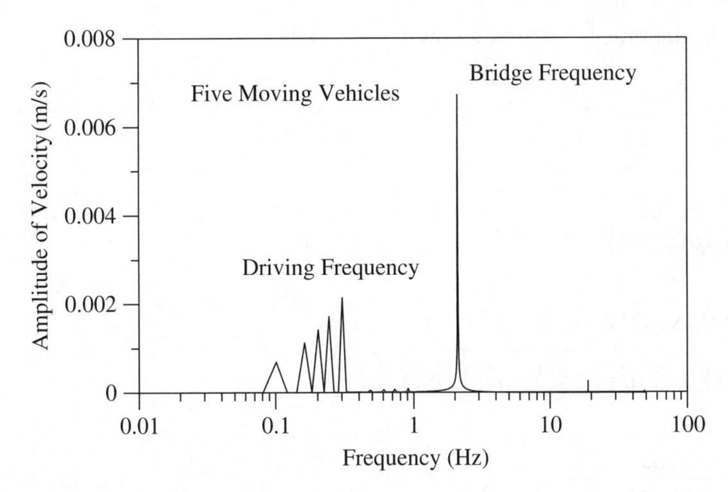


**Figure 3.11** Velocity response spectrum of beam subjected to five moving vehicles.

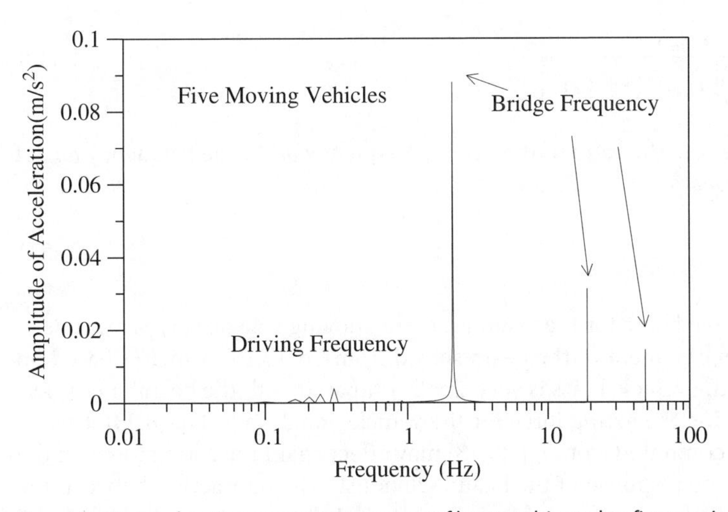


**Figure 3.12** Acceleration response spectrum of beam subjected to five moving vehicles.

The term denoted as $g(t)$ on the right-hand side of Eq. (3.24) represents the interaction effect between the bridge and moving vehicle, which is a function of time. The solution to Eq. (3.23) can be obtained by the Duhamel integral as

$$q_v(t) = \frac{1}{\omega_v} \int_0^t g(t) \cdot \sin \omega_v (t - \tau) d\tau \tag{3.24}$$

Using the expression for $g(t)$ in Eq. (3.23), the displacement response of the vehicle can be integrated as

$$\begin{aligned} q_v(t) = \sum_{n=1}^{\infty} \Bigg[ & A_{1n} + A_{2n} \cos\left(\frac{2n\pi v}{L}\right)t + A_{3n} \cos(\omega_v t) \\ & + A_{4n} \cos(\omega_{bn} - n\pi v/L)t + A_{5n} \cos(\omega_{bn} + n\pi v/L)t \Bigg] \end{aligned} \tag{3.25}$$

where the coefficients in the brackets are as follows:

$$A_{1n} = \frac{\Delta_{stn}}{2\left(1 - S_n^2\right)} \tag{3.26}$$

$$A_{2n} = \frac{\Delta_{stn}}{2\left(1 - S_n^2\right)\left(4\mu_n^2 S_n^2 - 1\right)} \tag{3.27}$$

$$A_{3n} = \frac{2\Delta_{stn} S_n^2 \mu_n^4 \left(2 + \mu_n^2 S_n^2 - \mu_n^2\right)}{\left(4\mu_n^2 S_n^2 - 1\right)\left[1 - 2\mu_n^2\left(1 + S_n^2\right) + \mu_n^4\left(1 - S_n^2\right)^2\right]} \tag{3.28}$$

$$A_{4n} = -\frac{\Delta_{stn} S_n}{2\left(1 - S_n^2\right)\left[1 - \mu_n^2\left(1 - S_n\right)^2\right]} \tag{3.29}$$

$$A_{5n} = \frac{\Delta_{stn} S_n}{2\left(1 - S_n^2\right)\left[1 - \mu_n^2\left(1 + S_n\right)^2\right]} \tag{3.30}$$

We shall use $\mu_n$ to denote the ratio of the vehicle frequency $\omega_v$ to the frequency $\omega_{bn}$ of the $n$th mode of the beam:

$$u_n = \frac{\omega_{bn}}{\omega_v} \tag{3.31}$$

The solution as derived here for the response of the moving vehicle is *approximate* in the sense that the displacement of the beam solved from Eq. (3.13), which is based on the assumption that the vehicle mass is very small compared with the beam's mass, has been used in computing the acting force for the vehicle equation in Eq. (3.23). In fact, the vehicle response computed from Eq. (3.23) may affect again the acting force on the beam, and therefore the response of the beam. Obviously, the interaction between the beam and moving vehicle is an issue of iterative nature, but for our purposes, only the first cycle of iteration was considered.

In the numerical study, it will be demonstrated that even with such an approximation, the closed-form solutions so derived appear to be quite accurate compared with the finite element solution that takes into account the full effect of interaction. Compared with the numerical approaches, the present approximate approach has the advantage that the *key parameters* involved in each phenomenon can be clearly identified and physically interpreted. It should be added that the solution presented in Eq. (3.25) reduces to the one presented in Chapter 2, where only the first mode of vibration of the beam is considered.

As can be seen from Eq. (3.25), the number of terms involved in the displacement response of the vehicle increases as there are more vibration modes of the beam considered. Five types of frequencies can be identified for the vehicle response, which can further be categorized into three groups as: (i) driving frequencies $n\pi v/L$; (ii) vehicle frequency $\omega_v$; and (iii) bridge-related frequencies, $\omega_{bn} - n\pi v/L$ and $\omega_{bn} + n\pi v/L$, where the index $n$ denotes the sequential number of mode of vibration of the beam. In particular, the third group reveals the fact that the frequencies $\omega_{bn}$ of the beam are shifted by an amount related to the driving frequency of the vehicle, $\pm n\pi v/L$, due to the Doppler effect.

We are particularly interested in the terms containing the frequencies of the beam with shift, i.e. $\omega_{bn} - n\pi v/L$ and $\omega_{bn} + n\pi v/L$, in Eq. (3.25), as they provide clues for *extracting the bridge frequencies* from the dynamic (displacement) response of a passing vehicle, at least from the theoretical point of view. Before this can be realized, however, we would like to have some idea about the relative influence of the terms containing $\omega_{bn} - n\pi v/L$ and $\omega_{bn} + n\pi v/L$ in the displacement response of the vehicle in Eq. (3.25) with respect to the remaining terms, and particularly to see if they are *practically visible* from the spectrum of the associated response.

Table 3.2 lists the magnitudes of all the coefficients $A_{1n}, A_{2n}, \ldots, A_{5n}$. It can be observed that the coefficient $A_{5n}$, i.e., the term associated with the frequency $\omega_{bn} + n\pi v/L$, remains generally the largest for the practical ranges of $\mu$ and $S$ considered, for $0 \leqq \mu \leqq 5$ and for $0 \leqq S \leqq 0.5$. Such a fact offers us a theoretical means to *extract* the bridge frequencies from the displacement response of the vehicle, if due account is taken of the shifting effect or if the vehicle speed is not too large. Furthermore, the magnitude for the second mode of the beam is about one quarter that of the first mode, and even smaller for higher modes, indicating that modes higher than the first one can generally be neglected in computing the displacement response of the vehicle, as will be demonstrated later on. In other words, we may not be able to extract the second and higher frequencies of the beam in a manner as easily as the first frequency from the displacement response of the moving vehicle. The above conclusion may need to be adjusted if the velocity and acceleration responses of the moving vehicle, rather than the displacement response, are to be used for extracting the bridge frequencies, since they are dominated by medium and higher frequencies.

By setting the denominator for the coefficient $A_{5n}$ equal to 0 for $n$ = 1, one can obtain the condition for the *resonance* to occur on the vehicle, as indicated by the vehicle-bridge frequency ratio $\mu$, which turns out to be equal to 1 for small speed parameter $S$, but slightly less than 1 for large speed parameter $S$. Under the condition of resonance,

**Table 3.2** Amplitude of vehicle acceleration response.

| Frequency | Amplitude | First Mode $n = 1$ | Second Mode $n = 2$ |
|---|---|---|---|
| $\frac{2n\pi v}{L}$ | $\frac{\Delta_{stn}}{2\left(1-S_n^2\right)\left(4\mu_n^2 S_n^2-1\right)}$ | $\frac{\Delta_{st1}}{2\left(1-S_1^2\right)\left(4\mu_1^2 S_1^2-1\right)}$ | $\frac{\Delta_{st1}}{16\left(1-S_1^2\right)\left(16\mu_1^2 S_1^2-1\right)}$ |
| $\omega_{bn}-\frac{n\pi v}{L}$ | $\frac{\Delta_{stn} S_n}{2\left(1-S_n^2\right)\left[1-\mu_n^2\left(1-S_n\right)^2\right]}$ | $\frac{\Delta_{st1} S_1}{2\left(1-S_1^2\right)\left[1-\mu_1^2\left(1-S_1\right)^2\right]}$ | $\frac{\Delta_{st1} S_1}{64\left(1-\frac{S_1^2}{4}\right)\left[1-16\mu_1^2\left(1-\frac{S_1}{2}\right)^2\right]}$ |
| $\omega_{bn}+\frac{n\pi v}{L}$ | $\frac{\Delta_{stn} S_n}{2\left(1-S_n^2\right)\left[1-\mu_n^2\left(1+S_n\right)^2\right]}$ | $\frac{\Delta_{st1} S_1}{2\left(1-S_1^2\right)\left[1-\mu_1^2\left(1+S_1\right)^2\right]}$ | $\frac{\Delta_{st1} S_1}{64\left(1-\frac{S_1^2}{4}\right)\left[1-16\mu_1^2\left(1+\frac{S_1}{2}\right)^2\right]}$ |

the coefficient $A_{5n}$ reaches a local maximum. The same is also true for the coefficient $A_{4n}$, except that the vehicle-bridge frequency ratio $\mu$ is equal to 1 for small speed parameter $S$, but slightly greater than 1 for large speed parameter $S$.

## 3.5 Numerical Verification

In order to evaluate the effect of the approximation made in deriving the solution to Eqs. (3.10) and (3.11), i.e. by assuming the mass of the vehicle to be very small compared with that of the bridge, an independent finite element analysis (Yang and Yau 1997) that does not rely on the said assumption will be conducted. The following properties are adopted for the bridge: cross-sectional area $A = 2.0\,\text{m}^2$, moment of inertia $I = 0.12\,\text{m}^4$, elastic modulus $E = 2.75 \times 10^{10}\ \text{N/m}^2$, length $L = 25\,\text{m}$, and mass per-unit-length $\bar{m} = 4{,}800\,\text{kg/m}$. The vehicle is assumed to have a mass $m_v = 1200\,\text{kg}$ and a spring constant $k_v = 500\,000\,\text{N/m}$, which implies a vibration frequency of $\omega_v = 20.41$ rad/s (= 3.25 Hz). The beam is divided into 20 beam elements for the present analysis. The first three eigenvalues $\omega_b$ computed for the beam are: 13.09 rad/s (= 2.08 Hz), 52.36 rad/s (= 8.33 Hz), and 117.81 rad/s (= 18.75 Hz).

The time-history responses computed for the displacement, velocity, and acceleration of the vehicle passing through the beam at the speed of 10 m/s considering either a single mode or two modes have been plotted in Figures 3.13–3.15, respectively, together with those by the finite element analysis. The first observation is that the solution obtained by considering only the first mode of vibration of the beam is very close to that obtained by considering the first two modes, indicating that the single-mode approach is reliable for simulating the vehicle response, and that high modes of the beam can generally be neglected. The discrepancy between the present analytical solutions and

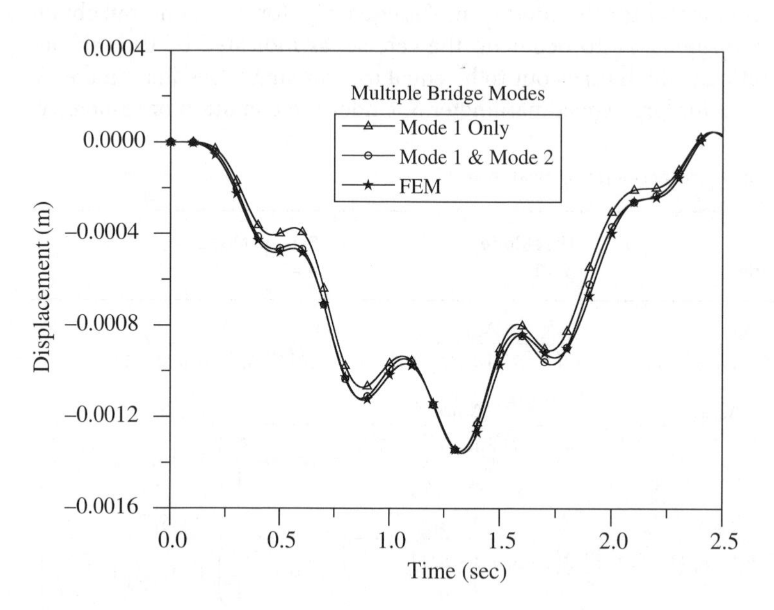


**Figure 3.13** Displacement response of vehicle: analytical and numerical.

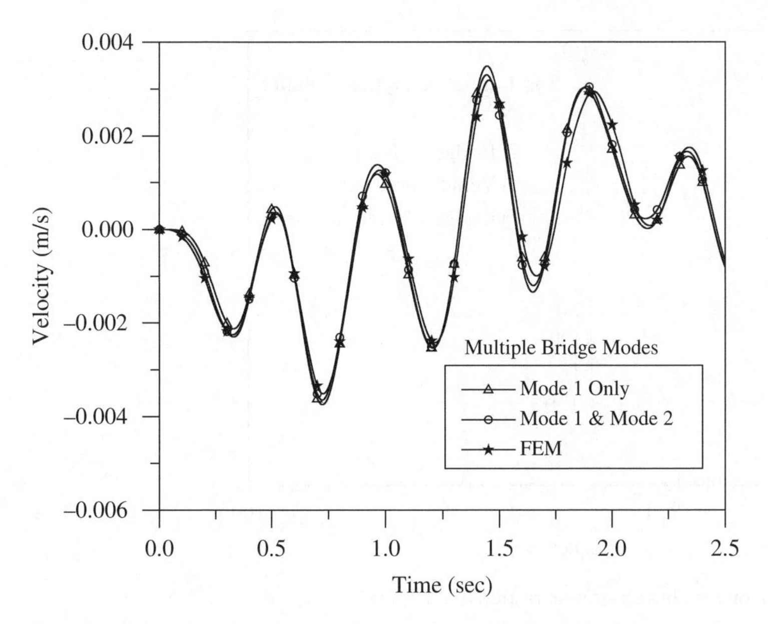


**Figure 3.14** Velocity response of vehicle: analytical and numerical.

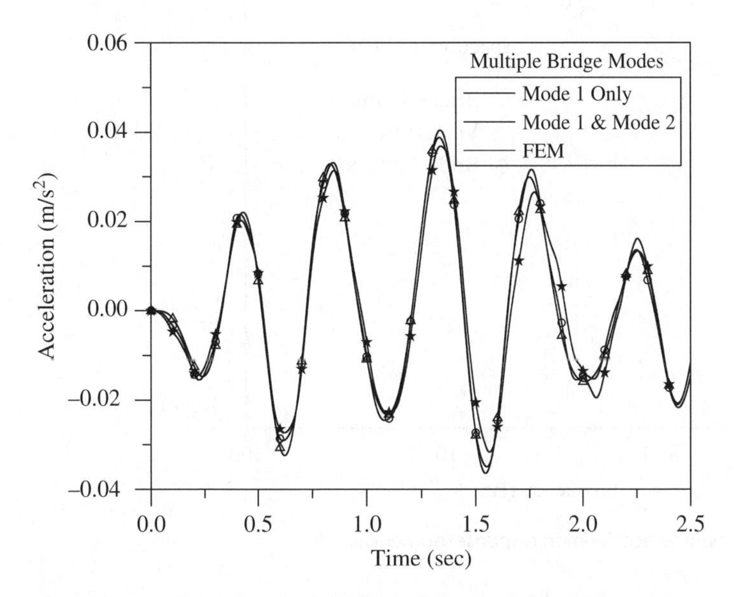


**Figure 3.15** Acceleration response of vehicle: analytical and numerical.

the finite element solutions is owing to the fact that iteration was not performed for updating the interaction forces existing between the moving vehicle and the bam, in addition to adoption of the assumption of small mass ratios for the vehicle. Such a discrepancy is generally negligible when observed in the frequency domain, as will be discussed below.

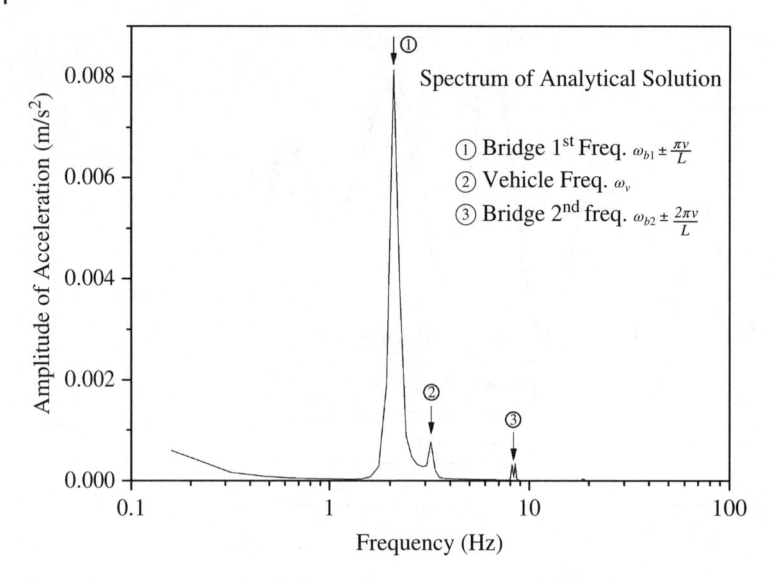


**Figure 3.16** Spectrum of vehicle acceleration response: analytical.

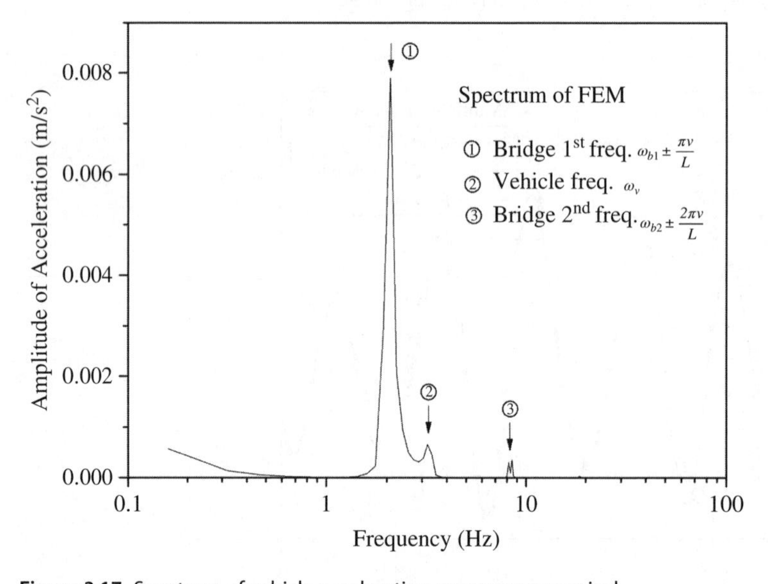


**Figure 3.17** Spectrum of vehicle acceleration response: numerical.

The Fourier transforms of the acceleration response of the vehicle obtained by the present analytical approach and the finite element method are plotted in Figures 3.16 and 3.17 with vehicle speed 4 m/s, respectively, in which several peaks can be clearly seen. An observation herein is that the frequencies corresponding to all the peaks in the two figures appear to be in good agreement, indicating that the use of the single-mode approach to solve for the vehicle response can yield sufficiently accurate results, particularly when the frequency content is of concern.

In each figure, the first peak (counted from the left-hand side) relates to the driving frequency $\pi v/L$. The frequencies associated with the second and third peaks relate to the fundamental frequency of the beam with shift, i.e., $\omega_b - \pi v/L$ and $\omega_b + \pi v/L$. By removing the shifting effect, the fundamental frequency of the beam can be recovered as $\omega_b$. The frequency identified from the figures for the fourth peak is actually the vehicle frequency, i.e. $\omega_v$ = 3.25 Hz. Of interest are the two small peaks on the right-hand side of the figures. They are associated with the second frequency of the beam with shift.

In each figure, the first peak relates to the bridge first frequency, $\omega_{b1}$ = 2.08 Hz. It should be noted that the Doppler effect related driving frequency $\pi v/L$ = 0.08 Hz, which makes it is difficult to distinguish the two frequencies due to the Doppler shifting. This explain why there is only one peak in the location of the bridge's first frequency in Figures 3.16 and 3.17. The frequency identified from the figures for the second peak is actually the vehicle frequency, i.e. $\omega_v$ = 3.25 Hz. Of interest are the two small peaks on the right-hand side of the figures. They are associated with the second frequency of the beam with shift. By removing the shifting effect, the second frequency of the beam can be recovered as $\omega_{b2}$ = 8.33 Hz.

It should be noted that the results presented in Figures 3.16 and 3.17 are based on the assumption of zero damping and smooth pavement surface for the bridge. Should these two factors be taken into account, it is likely that some of the peaks, particularly those of low magnitudes, will be nullified, while some new peaks will be generated by irregularities in the pavement. Under such a situation, we suspected that the third peak associated with the fundamental frequency of the beam, $\omega_{b1} + \pi v/L$ = 2.28Hz, could still be observed. But it is hard to say anything about the second frequency of the beam. Further research is required in this regard, since it presents lots of advantages if the *bridge* frequency can be extracted indirectly from the acceleration response of a passing test *vehicle*, compared with the conventional approaches of measuring the bridge frequencies directly based on the bridge response.

## 3.6 Concluding Remarks

In this chapter, a modal superposition approach was adopted to derive the solutions for a vehicle-bridge system, with the vehicle simulated as a sprung mass and the bridge as a simply supported beam. Such an approach is approximate in that no iteration was conducted for updating the interaction force existing between the two subsystems. The solutions obtained, however, were shown to be quite accurate compared with those obtained by the finite element method. For both the beam and vehicle responses, we concluded that sufficiently accurate solutions can be obtained by considering only the first mode of vibration of the beam. As for the beam, the displacement response is dominated by the driving frequency component; the velocity is equally governed by both the driving frequency and frequency components of the beam; and the acceleration is dominated by the frequency component of the beam. There exists a possibility, at least theoretically, that the vehicle speed can be identified from the displacement or velocity spectrum of the beam in the form of driving frequency, since it appears as a low-frequency peak. Further investigation is required in this regard. There are three groups of frequencies involved in vehicle response: the driving frequency, vehicle frequency, and beam frequency with shift. For the case where zero damping and smooth pavement surface are assumed for the beam, all the three groups of frequencies can be identified from the vehicle response. Of interest is the fact that the shifted fundamental frequency of the beam appears as the highest peak in the spectrum. If the effects of damping and irregular pavement are taken into account, we suspect that the fundamental frequency of the beam can still be identified from the vehicle response. Both further theoretical and experimental studies are required in this regard

# 4

# Vehicle Scanning of Bridge Frequencies

## Experiment

Following the theoretical development in the preceding two chapters, a single-axle, two-wheel cart towed by a light truck is used to scan the frequencies of vibration of the sustaining bridge, as a feasibility test of the idea presented. The truck excites the bridge through its movement at a constant speed, thereby serving as an *exciter* to the bridge. The vibration of the bridge is transmitted via the contact point and *scanned* by an accelerometer mounted inside the test cart. The bridge that is set into motion by the truck will dominate the vertical dynamic response of the test cart in movement, which, in turn, serves as a *scanner* or *message receiver* of the vertical oscillation or dynamic properties, i.e., frequencies, of the sustaining bridge. The response recorded by the accelerometer mounted on the test cart during its passage over the bridge is then processed by the fast Fourier transform (FFT) to extract the frequencies of excitation or vibration transmitted from the bridge. The feasibility in scanning the fundamental frequency of the bridge using the moving test cart is confirmed by the field tests conducted in a highway bridge located in northern Taiwan. The methodology developed for the first time herein is simple and efficient, and can be promisingly applied to scanning a wide range of bridge structures, if only the fundamental frequency is of interest. The materials presented in this chapter are based primarily on the paper by Lin and Yang (2005).

## 4.1 Introduction

Frequencies of vibration are crucial information for bridges, as they relate closely to the stiffness and overall health of the bridge. In some applications, the bridge frequencies of vibration, especially of the fundamental mode, have been used as an index of structural integrity or health condition (Ward 1984; Kato and Shimada 1986; Mazurek and DeWolf 1990). It is well known that a drop in the frequency of vibration implies a deterioration in the stiffness of the bridge, whether it is caused by a damage or failure in any component, joint, or support of the bridge, or by a loss in tendon or cable forces, depending on the type of bridges considered. The indirect cause for structural deterioration may be aging in materials, overloading by trucks heavier than allowable, previous earthquake shaking, scouring in bridge supports, or other damaging events. For the purpose of maintenance, it is desirable that the variation in frequencies of a bridge be monitored

*Vehicle Scanning Method for Bridges*, First Edition. Yeong-Bin Yang, Judy P. Yang, Bin Zhang and Yuntian Wu.

throughout its service life, such that precautions can be duly undertaken to ensure its safe and normal functioning.

Traditionally, a number of techniques have been developed for measuring the bridge frequencies, which differ on the use of different vibration sources, such as those based on ambient vibrations (Fujino et al. 2000; Chang et al. 2001), wind forces (Brownjohn et al. 1994; Xu et al. 2000), normal traffic loads (McLamore et al. 1971; Abdel-Ghaffar and Housner 1978; Farrar and James 1997; Ward 1984; Mazurek and DeWolf 1990), controlled traffic loads (Paultre et al. 1995; Conner et al. 1997), forced vibrations (Fujino et al. 2000; Farrar and James 1997), and impact forces (Douglas and Reid 1982; Huang et al. 1999a; Maragakis et al. 2001), among others (Salawu and Williams 1995). The selection of a specific vibration source depends mainly on the dynamic properties of the bridge of interest, the type of the bridge structure, the readiness of vibration source, the level of traffic control that can be exerted, and even the time, effort, and budget available for conducting the test. In general, a stronger vibration source should be adopted if the frequencies and vibration shapes of the higher modes are of interest.

All the aforementioned approaches for measuring the bridge frequencies will be referred to as the *direct approaches*, since they all require on-site installation of the measurement equipment and sensors directly on the bridge in order to record its vibration data. With direct approaches, one needs to mount a good number of vibration sensors, such as accelerometers, at different locations of the bridge and have them connected to a central PC-driven data acquisition system. Although a wide range of dynamic properties can be monitored for the bridge, the cost of installing and maintaining the sensors and data acquisition system is generally high. For this reason, such approaches have been applied mainly to bridges that are considered to be functionally important, or featured by some new attributes in span length, shape, materials, and/or in construction.

Unlike the traditional direct approaches already mentioned, the *indirect approach* for measuring the bridge frequencies, especially of the fundamental mode, was first proposed by Yang et al. (2004a) and described here in Chapter 2. In this theoretical study, a vehicle modeled as a sprung mass moving over the bridge was used as the *exciter* to the bridge and simultaneously as a *receiver* of the vibration transmitted from the bridge. The frequencies of the bridge were then extracted indirectly from the dynamic response recorded off the vehicle during its passage over the bridge, rather than from that of the bridge. This chapter can be regarded as an experimental realization of the idea proposed in the paper by Yang et al. (2004a) or in Chapter 2. Remarks will be given concerning the advantages and limitations of the indirect approach.

## 4.2 Objective of This Chapter

The objective of this chapter is to experimentally demonstrate the feasibility of measuring the fundamental bridge frequency from the dynamic response recorded of a vehicle during its passage over the bridge. Since the experiment is carried out with no preconditions on the bridge, that is, no sensors or equipment need to be installed the bridge, neither is the detour of ongoing traffic, this technique, when fully developed, will be applicable to extracting frequencies from virtually all bridges. Although the health condition of a bridge cannot be assessed solely by monitoring its fundamental frequency, a low-cost routine monitoring of the variation of the fundamental frequency of the bridge can still provide useful information for the maintenance of the bridge. The low cost and high mobility make test vehicles extremely attractive options for regularly monitoring a number of bridges of concern in an area, and further investigation is thus warranted.

In Yang et al. (2004a), as presented in Chapter 2, a single moving vehicle (modeled as a sprung mass) was adopted, which offers the dual function of an *exciter* to the bridge and a *receiver* of the bridge response. In the study being discussed here, however, a small *tractor-trailer* system will be adopted. The tractor is a two-axle, four-wheel commercial light truck, which, when moving over the bridge, serves to excite the bridge into motion, thereby playing the role of an exciter to the bridge. The trailer is a small one-axle, two-wheel test cart, which will be excited via the contact point by the bridge that is already set in vibration, thereby serving as a *scanner* or message *receiver* of the underlying bridge motions. Theoretically, the bridge frequencies will be transmitted to the test cart and dominate its dynamic response, as they are the frequencies of the source (Yang et al. 2004a). Naturally, by recording the dynamic response of the test cart during its passage over the bridge, we can extract the bridge frequencies from such a response, if they can be separated from the frequencies of the test cart itself. To facilitate comparison with the theoretical study presented in Yang et al. (2004a), the single-axle test cart used is modeled as a *sprung mass* with the effect of suspension damping ignored, and the bridge as a simply supported beam. The assumption of a sprung mass for the test cart is good, since it is small and light compared with the bridge under test.

## 4.3 Description of the Test Bridge

The bridge selected is the Da-Wu-Lun Bridge, completed in 2000, which forms part of the Taiwan Provincial Highway No. 2 near the northern coast of Taiwan. The entire bridge, slightly curved, is composed of six spans of prestressed concrete girders resting on columns of various depths, as shown in Figure 4.1a. However, the test span selected is straight and simply supported. The highway is an important link for heavy loaded cargo trucks and tractor-trailers with their destination set for the Keelung Harbor in the north end. For the sake of long-time maintenance, the dynamic properties of the entire link, including the bridge under test, must be monitored.

As shown in Figure 4.1, the bridge unit is composed of six identical prestressed I-girders of span length of 30 m, placed at a center-to-center distance of 2.8 m. The cross-sectional area and moment of inertia of each I-girder are $0.64\,m^2$ and $0.2422\,m^4$, respectively. On top of the I-girders is a concrete deck slab with a thickness of 20 cm, which is covered by AC pavement of 5 cm in thickness. The total width of the bridge's cross section is 16.5 m. The concrete of the bridge has an elastic modulus of 29 GPa and density of $2400\,kg/m^3$. The two ends of the bridge unit considered have an elevation difference of 1.91 m, equivalent to a longitudinal slope of 6.4%.

## 4.4 Description of the Test Vehicle

The vehicle system mainly used for measurement is a tractor-trailer, as shown in Figure 4.2a. The tractor is a two-axle commercial light truck (Ford model Econovan 2A 2.0 c.c.) weighing 1.4 tons with an axle spacing of 2.4 m. The trailer is a single-axle,

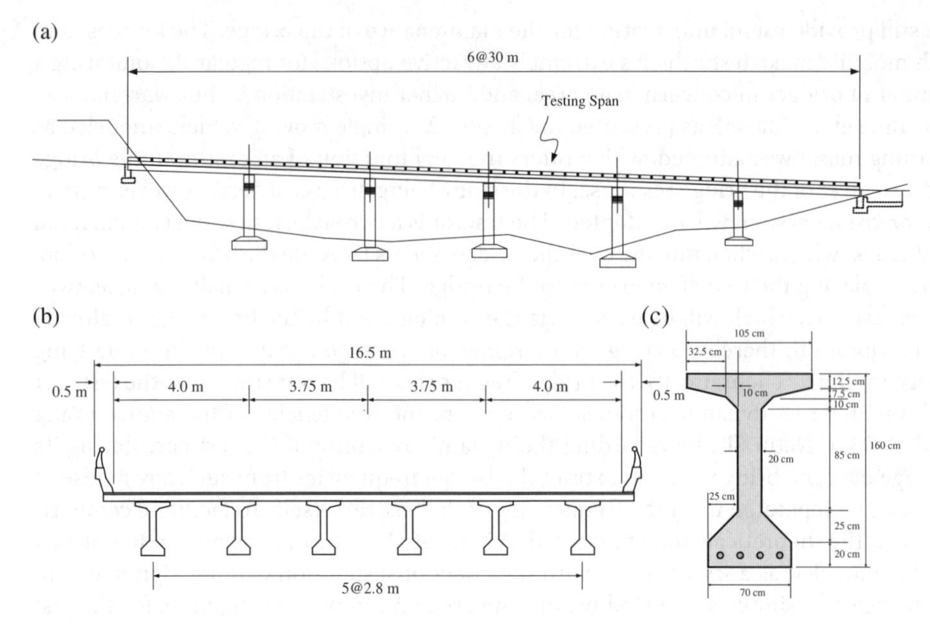


**Figure 4.1** Bridge under test: (a) elevation; (b) cross section; (c) girder.

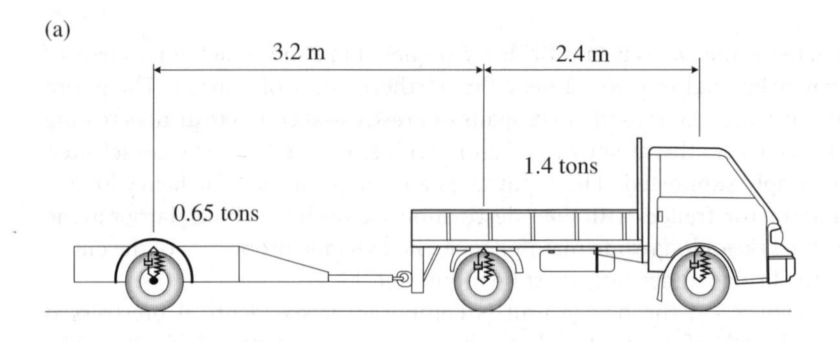


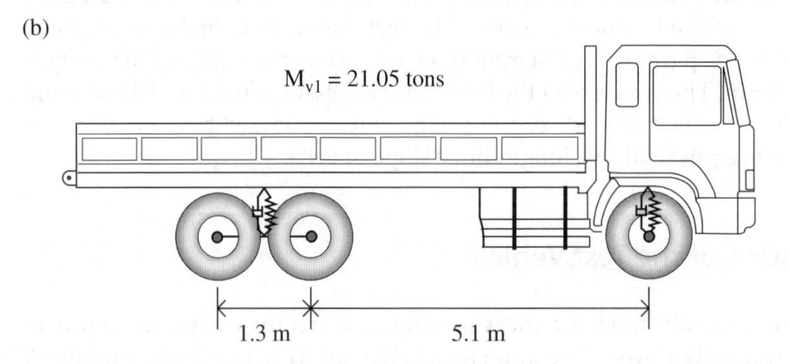


**Figure 4.2** Vehicles used in the test: (a) tractor-trailer; (b) heavy truck.

two-wheel test cart of 0.65 tons. The latter can be considered as a single-degree-of-freedom (DOF) sprung mass. Free from the engine disturbance, the vertical vibration of the test cart can be ideally regarded as measured from its own suspension system. The linkage connecting the truck and the test cart is free to rotate in all directions, meaning that no bending moment will be transmitted between the two parts.

During the field test, a heavy truck of the ISUZU F-series FWR34Q with a net weight of 21.05 tons was employed. It has an axle spacing of 5.1 m between the front and first rear axles and 1.3 m between the two rear axles, as shown in Figure 4.2b. This heavy truck played the role of a moving load in a comparison test designed to measure the bridge frequencies directly from the bridge response. It also played the role of ongoing traffic in one of the tests when the tractor-trailer was moving over the bridge.

## 4.5 Instrumentation

Two kinds of vibration transducers were used in the field test. One is the acceleration-type model AS-1GA made by Kyowa Inc. with a size of 1.4 × 1.4 × 1.8 cm, as shown in Figure 4.3a. The specifications of this subminiature low-capacity acceleration transducer include rated capacity ±9.807 m/s$^2$, response frequency range from DC to 40 Hz, and resonance frequency 70 Hz. The other is the velocity-type transducer model VSE-15D made by Tokyo Shokushin Co., of which the specifications are as follows: response frequency range from 0.2 to 70 Hz and rated capacity range from 150 $\mu$ kine to 10 kine (1 kine = 1 cm/s), depending on the setting by software. The velocity transducer was used only in the ambient vibration test for comparison.

A portable ambient vibration monitoring system called the SPC51 system and provided by the Tokyo Shokushin Co. was used in all field tests (see Figure 4.3b). This system consists of a laptop computer, data processing unit, dual power supply switcher, and amplifiers and high-frequency data scanner capable of dealing with 16 channels simultaneously.

In measuring the dynamic response of the test cart during its movement, the vibration transducer was installed vertically at a position near the gravity center of the test cart, while the data acquisition system was placed in the front seats of the light truck, i.e., the tractor, for which the power supply was made available through a battery (see Figure 4.3b). For the purpose of comparison, an ambient vibration test was conducted for the bridge. In this test, the vibration transducer was installed vertically at the center of the surface deck of the bridge of concern.

## 4.6 Testing Plan

Before conducting the cart response test, an ambient vibration test was conducted for the bridge free of any traffic for directly measuring the bridge frequencies, which will be used as the reference for comparison. In addition, a free vibration test was conducted for the test cart (trailer) at rest to determine its dynamic properties, including the vertical vibration frequency and damping coefficient. Four key items are included in this experiment:

1) Record the dynamic response of the bridge to the moving action of the heavy truck. Such a response allows us to measure the bridge frequencies directly from the bridge response that serve as the basis of comparison.

(a)

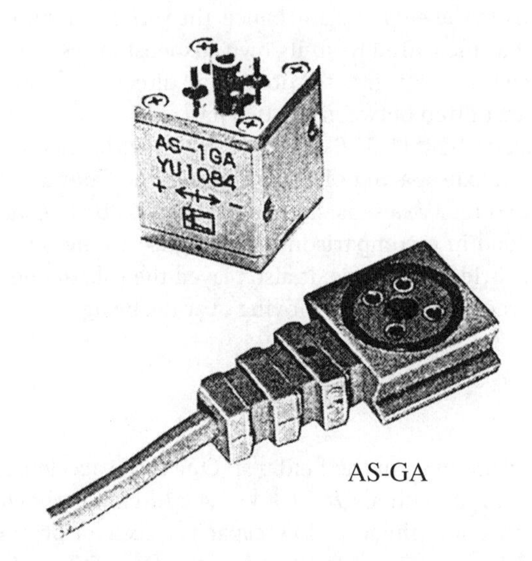


(b)

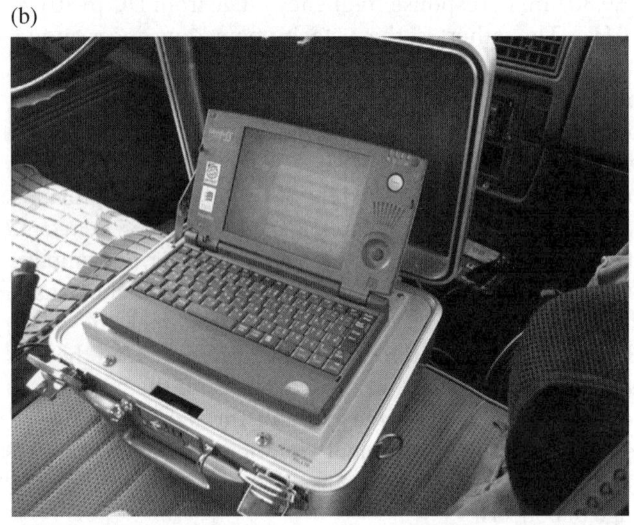

**Figure 4.3** Instruments: (a) acceleration transducer; (b) central control system.

2) During the passage of the heavy truck over the bridge, record the vertical response of the test cart placed at rest at the midspan of the bridge. The bridge frequencies extracted from the recorded cart response provides a means for evaluating the effect of the cart structure in transmitting the bridge frequencies.
3) Record the dynamic response of the test cart towed by the tractor moving over the bridge. The bridge frequencies extracted from the test cart response are exactly the ones desired.
4) Same as item 3, but in the meantime, let the heavy truck move simultaneously over the bridge to play the role of ongoing traffic. From the test cart response recorded herein, the influence of ongoing traffic on the extracted bridge frequencies can be evaluated.

In each of the above tests, four different vehicle speeds are considered, so as to evaluate the effect of vehicle speed on the measured results.

## 4.7 Eigenvalue Analysis Results

As shown in Figure 4.1, the bridge deck is supported by six girders. An eigenvalue analysis was conducted for a single girder with simple supports to obtain the first vertical frequency as 3.761 Hz. The bridge was then model as a gird structure simply supported at the two longitudinal ends, in which the bridge deck was represented by three sets of horizontal bracings to account for its transverse stiffness contribution. The first five frequencies and vibration modes computed for the grid structure were shown in Figure 4.4, from which the first two frequencies, 3.732 and 7.571 Hz, are identified to be associated with the first vertical and lateral modes, respectively.

## 4.8 Experimental Results

### 4.8.1 Ambient Vibration Test

Figure 4.5 shows the vertical ambient vibration responses of the bridge, recorded at the midspan of the bridge along the centerline, using the acceleration and velocity-type transducers for parts (a) and (b), respectively. The total length of each record is 200 seconds with a sampling rate of 100 Hz. Figure 4.6 shows the fast Fourier transform (FFT) of the recorded data, from which the peaks associated with the dominant frequencies of the bridge are identified. In Figure 4.6a, the lowest frequency can be read as 3.76 Hz, which is the same as the one in Figure 4.6b. However, the peak associated with the second frequency 7.60 Hz in the acceleration spectrum of part (a) was not well matched by the velocity spectrum of part (b) in terms of magnitude. This is due to the fact that the acceleration sensor was not mounted in a perfectly vertical position, because it was

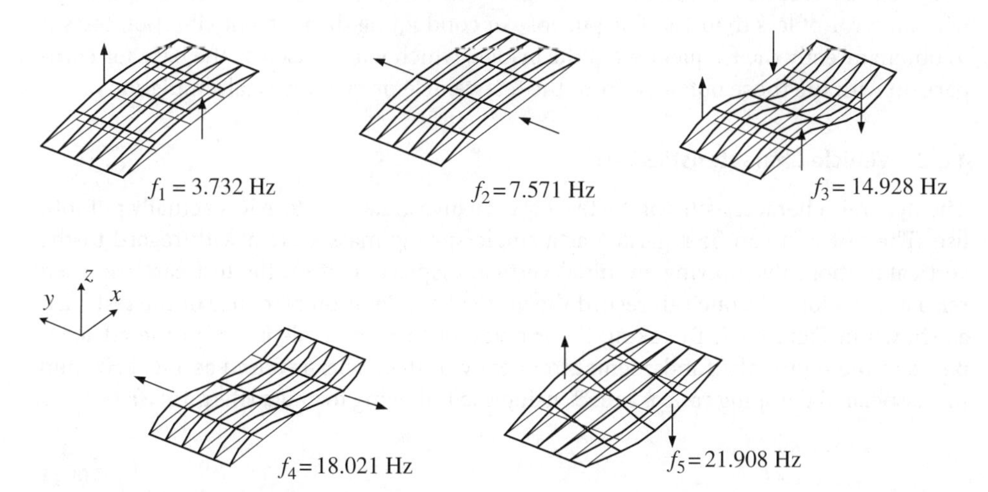


**Figure 4.4** Frequencies and modal shapes of the test span of the bridge.

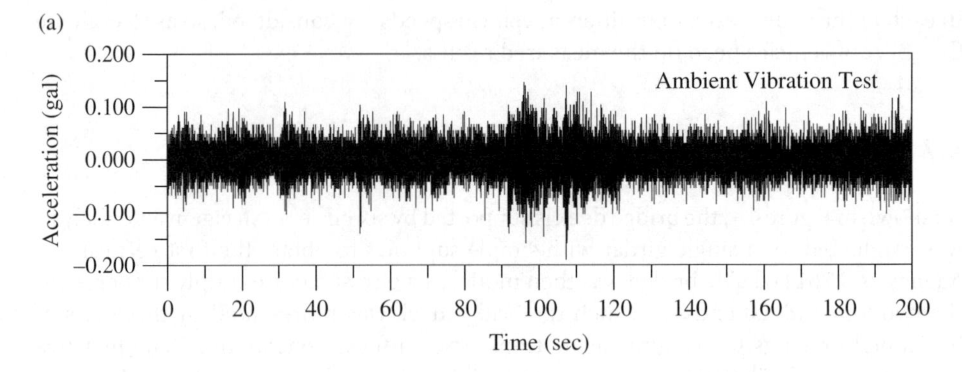


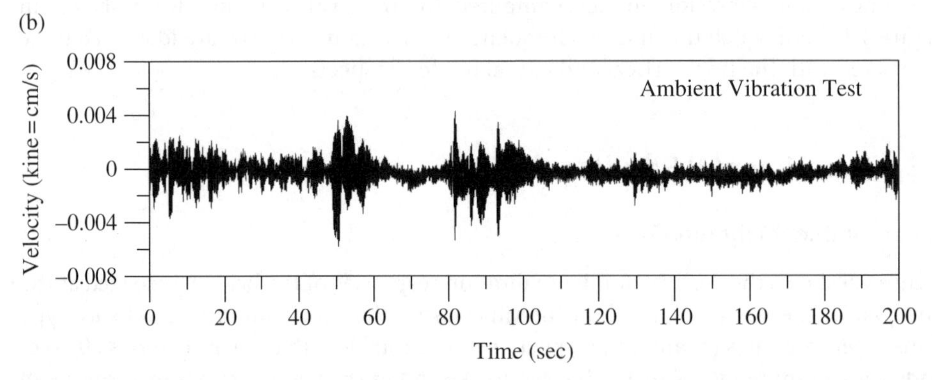


**Figure 4.5** Time-history record from ambient vibration test: (a) acceleration; (b) velocity.

directly attached to the pavement surface of the bridge, which, as mentioned previously, is not horizontal. However, the velocity sensor was installed on a supporting base with adjustment to ensure that the sensor is fully vertical.

The fundamental vertical and lateral frequencies measured from the ambient vibration tests are identical to those of the eigenvalue analysis results shown in Figure 4.4, with an error of less than 1%. The purpose of conducting the ambient vibration tests is to obtain directly the frequencies of the bridge, which will be used as the base for comparison with the those obtained from the test vehicle or by numerical simulations.

### 4.8.2 Vehicle Characteristics Test

The dynamic characteristics of the test cart are investigated before it is actually put into use. The test cart can be regarded as a single sprung mass system with regard to the vertical motion. By applying an initial vertical displacement on the test cart and then releasing it suddenly, one can record the vertical acceleration response of the test cart, as shown in Figure 4.7. By taking the average of four intervals between the adjacent peaks in the figure, the fundamental frequency of the trailer is found as 1.814 Hz, and the associated damping ratio $\xi$ as 13% using the following formula (Berg 1988):

$$\xi = \frac{\delta / 2\pi}{\left[1 + \left(\delta / 2\pi\right)^2\right]^{1/2}} \tag{4.1}$$

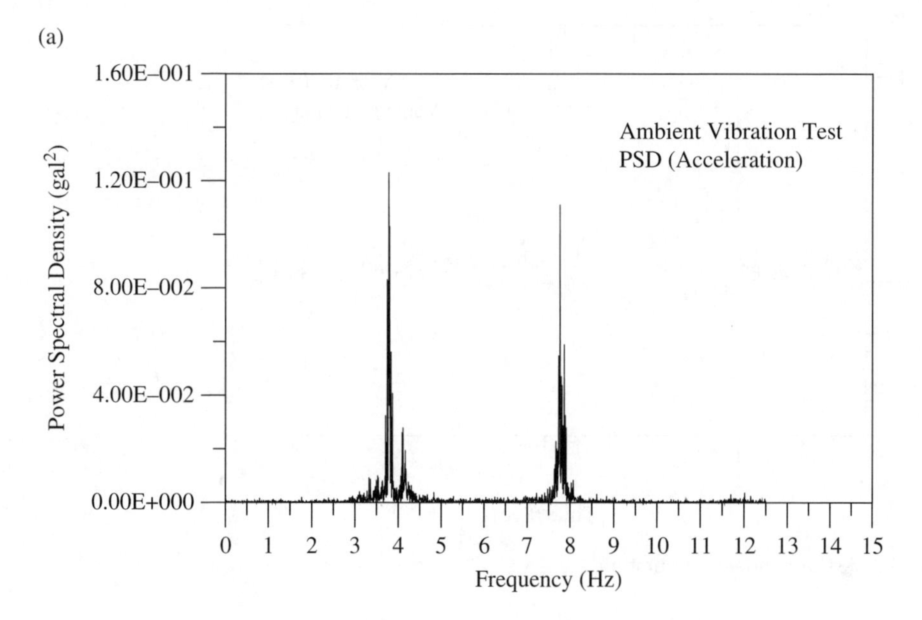


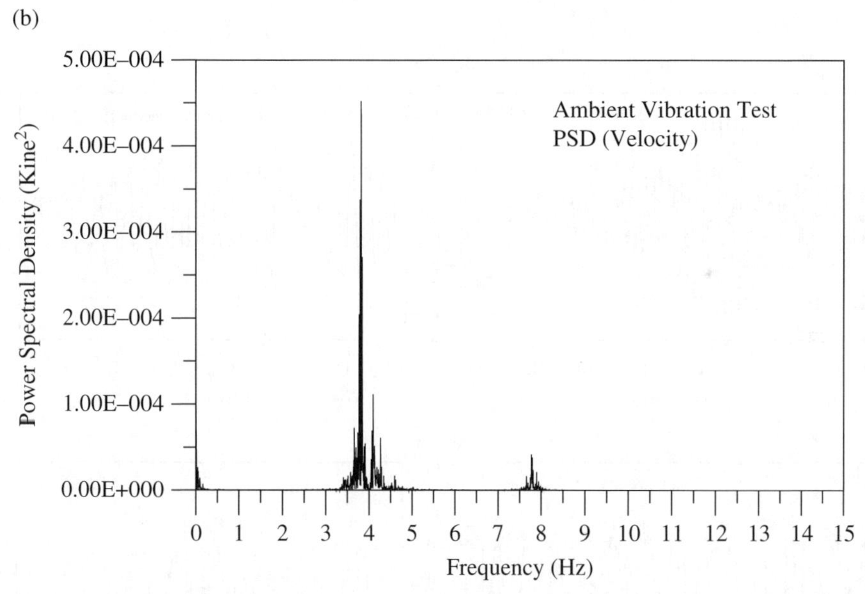


**Figure 4.6** Power spectral density of ambient vibration records: (a) acceleration; (b) velocity.

where $\delta$ denotes the logarithmic value of the ratio of two adjacent peaks in the free-vibration response.

### 4.8.3 Bridge Response to the Moving Truck

For comparison, the frequencies of vibration of the bridge will be measured directly from its response to the moving heavy truck. In Figure 4.8a–d, the acceleration responses of the bridge to the heavy truck moving at the four speeds 10.21, 18.97, 37.29, and

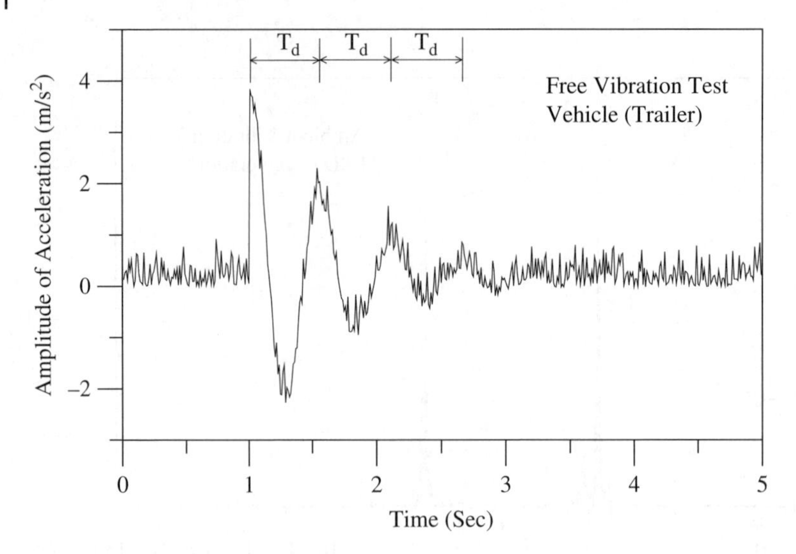


**Figure 4.7** Test cart under free vibration.

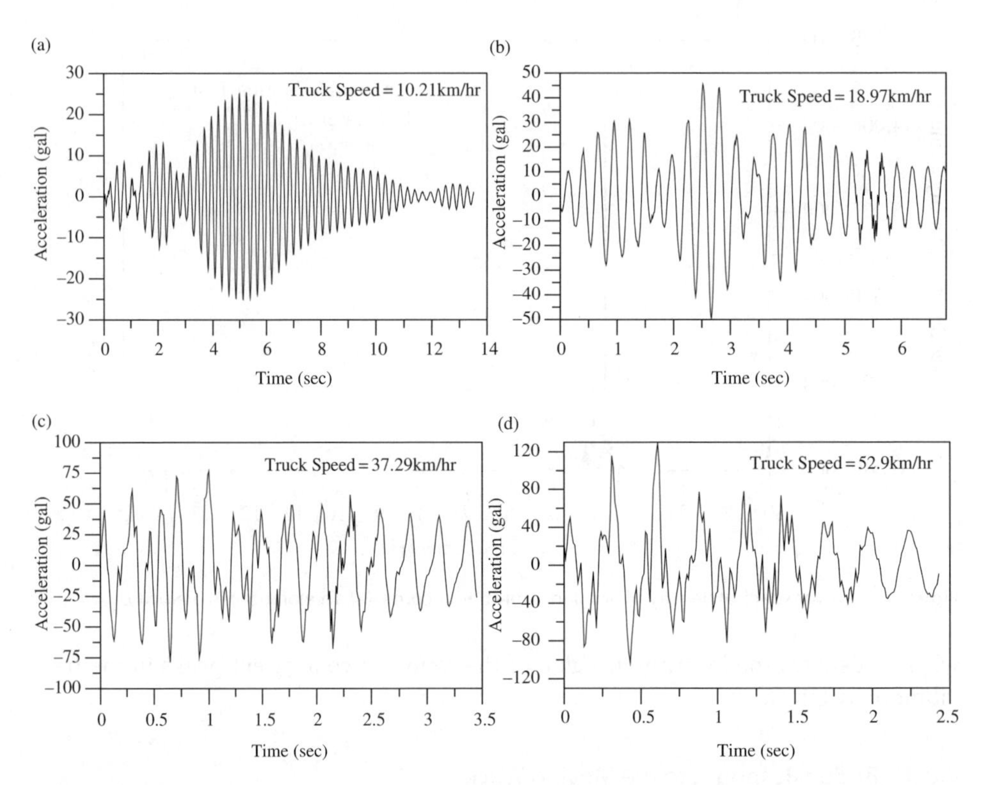


**Figure 4.8** Bridge response to the heavy truck moving at the speed of: (a) 10.21 km/h; (b) 18.97 km/h; (c) 37.29 km/h; (d) 52.9 km/h.

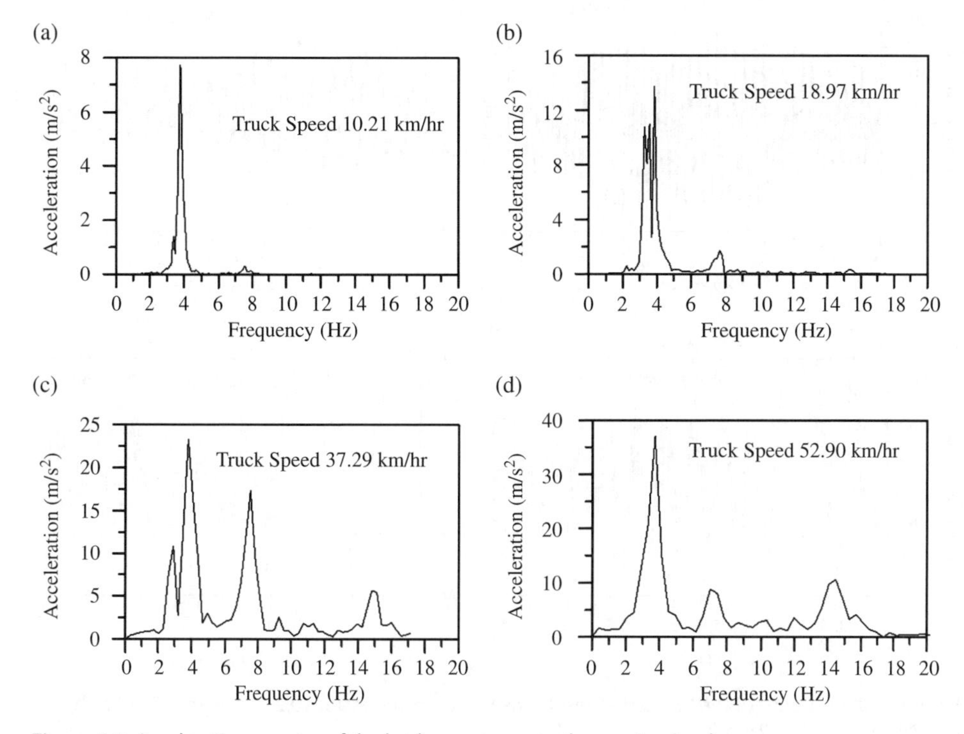


**Figure 4.9** Acceleration spectra of the bridge response to the moving truck.

52.9 km/h were plotted. As can be seen, the maximum response of the bridge increases as the truck speed increases, a trend generally consistent with the theoretical prediction in Yang and Lin (1995). From Figure 4.8, the peak responses can be read as 22, 45, 76, and 128 gal for the four speeds considered. The FFTs of the responses in Figure 4.8 have been plotted in Figure 4.9, from which the first and second bridge frequencies are identified to be around 3.7 and 7.5 Hz, respectively, for all the speeds considered, which agree well with those computed from the eigenvalue analysis. The damping ratio of the bridge can be estimated using Eq. (4.1) from the free decay response after the truck left the span in Figure 4.8, which is around 3%.

### 4.8.4 Response of the Test Cart Resting on the Bridge to the Moving Truck

It is essential to examine how the test cart responds when it stays on the bridge under the action of other moving loads. To this end, the test cart was placed at the midspan of the bridge, when the heavy truck moved at different speeds over the bridge. The vertical acceleration responses recorded of the test cart during the passage of the heavy truck at the four speeds: 10.21, 18.55, 37.8, and 52.32 km/h, were plotted in Figure 4.10a–d, respectively. As can be seen from the FFT plotted in Figure 4.11, the test cart response was dominated primarily by the dynamic property of the bridge. Because of the damping effect of the test cart, only the first frequency of the bridge, 3.76 Hz, can be observed from the response spectra of the test cart.

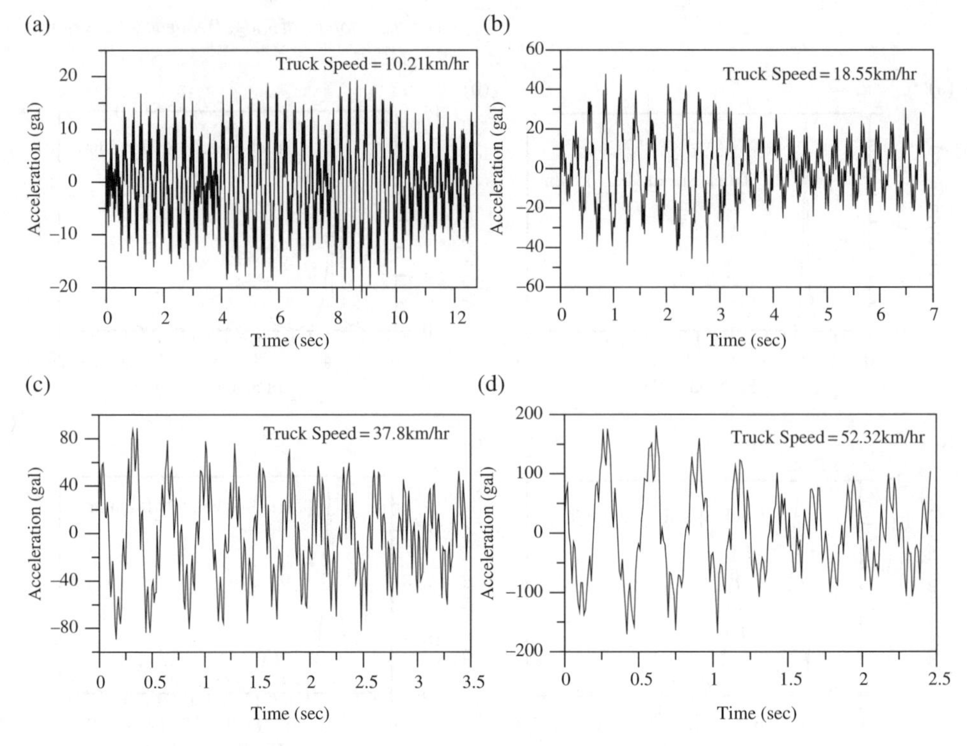


**Figure 4.10** Response of the test cart at rest to the truck moving at: (a) 10.21 km/h; (b) 18.55 km/h; (c) 37.8 km/h; (d) 52.32 km/h.

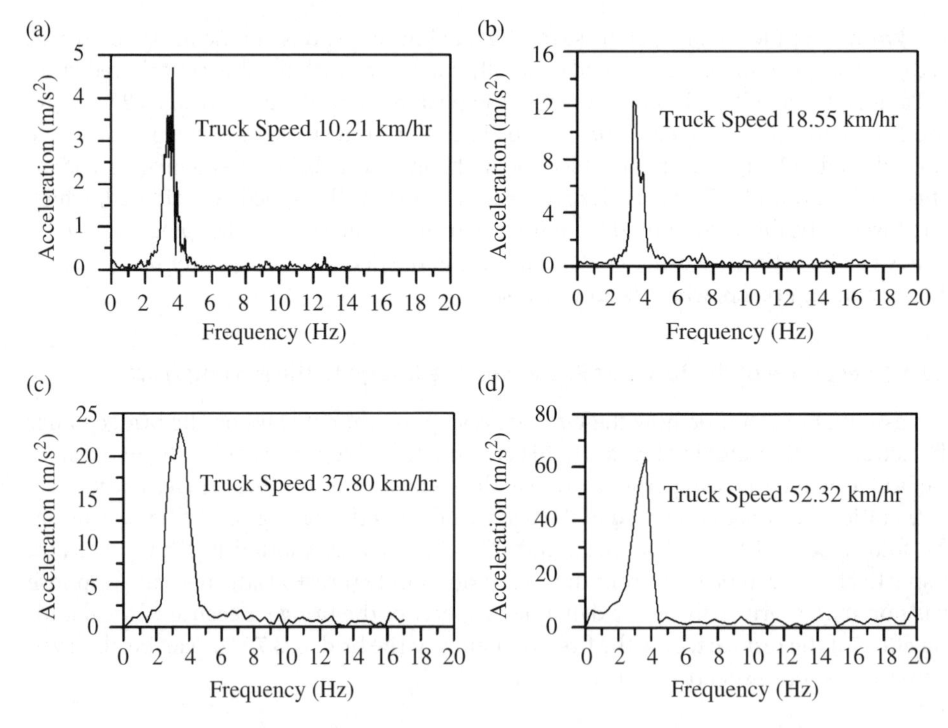


**Figure 4.11** Acceleration spectra of the response of the test cart at rest to the moving truck.

From Figure 4.11, it can also be observed that the peak amplitude associated with the bridge frequency 3.76 Hz increases as the truck speed increases. By comparing Figure 4.11 with Figure 4.9, one observes that the bridge responses measured from the test cart far exceed those measured directly from the bridge deck. The reason is that the response of the test cart, a single DOF system, will be amplified when the ratio of the excitation frequency to the structural frequency of the test cart is greater than 0.7 for a damping ratio of 10% (Berg 1988), which is 3.76/1.81 = 2.08 for the case considered. Clearly, the frequency ratio of the bridge (source) to the test cart (receiver) is a parameter that must be considered in development of the indirect measurement technique for bridges.

### 4.8.5 Response of the Moving Test Cart with No Ongoing Traffic

Here, we shall proceed to examine whether the cart response still contains the bridge frequency, if it is allowed to move over the bridge at certain speeds, actually towed by the tractor as already mentioned. To mark the beginning and ending time of the test cart's movement over the bridge, two ropes were placed transversely across the bridge deck, each at the entrance and exit of the span considered. When the test cart rolls over the rope, a jump in the test cart response will be recorded, which has been used to trim off the responses in Figure 4.12 for the four speeds 13.0, 17.28, 35.13, and 51.84 km/h.

For the test cart moving at speed 13.0 km/h, three to four peaks occurred within each time unit, as in Figure 4.12a. This implies that the test cart response is dominated by the

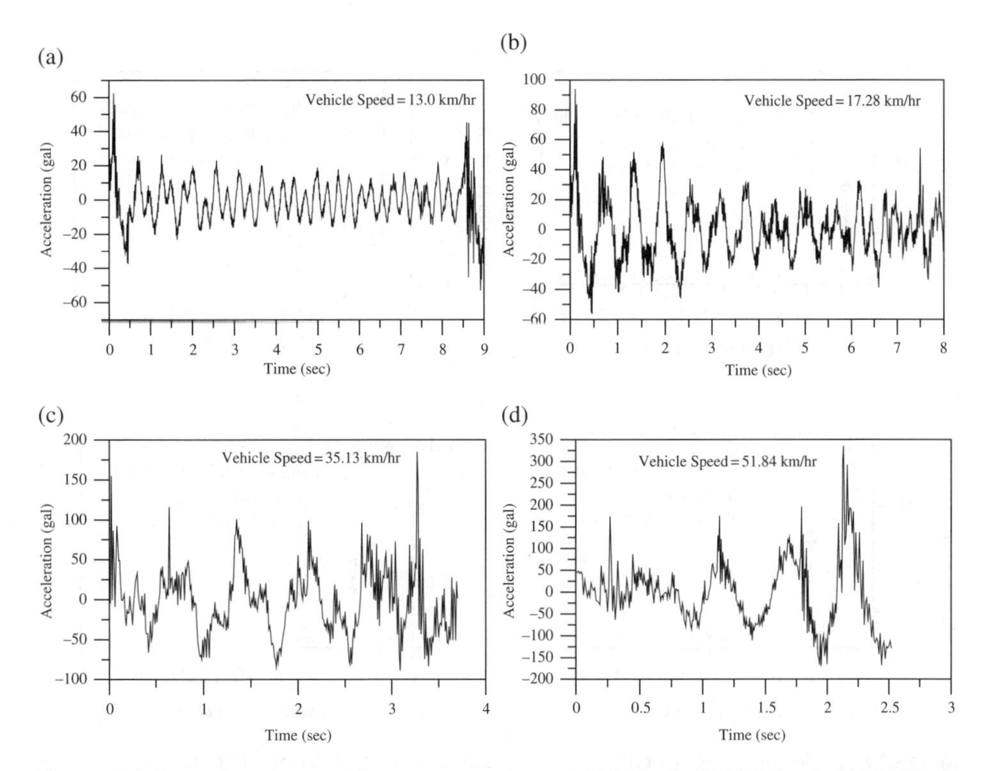


**Figure 4.12** Dynamic response of the test cart moving at: (a) 13.0 km/h; (b) 17.28 km/h; (c) 35.13 km/h; (d) 51.84 km/h.

bridge frequency, i.e., the source frequency to the test cart. In contrast, the response shown in Figure 4.12d seems to be dominated by some high-frequency components, as the number of peaks occurring within each time unit cannot be visually counted.

Figures 4.12a–d reveal that as the vehicle speed increases, the period of the dominant vibration component becomes longer. This seems to be natural, since the dynamic properties, especially the fundamental frequency, of the test cart tend to be excited and reflected in the response when traveling at high speeds. On the other hand, one also observes from Figures 4.12a–d that as the vehicle speed increases, there is an increasing influence of the high-frequency components in the response, mainly due to contributions associated with the mechanical parts of the cart, roughness in pavement, or minor collision in the connector between the tractor and trailer.

The FFT of the test cart responses in Figures 4.12a–d are plotted in Figure 4.13. As was expected, the lowest peaks in this figure indicate the fundamental vibration frequency of the test cart. Obviously, the peaks representing the fundamental frequency of the bridge can be identified from the results for the lowest three speeds. However, the same is not true for the highest speed, 51.84 km/h, for which the bridge frequency seems to be blurred by the same high-frequency components, possibly related to the pavement roughness and mechanical parts of the test cart, which were excited by the relatively large amount of energy carried by the vehicle at high speeds. For the vehicle moving at high speeds, one also observes that the bridge frequency is not as obvious in the test cart response, because the energy transmitted directly from the tractor is

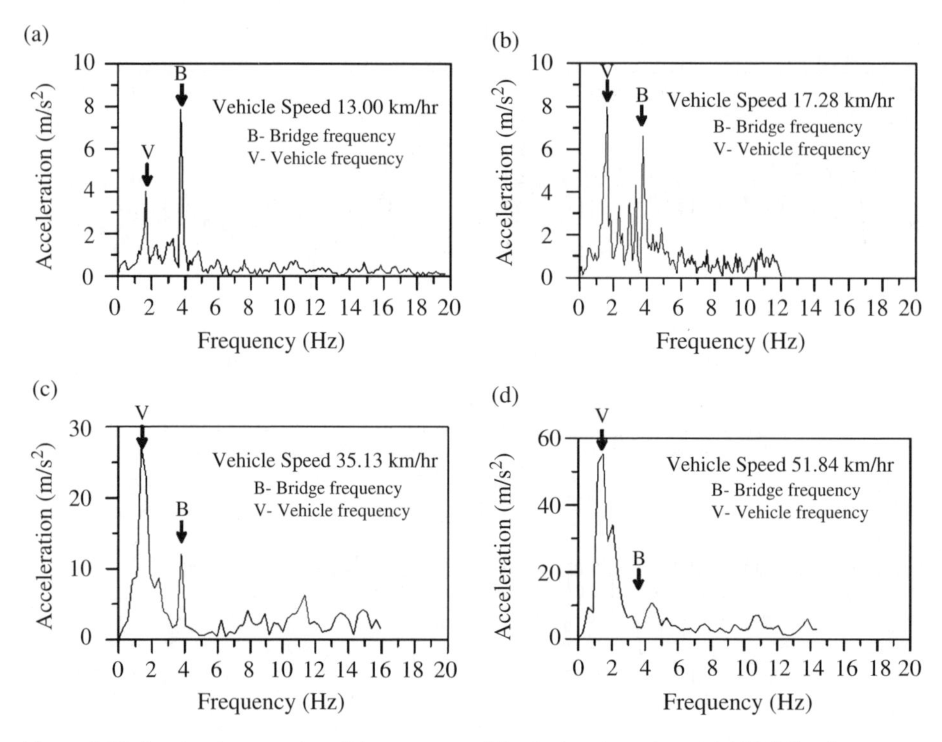


**Figure 4.13** Acceleration spectra of the response of the test cart moving at: (a) 13.00 km/h, (b) 17.28 km/h; (c) 35.13 km/h; (d) 51.84 km/h.

higher than that from the bridge. Therefore, to make the bridge frequency more visible in the test cart response, the tractor-trailer should not be allowed to move at too high a speed.

As a side remark, the second frequency of the bridge that was identified by the ambient vibration test in Figure 4.6 is invisible from the test cart response spectra in Figure 4.13, as it was mixed with other high-frequency components.

### 4.8.6 Response of the Moving Test Cart with Ongoing Traffic

One essential question is whether the ongoing traffic can affect the dynamic response of the test cart hauled by the tractor. To answer this question, the heavy truck is allowed to proceed at a speed close to that of the tractor-trailer to simulate the effect of ongoing traffic. The response recorded of the test cart at the four speeds: 9.79, 15.41, 38.56, and 46.94 km/h are plotted in Figures 4.14a–d, respectively. Compared with those in Figure 4.13, one observes that under similar vehicle speeds, the existence of an on-going moving truck has resulted in significant increase in the magnitude of the test cart response. Such a phenomenon is considered beneficial and also reasonable, as the bridge (the source to the test cart) was excited to a larger extent by the on-going truck. From the FFT plotted in Figure 4.15, one observes that for the speeds less than 40 km/h, the bridge frequency can be easily identified from the test cart response. However, for the speed at 46.94 km/h, the test cart response was dominated primarily by its own

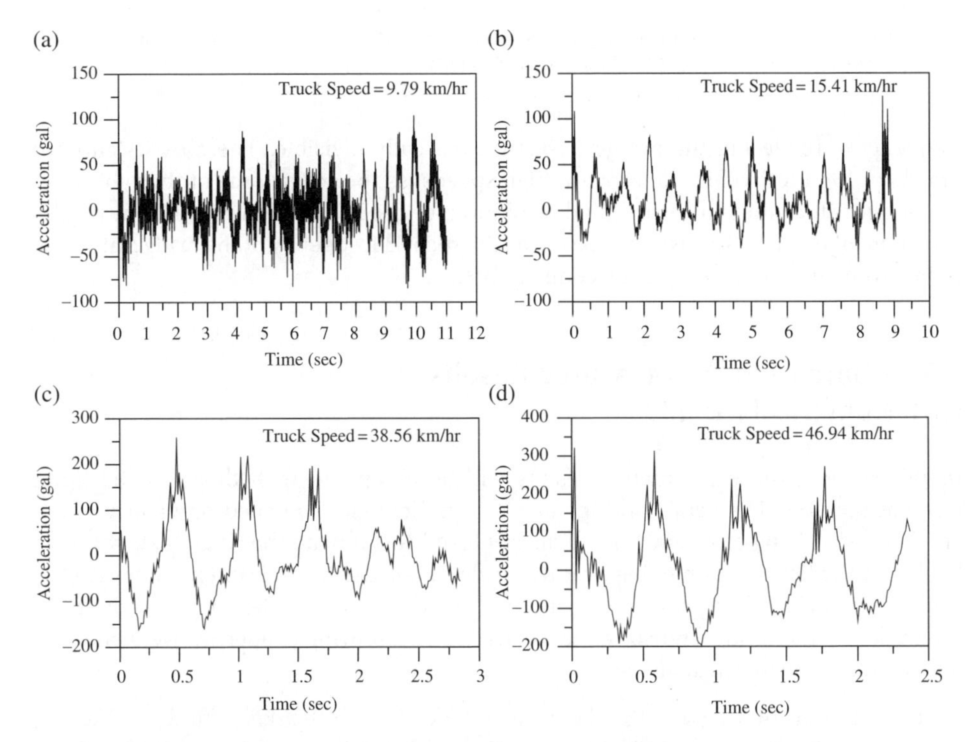


**Figure 4.14** Dynamic response of the test cart (plus the truck) moving at: (a) 9.79 km/h; (b) 15.41 km/h; (c) 38.56 km/h; (d) 46.94 km/h.

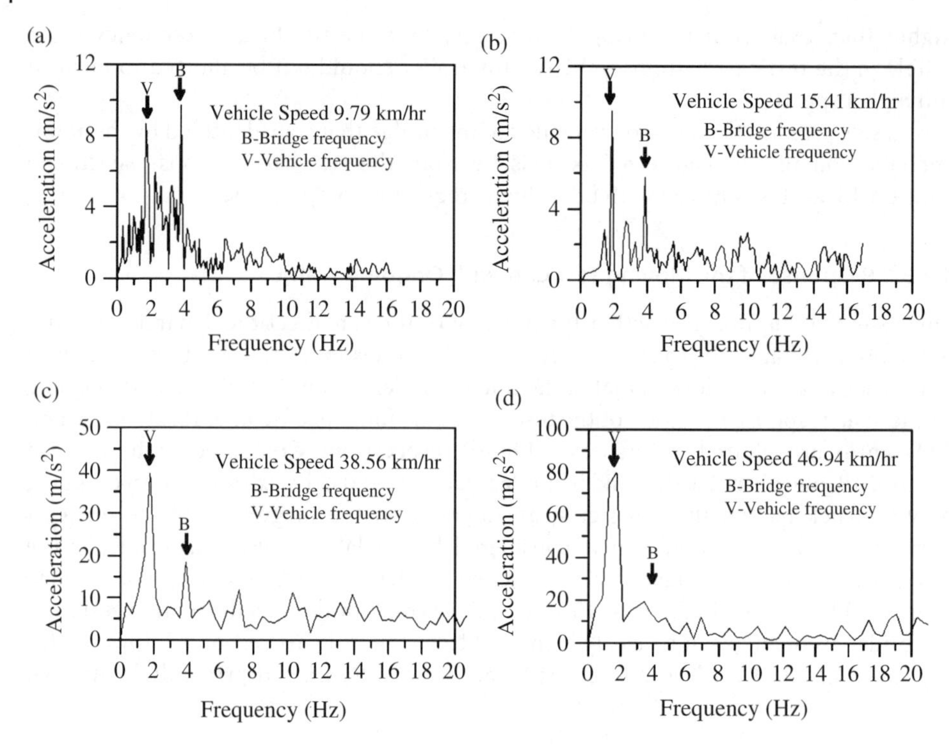


**Figure 4.15** Acceleration spectra of the response of the test cart (plus the truck) moving at: (a) 9.79 km/h; (b) 15.41 km/h; (c) 38.56 km/h; (d) 46.94 km/h.

frequencies, rendering the bridge frequency basically invisible. There exists another drawback for a test vehicle to move at high speeds, as the acting time of the vehicle on the bridge is too short to ensure good resolution for the FFT. Meanwhile, for the results shown in Figure 4.15, no visibility exists for the second frequency of the bridge, also due to involvement of high-frequency components.

## 4.9 Comparing the Measured Results with Numerical Results

In this section, the experimental results will be compared with those by the finite element method. The vehicle-bridge interaction element developed previously (Yang and Yau 1997; Yang et al. 2004a) will be employed to simulate the single-axle test cart hauled by the light truck moving over the bridge, considering their interaction with the bridge. The bridge is represented by three-dimensional beam elements and the tractor-trailer by the model shown in Figure 4.16. The following properties were adopted for the tractor-trailer in analysis:

a) Tractor: $M_t = 1450\,\text{kg}$, $I_t = 1865\,\text{kg}\cdot\text{m}^2$, $k_f = 49\,\text{kN/m}$, $c_f = 0.26\,\text{kN}\cdot\text{s/m}$, $k_v = 63\,\text{kN/m}$, $c_v = 0.22\,\text{kN}\cdot\text{s/m}$, $m_{wf} = 110\,\text{kg}$, $m_{wr} = 125\,\text{kg}$, $d_1 = 0.8\,\text{m}$, $d_2 = 1.6\,\text{m}$, and $d_3 = 3.2\,\text{m}$.
b) Trailer: $m_v = 650\,\text{kg}$, $k_v = 84\,\text{kN/m}$, $c_v = 2.85\,\text{kN}\cdot\text{s/m}$, and $m_{wv} = 50\,\text{kg}$.

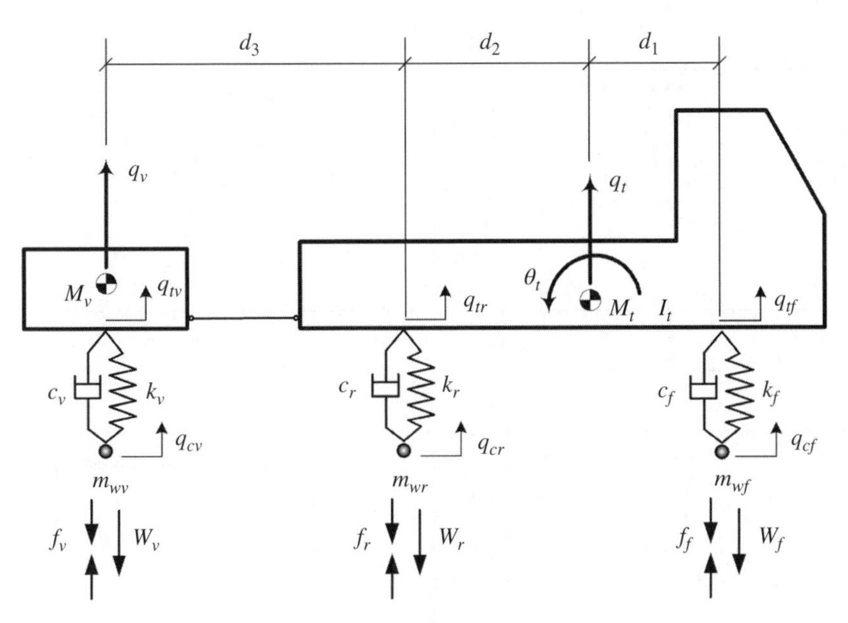


**Figure 4.16** Mathematical model for the tractor-trailer.

The Newmark's $\beta$ method is adopted for solving the finite element equations, with the control parameters set as $\beta = 0.25$ and $\gamma = 0.5$ for unconditional numerical stability. No consideration is made for pavement roughness.

The numerical results obtained for the test cart response by the finite element method have been plotted in Figures 4.17a–d for the four speeds 13.0, 17.28, 35.13, and 51.84 km/h, respectively. Similarly, the maximum response of the vehicle increases as the speed increases. Compared with the experimental results in Figure 4.12, the maximum vehicle responses computed by the finite element simulation appear to be generally lower, except for the speed of 13.0 km/h, as indicated by Table 4.1.

On the other hand, significant differences exist in the phase angles between the experimental and numerical results, due to the fact that the initial conditions for both the bridge and vehicles were not taken into account in the numerical simulation. Figure 4.18 shows the FFT of the test cart responses shown in Figure 4.17. Obviously, both the vehicle and bridge frequencies can be identified from the test cart response, especially for the cases with higher speeds. From Table 4.2, we observe that the fundamental frequencies of the bridge obtained either numerically or by scanning agree with each other to the first decimal digit.

## 4.10 Concluding Remarks

The feasibility of using a passing vehicle to scan the bridge frequency was tested using a tractor-trailer system. The tractor is a two-axle light truck that serves to tow the single-axle test cart as the trailer, while exciting the bridge. The test cart serves as the receiver or scanner of the bridge vibration. The feasibility of employing the tractor-trailer system to scan the bridge frequencies has been verified through comparison of

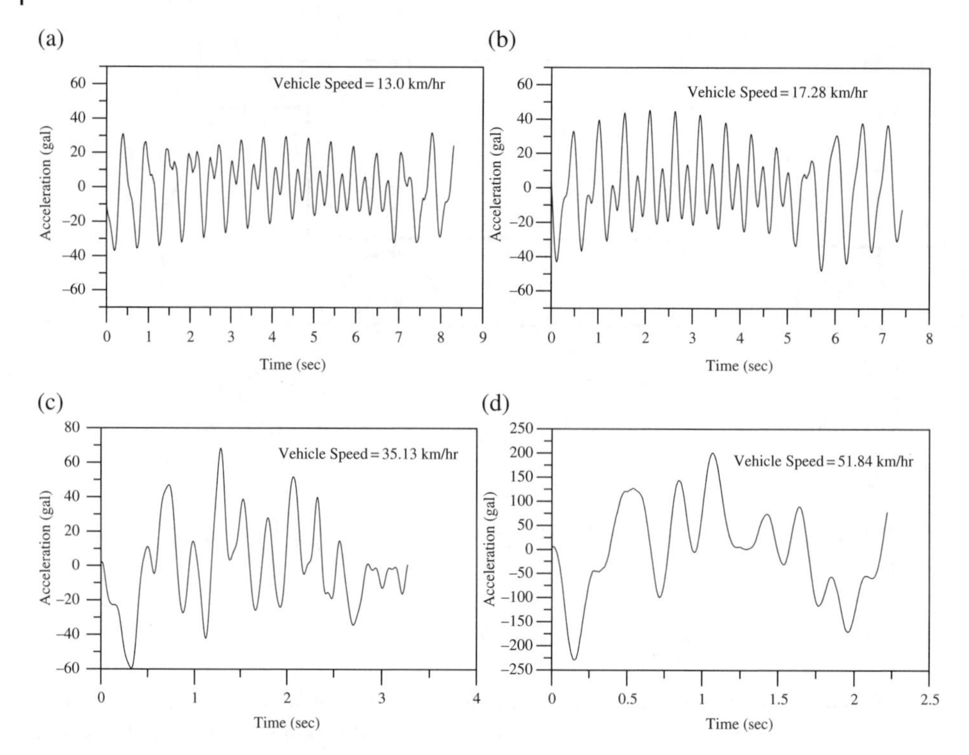


**Figure 4.17** Test cart response by numerical simulation for speed of: (a) 13.0 km/h; (b) 17.28 km/h; (c) 35.13 km/h; (d) 51.84 km/h.

**Table 4.1** Maximum test cart response for various speeds (gal).

| | **Vehicle Speed (km/h)** | | | |
|---|---|---|---|---|
| **Methods** | **13.0** | **17.28** | **35.13** | **51.84** |
| Numerical Simulation | 34.5 | 44.3 | 70.8 | 210.9 |
| Experiment Results | 28.7 | 50.2 | 98.1 | 215.4 |

the measured results with those from the comparative dynamic tests and numerical studies.

Based on the field tests, for the tractor-trailer moving at speeds less than 40 km/h, the bridge frequency can be easily identified from the response spectrum of the test cart. However, as the speed increases, the bridge frequency will be blurred due to pollution of high-frequency components, resulting from the cart structure or pavement roughness. The existence of ongoing traffic is considered beneficial because it tends to intensify the bridge vibration, which will be reflected in the test cart response.

For application of the present technique to practical problems, the following are suggested: (i) The dynamic properties of the test cart should be identified first; (ii) The vehicle speed should be kept lower for poorer road conditions in order to get better

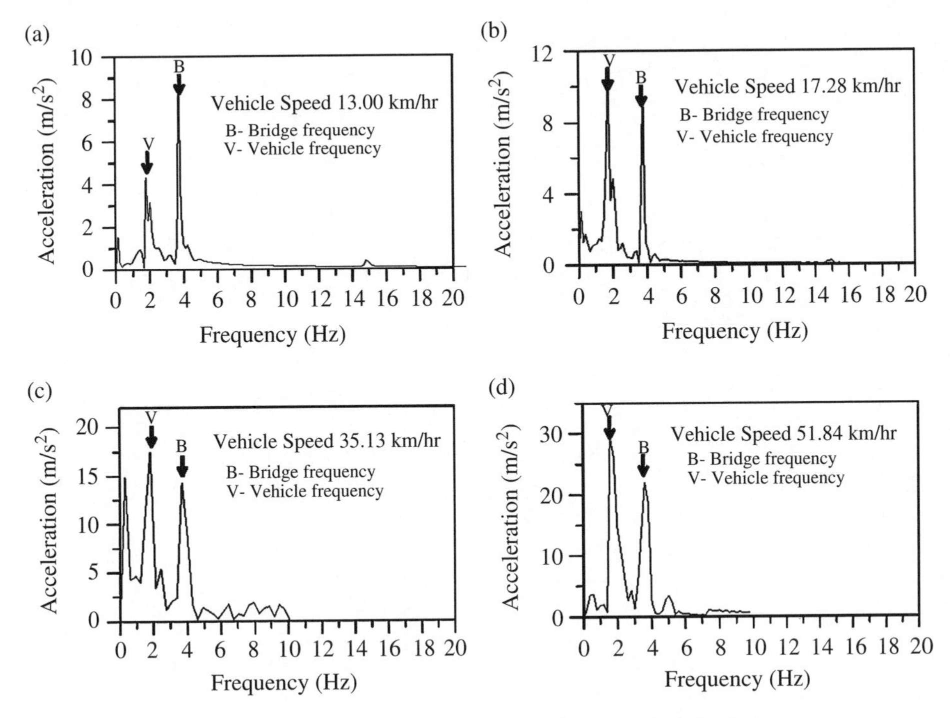


**Figure 4.18** Acceleration spectra of the test cart response by numerical simulation.

**Table 4.2** Bridge frequency extracted from test cart response for various speeds (Hz)[a].

| | Vehicle Speed (km/h) | | | |
|---|---|---|---|---|
| **Methods** | **13.0** | **17.28** | **35.13** | **51.84** |
| Numerical Simulation | 3.741 | 3.748 | 3.750 | 3.752 |
| Experiment Results | 3.728 | 3.727 | 3.734 | N/A |

[a] *Note:* Reference from ambient vibration test: 3.76 Hz.

resolution in the frequency domain; (iii) At least three runs of tests with different speeds should be conducted before the bridge frequency is ascertained from the test cart spectrum.

The technique presented herein can be used for scanning the first bridge frequency. To scan the second and higher frequencies of the bridge requires further improvement in the level of resolution for the FFT and the precision of the instrumentation in collecting the data. It should be admitted that all the conclusions made in this chapter are strictly valid for the conditions, assumptions, and instrumentation adopted in the test. Further studies should be carried out to identify the most suitable range of application of the approach proposed herein. In later chapters, we will discuss techniques for removing the pollution from high frequency components.

# 5

# EMD-Enhanced Vehicle Scanning of Bridge Frequencies

The method previously proposed for extracting the bridge frequencies from a passing vehicle works mainly for the first frequency. In order to extract bridge frequencies of higher modes, the vehicle response will first be processed by the *empirical mode decomposition* (EMD) to generate the *intrinsic mode functions* (IMFs), and then by the fast Fourier transform (FFT) to extract the bridge frequencies. One feature with the EMD technique is that frequencies of higher modes can be made more visible by the sifting process. To verify the feasibility of such an approach, the vehicle response generated by the finite element simulations will first be analyzed, with results compared with the analytical ones. Then the same procedure will be adopted to extract bridge frequencies from the recorded response of a passing vehicle, and the results will be compared with those from an ambient vibration test. It was demonstrated that using the IMFs computed from the vehicle response, rather than the original vehicle response, bridge frequencies of higher modes can be successfully extracted. The materials presented in this chapter are based primarily on Yang and Chang (2009a), but with slight modifications according to Chapter 3.

## 5.1 Introduction

The frequency of vibration is an important parameter for bridges, since it relates closely to the overall stiffness of the bridges. A drop in the frequency of vibration reflect on the structure's health, such as a deterioration in bridge stiffness, whether it is due to damages in any member or joint, settlement in support, or due to aging of materials of the bridge. For the purpose of structural control, the frequencies of a bridge are key parameters in determining the feedback forces that should be generated to counterbalance the forces induced on the bridge by external excitations, such as earthquakes. For a newly completed bridge, the frequencies of vibration are measured generally for two reasons. First, they serve as a reference for comparison with the frequencies used in design, by which the appropriateness of the design model can be assessed. Second, they represent the dynamic signature or baseline for monitoring the long-term variations of bridge behaviors. For the reasons stated above and beyond, the measurement of frequencies of vibration is often regarded as an essential part of the maintenance work for bridges during their service life.

*Vehicle Scanning Method for Bridges*, First Edition. Yeong-Bin Yang, Judy P. Yang, Bin Zhang and Yuntian Wu.

Conventionally, the bridge frequencies are measured directly from the dynamic response of the bridge, using vibration sensors directly mounted on the bridge. A variety of direct approaches have been summarized by Salawu and Williams (1995) according to the sources of vibration exerted on the bridge. One drawback with the conventional approaches for measuring bridge frequencies is that the deployment and maintenance of the vibration sensors and data loggers are generally costly, labor-intensive, and time-consuming, even though wireless data transmission is becoming popular nowadays. As was noted previously, all these drawbacks can be circumvented if the bridge frequencies are extracted indirectly from the recorded response of a moving test vehicle during its passage over the bridge. This has been the major advantage of the vehicle scanning method, since it does not require any instrumentation to be deployed on the bridge.

The key idea of the indirect approach is that the passing vehicle plays the dual role of vibration exciter to the bridge and message receiver of the bridge response. As a vibration exciter, the vehicle will set the bridge in vibration via its horizontal movement. Meanwhile, the vehicle will be excited by the vertical motion of the bridge. Since the vehicle is very small compared with the bridge in terms of mass or weight, the vertical motion of the vehicle will be dominated primarily by the bridge frequencies. As a result, the response recorded of the vehicle during its passage over the bridge can be processed to yield the bridge frequencies. This has been the idea behind the vehicle scanning method for extracting the bridge frequencies using the passing vehicle, as was theoretically and experimentally studied by Yang et al. (2004a), Yang and Lin (2005), and Lin and Yang (2005). Nevertheless, due to limitations in the resolution of the instruments and processing methods used, the proposed approach works mainly for extraction of the first frequency of the bridge.

In order to extend applicability of the vehicle scanning method to extracting the higher frequencies of the bridge, the EMD technique proposed by Huang et al. (1998, 1999b) are adopted herein to preprocess the data recorded for the vehicle during its passage over the bridge. The EMD is a newly developed signal processing technique, which is especially suitable for the processing of nonlinear and nonstationary signals. By the EMD procedure or by a sifting process, any complicated dynamic response can be decomposed into a set of *intrinsic mode functions* (IMFs), generally arranged in order from high to low frequencies. Since the sifting process can recover low-amplitude riding waves with repeated siftings (Huang et al. 1998), it is expected that bridge frequencies of higher modes, which appear to be small in amplitudes in the vehicle spectrum with no EMD processing, can be made more visible in the first few IMFs associated with high-frequency components.

For a vehicle moving over the bridge, the signals recorded from the passing vehicle are nonlinear and nonstationary in nature, which is the case where the EMD can find its application. In this chapter, the response recorded (or computed) for a vehicle traveling over a bridge will first be preprocessed by the EMD to generate the IMFs. The FFT will then applied to all the IMFs to extract the bridge frequencies. Such a procedure enables us to recover some of the higher modes missing from the original vehicle spectrum.

In order to unveil the key frequencies associated with the vehicle-bridge system, an analytical formulation will first be conducted, based on the assumption of small mass ratio for the vehicle relative to the bridge. Three examples will then be studied using the vehicle response generated from the finite element simulations. The bridge frequencies extracted in this regard will be compared with those obtained analytically. Furthermore, the procedure used will be extended to treating the response recorded of the test vehicle during its passage over a bridge. The bridge frequencies extracted will be compared with those obtained from an ambient vibration test.

## 5.2 Analytical Formulation of the Problem

To highlight the major dynamic characteristics of the coupled vehicle-bridge system, the simplified mathematical model used previously, as in Figure 5.1, will be adopted. The vehicle is modeled as a lumped mass $m_v$, supported by a spring of stiffness $k_v$ and traveling at speed $v$ across a simply supported beam of length $L$. The beam is assumed to be of the Bernoulli-Euler type with constant cross section and smooth pavement.

By neglecting the damping effects of both the bridge and vehicle, the equations of motion for the bridge and vehicle can be written as

$$\bar{m}\ddot{u} + EIu'''' = f_c(t)\delta(x - vt), \tag{5.1}$$

$$m_v\ddot{q}_v + k_v\left(q_v - u\big|_{x=vt}\right) = 0, \tag{5.2}$$

where for the beam, $\bar{m}$ denotes the mass per unit length, $E$ the elastic modulus, $I$ the moment of inertia, and $u(x, t)$ the vertical displacement of the beam, for the vehicle, $q_v$ denotes the vertical displacement, measured from the static equilibrium position of the vehicle, and a dot and a prime denote differentiation with respect to time $t$ and coordinate $x$, respectively, of the beam. The contact force $f_c$ between the vehicle and bridge can be expressed as the sum of the weight of the vehicle and the elastic force of the suspension system:

$$f_c(t) = -m_v g + k_v\left(q_v - u\big|_{x=vt}\right), \tag{5.3}$$

where $g$ is the acceleration of gravity. The analytical solution to the aforementioned bridge and vehicle equations presented in previous chapters will be summarized below, since it allows us to identify the key parameters involved in the vehicle-bridge interaction (VBI).

By the modal superposition method, the solution to Eq. (5.1) can be expressed in terms of the modal shapes, $\phi_n(x)$, and modal coordinates, $q_{b,n}(t)$. In addition, for the

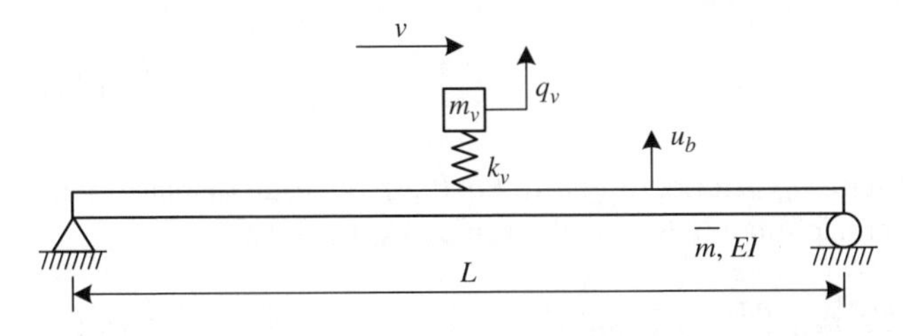


**Figure 5.1** Lumped sprung mass moving across a simply supported beam.

simple beam considered herein, the modal shapes of the beam satisfying the boundary conditions are of the sinusoidal form. Namely, the solution to Eq. (5.1) can be expressed as (Biggs 1964)

$$u(x,t) = \sum_n \phi_n(x) q_{b,n}(t) = \sum_n \sin\frac{n\pi x}{L} q_{b,n}(t). \tag{5.4}$$

where $\phi_n$ is the modal shape and $q_{b,n}$ the modal coordinate. Substituting Eq. (5.4) for the beam displacement $u(x, t)$ into Eq. (5.1) and using the orthogonality conditions for the modal shapes, along with the assumption that the vehicle mass $m_v$ is much less than the beam mass $\bar{m}L$, i.e., $m_v/\bar{m}L \ll 1$, the equation of motion for the beam in Eq. (5.1) can be expressed in terms of the modal coordinates as

$$\ddot{q}_{b,n} + \omega_{b,n}^2 q_{b,n} = -\frac{2m_v g}{\bar{m}L}\sin\frac{n\pi vt}{L}, \tag{5.5}$$

where $\omega_{b,n}$ is the natural frequency of the beam of the $n$th mode:

$$\omega_{b,n} = \frac{n^2\pi^2}{L^2}\sqrt{\frac{EI}{\bar{m}}}. \tag{5.6}$$

Meanwhile, the vehicle equation in Eq. (5.2) can be rewritten as

$$\ddot{q}_v + \omega_v^2 q_v = \omega_v^2 \left. u \right|_{x=vt}, \tag{5.7}$$

where $\omega_v$ denotes the frequency of vibration of the vehicle,

$$\omega_v = \sqrt{\frac{k_v}{m_v}}. \tag{5.8}$$

For zero initial conditions, the solution for the modal coordinate $q_{b,n}$ to the beam equation in Eq. (5.5) is

$$q_{b,n} = \frac{\Delta_{st,n}}{1-S_n^2}\left[\sin\frac{n\pi vt}{L} - S_n \sin\omega_{b,n}t\right], \tag{5.9}$$

where $\Delta_{st,n}$ is the static deflection of the $n$th mode of the beam induced by the vehicle,

$$\Delta_{st,n} = \frac{-2m_v g L^3}{n^4\pi^4 EI}, \tag{5.10}$$

and $S_n$ is a nondimensional speed parameter,

$$S_n = \frac{n\pi v}{L\omega_{b,n}}. \tag{5.11}$$

Substituting the modal coordinates $q_{b,n}$ in Eq. (5.9) into Eq. (5.4), the total displacement response of the beam to the vehicle moving at speed $v$ can be expressed as

$$u(x,t) = \sum_n \frac{\Delta_{st,n}}{1-S_n^2}\left\{\sin\frac{n\pi x}{L}\left[\sin\frac{n\pi vt}{L} - S_n\sin\omega_{b,n}t\right]\right\}. \tag{5.12}$$

With the substitution of the total displacement of the beam, $u(x, t)$, in Eq. (5.12), one can obtain by Duhamel's integral from Eq. (5.7) the vehicle displacement as follows:

$$q_v(t)=\sum_n\left[A_{1n}+A_{2n}\cos\left(\frac{2n\pi v}{L}\right)t+A_{3n}\cos(\omega_v t)\right.$$
$$\left.+A_{4n}\cos\left(\omega_{b,n}-\frac{n\pi v}{L}\right)t+A_{5n}\cos\left(\omega_{b,n}+\frac{n\pi v}{L}\right)t\right], \tag{5.13}$$

where the coefficients, i.e., the amplitudes, are given as follows:

$$A_{1n}=\frac{\Delta_{stn}}{2\left(1-S_n^2\right)}, \tag{5.14a}$$

$$A_{2n}=\frac{\Delta_{stn}}{2\left(1-S_n^2\right)\left(4\mu_n^2S_n^2-1\right)}, \tag{5.14b}$$

$$A_{3n}=\frac{2\Delta_{stn}S_n^2\mu_n^4\left(2+\mu_n^2S_n^2-\mu_n^2\right)}{\left(4\mu_n^2S_n^2-1\right)\left[1-2\mu_n^2\left(1+S_n^2\right)+\mu_n^4\left(1-S_n^2\right)^2\right]}, \tag{5.14c}$$

$$A_{4n}=-\frac{\Delta_{stn}S_n}{2\left(1-S_n^2\right)\left[1-\mu_n^2\left(1-S_n\right)^2\right]}, \tag{5.14d}$$

$$A_{5n}=\frac{\Delta_{stn}S_n}{2\left(1-S_n^2\right)\left[1-\mu_n^2\left(1+S_n\right)^2\right]}, \tag{5.14e}$$

and the nondimensional parameter $\mu_n$ is defined as the ratio of the $n$th mode natural frequency of beam $\omega_{b,n}$ to the vehicle frequency $\omega_v$:

$$\mu_n=\frac{\omega_{b,n}}{\omega_v}. \tag{5.15}$$

The solution given in Eq. (5.13) for the vehicle differs from that of Yang and Lin (2005) in that the amplitudes are given in a physically more meaningful sense.

The bridge displacement in Eq. (5.12) reveals that there are two groups of frequencies, i.e., the driving frequencies $n\pi v/L$ of the vehicle and the natural frequencies $\omega_{b,n}$ of the beam. The same property also holds for the velocity and acceleration of the beam (not shown here). This is exactly the idea behind the conventional direct approach for measuring the bridge frequencies from the bridge responses.

The acceleration response of the vehicle can be obtained by differentiating the displacement in Eq. (5.13) twice:

$$\ddot{q}_v(t)=\sum_n\left[\tilde{\tilde{A}}_{2n}\cos\left(\frac{2n\pi v}{L}\right)t+\tilde{\tilde{A}}_{3n}\cos(\omega_v t)\right.$$
$$\left.+\tilde{\tilde{A}}_{4n}\cos\left(\omega_{b,n}-\frac{n\pi v}{L}\right)t+\tilde{\tilde{A}}_{5n}\cos\left(\omega_{b,n}+\frac{n\pi v}{L}\right)t\right], \tag{5.16}$$

**Table 5.1** Coefficients for vehicle's acceleration response.

| Coefficient | $\tilde{\tilde{A}}_{2n}$ | $\tilde{\tilde{A}}_{3n}$ | $\tilde{\tilde{A}}_{4n}$ | $\tilde{\tilde{A}}_{5n}$ |
|---|---|---|---|---|
| Definition | $-A_{2n}(2\omega_{b,n}S_n)^2$ | $-A_{3n}\omega_v^2$ | $-A_{4n}\omega_{b,n}^2(1-S_n)^2$ | $-A_{5n}\omega_{b,n}^2(1+S_n)^2$ |

where the coefficients are listed in Table 5.1.

From Eqs. (5.13) and (5.16), one observes that there exist four different frequency components, which can be categorized into three groups as: (i) driving-related frequencies, i.e., $2n\pi v/L$ with amplitude $A_{2n}$; (ii) vehicle frequency $\omega_v$ with amplitude $A_{3n}$; and (iii) bridge-related frequencies, including the left-shifted ones, $\omega_{bl,n} = \omega_{b,n} - n\pi v/L$, and right-shifted ones, $\omega_{br,n} = \omega_{b,n} + n\pi v/L$, with amplitudes $A_{4n}$ and $A_{5n}$, respectively, with $n$ indicating the sequential mode number. Such a result offers a clue for extracting the bridge frequencies from the vehicle responses, which is known to be the theoretical basis for the vehicle scanning method.

Whether each frequency of the bridge can be successfully extracted from the vehicle response depends on its amplitude relative to the others, the vehicle passing speed, the sampling rate for taking the data, and the numerical method of data processing. In practice, the acceleration response of the vehicle can be measured using the seismometers commonly available. For this reason, only the acceleration response of the vehicle will be discussed in this chapter.

## 5.3 Finite Element Simulation of the Problem

Before the idea of preprocessing the vehicle response by the EMD to extract the bridge frequencies is physically tested in the field, it will first be numerically tested for feasibility. To this end, the finite element procedure to be adopted for generating the vehicle response is summarized, by which virtually no assumption is made, compared with the analytical approach presented in Section 5.2.

The beam element and the sprung mass model shown in Figure 5.2 have been referred to as the VBI element, of which the relevant matrices based on Chang et al. (2010) are summarized in the Appendix, as a modification from the ones by Yang and Yau (1997).

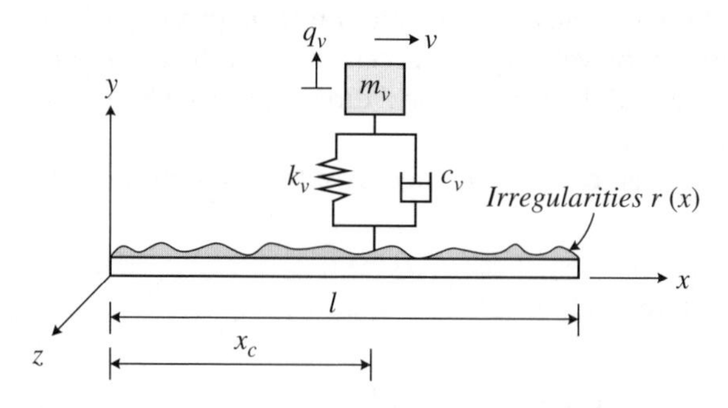


**Figure 5.2** Vehicle-bridge interaction (VBI) element.

In this figure, $r(x)$ denotes the surface irregularity, $x_c$ position of the contact point, $m_v$ sprung mass, and $k_v$ and $c_v$ the stiffness and damping coefficients of the suspension unit, respectively. The beam is modeled as a 12-degree-of-freedom (DOF) system and the sprung mass as a single DOF system. The equations of motion for the sprung mass $m_v$ and the bridge element directly in contact can be written as follows:

$$\begin{bmatrix} m_v & 0 \\ 0 & [m_b] \end{bmatrix} \begin{Bmatrix} \ddot{q}_v \\ \{\ddot{u}_b\} \end{Bmatrix} + \begin{bmatrix} c_v & -c_v\{N\}_c^T \\ -c_v\{N\}_c & [c_b]+c_v\{N\}_c\{N\}_c^T \end{bmatrix} \begin{Bmatrix} \dot{q}_v \\ \{\dot{u}_b\} \end{Bmatrix}$$
$$+\begin{bmatrix} k_v & -vc_v\{N'\}_c^T - k_v\{N\}_c^T \\ -k_v\{N\}_c & [k_b]+vc_v\{N\}_c\{N'\}_c^T + k_v\{N\}_c\{N\}_c^T \end{bmatrix} \begin{Bmatrix} q_v \\ \{u_b\} \end{Bmatrix} \quad (5.17)$$
$$= \begin{Bmatrix} 0 \\ -m_v g\{N\}_c \end{Bmatrix}.$$

where $\{u_b\}$ denotes the displacement vector of the bridge element, $[m_b]$, $[c_b]$, $[k_b]$ the mass, damping, stiffness matrices of the bridge element, $\{N\}$ is a vector containing the cubic Hermitian interpolation functions, and $\{N\}_c$ represents the vector $\{N\}$ evaluated at the coordinate position of the contact point.

By a dynamic condensation procedure, the vehicle displacement $q_v$ in the first line of Eq. (5.17) can be condensed into the displacement vector $\{u_b\}$ of the element in contact, resulting in the so-called VBI element (Yang and Yau 1997). The VBI element so derived can then be assembled with equations of other bridge elements not directly in contact with the sprung mass to yield the equations of motion as

$$[M]\{\ddot{u}\}+[C]\{\dot{u}\}+[K]\{u\}=\{F\}, \quad (5.18)$$

where $\{u\}$ denotes the system displacement vector, $[M]$, $[C]$, $[K]$ the mass, damping, and stiffness matrices of the system, and $\{F\}$ the external force vector.

The VBI element just mentioned not only has the same number of DOFs as the parent element but also possesses the property of symmetry in element matrices. Consequently, the procedure for solving Eq. (5.18) is identical to that conventionally used for solving structures subjected to dynamic loads, except that the VBI element matrices should be updated at each time step. In this chapter, Eq. (5.18) will be solved by Newmark's $\beta$ method with constant average acceleration (i.e., with $\beta = 0.25$ and $\gamma = 0.5$). Once the displacement, velocity, and acceleration responses of the bridge are made available, the vehicle response can be computed by backward substitution.

## 5.4 Empirical Mode Decomposition

In general, the data collected from the vehicle passing over a bridge are of finite duration, nonstationary, and nonlinear. They may also be polluted by the sensing and numerical procedures used. Thus, a direct processing of the data by the Fourier transform is not physically sound, since the latter is strictly valid to linear and stationary systems. The EMD method proposed by Huang et al. (1998) was specially tailored for treating nonstationary and nonlinear data. In this chapter, the EMD will be adopted to decompose the data collected (or computed) for the vehicle into IMFs that admit well-behaved Hilbert transforms (1998). The essence of the EMD is to identify the intrinsic oscillatory modes by their characteristic time scales in the data empirically, and then to decompose the data accordingly. Generally, the finest vibration mode or component of the shortest period at each instant will be identified and decomposed into the first IMF, and the components of longer periods will be identified and decomposed into the following IMFs in sequence. Therefore, the first bridge frequency, with a longer period, may not appear in the first few IMFs.

Each IMF represents a simple oscillation mode embedded in the data, which satisfies two conditions:

1) It possesses the same or nearly the same numbers of extrema and zero-crossings.
2) It is symmetric with respect to the local zero mean.

In numerical processing, we shall demonstrate that a better resolution can be achieved for higher frequencies of the bridge if the FFT is performed on the IMFs rather than on the original data.

The procedure of the EMD is to construct the upper and lower envelopes of the data set by spine fitting, and to compute the mean of both envelopes. Then, the data set is subtracted from the mean, referred to as the sifting process. By repeating the sifting process until the resulting data set satisfies the aforementioned two conditions, then it is treated as an IMF. The original data can then be subtracted from the IMF, and the process repeated on the remaining data set to obtain another IMF. The IMF obtained in each cycle by the sifting process involves only one mode of oscillation, with no complex riding waves allowed. As a counterpart to the well-known simple harmonic function, an IMF represents a much more general simple oscillatory mode. It may have a variable amplitude and frequency as functions of time, instead of constant amplitude and frequency in a simple harmonic function.

Given a set of measured data $X(t)$, the algorithm of the EMD, characterized by the sifting process, can be briefly described as follows:

1) Identify all the local maxima and minima of the data $X(t)$, and then form the upper and lower envelopes by interpolating the local maxima and minima, respectively, by cubic spline curves. All the extrema should be covered in these two envelopes. Let $m_1$ denote the mean of the upper and lower envelopes. The difference between the data and the mean $m_1$ is

$$h_1 = X(t) - m_1 \tag{5.19}$$

2) Ideally, $h_1$ should be the first IMF component. If $h_1$ does not satisfy the IMF requirements, treat it as the original data and repeat the first step until the requirements are satisfied. The first IMF component obtained is designated as $c_1$.
3) By subtracting $c_1$ from the original data, one obtains the residue $r_1$ as

$$r_1 = X(t) - c_1. \tag{5.20}$$

If $r_1$ still contains information of other period components, treat it as the new data and repeat the above sifting processes to obtain the next IMF $c_2$. The process is repeated until the following predetermined criteria are met: either when the component, $c_n$, or the residue $r_n$ becomes too small to be physically meaningful or when the residue $r_n$ becomes a monotonic function from which no more IMF can be extracted. Consequently, the data set $X(t)$ is decomposed into $n$ IMFs, $c_1$ to $c_n$, plus the final residue, $r_n$:

$$X(t) = \sum_{i=1}^{n} c_i + r_n. \tag{5.21}$$

As a whole, the first IMF $c_1$ should contain the finest vibration mode or component of the shortest period, and the following IMFs contain components of longer periods in sequence. As for the final residue $r_n$, it can be either the mean trend or a constant.

## 5.5 Extraction of Bridge Frequencies by Numerical Simulation

The following are the properties adopted for the simple beam: length $L$ = 25 m, elastic modulus $E = 2.75 \times 10^{10}\,\text{N/m}^2$, mass per unit length $\bar{m}$ = 4800 kg/m, and moment of inertia $I$ = 0.12 m$^4$. The vehicle has a lumped mass $m_v$ = 500 kg, spring constant $k_v$ = 500 kN/m, and frequency $\omega_v$ = 5.03 Hz. The beam is discretized into 20 elements, for which the first three natural frequencies are: $\omega_{b,1}$ = 2.08 Hz, $\omega_{b,2}$ = 8.33 Hz, $\omega_{b,3}$ = 18.75 Hz. The time step used in each simulation is 0.001 second.

Figure 5.3 shows the acceleration response of the vehicle traveling at speed $v$ = 4 m/s over the beam, obtained both by the finite element and analytical methods. The slight deviation of the analytical result from the finite element one indicates that the

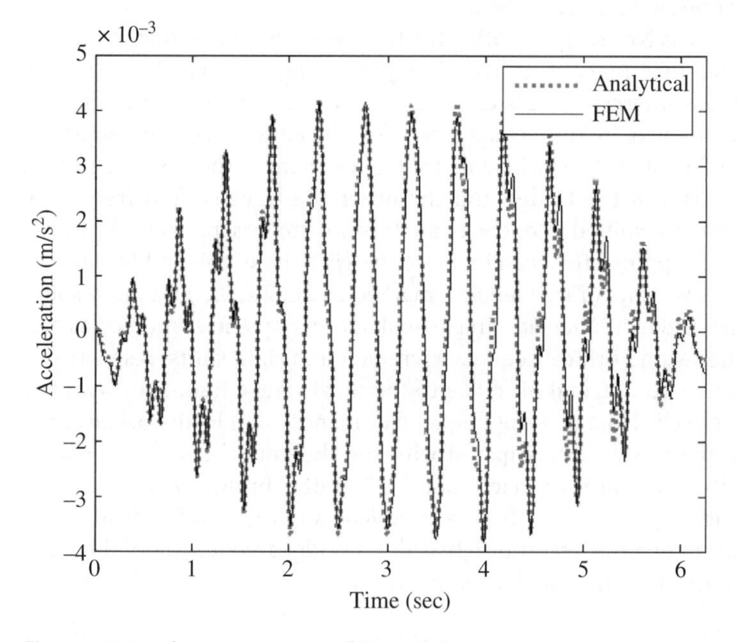


**Figure 5.3** Acceleration response of the vehicle.

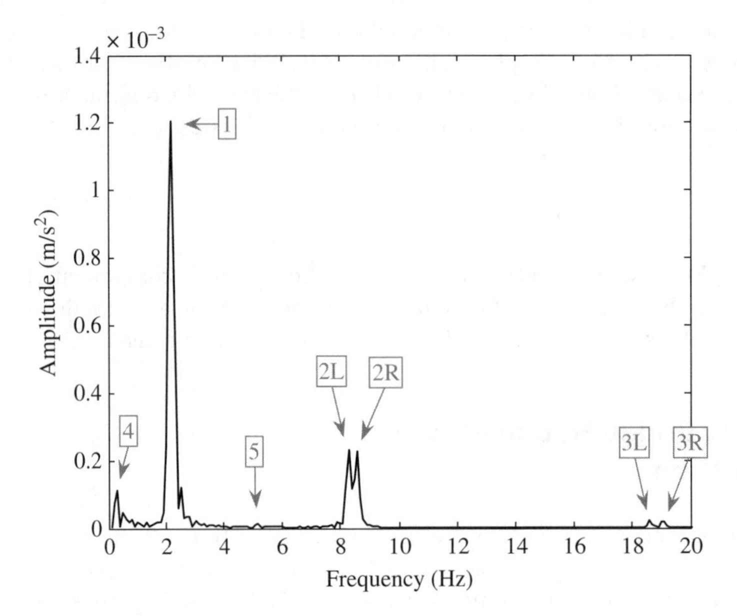


**Figure 5.4** Acceleration spectrum of the vehicle.

assumption of a small mass ratio for the vehicle to the bridge is acceptable. In order to examine the frequency contents in the acceleration response of the vehicle, the Fourier transformation is performed on the result obtained from the finite element analysis. As can be seen from the spectrum given in Figure 5.4, several frequencies associated with the peak amplitudes can be identified. All of them can be interpreted using the analytical theory previously presented, as given below.

The frequency marked as No. 4, representing the first right-shifted driving frequency ($\pi v/L$ = 0.08 Hz), and that marked as No. 5, representing the vehicle frequency $\omega_v$ = 5.03 Hz, are of less interest. In this chapter, we are concerned mainly with the bridge frequencies contained in the vehicle response. Theoretically, the frequency marked as No. 1 should contain two individual frequencies: one is the first left-shifted frequency ($\omega_{bl,1}$ = 2.0 Hz) of the bridge and the other the right-shifted frequency ($\omega_{br,1}$ = 2.16 Hz). But the two shifted frequencies are too close to be separately identified. In practice, such an overlapping effect can increase the visibility of the first bridge frequency. The following two pairs of frequencies, marked as 2L, 2R, 3L, and 3R, respectively, also relate to the bridge frequencies. In particular, the frequencies marked as No. 2L and 2R relate to the second bridge frequency with left and right shifts, respectively, while the ones marked as No. 3L and 3R relate to the third bridge frequency with left and right shifts, respectively. Note that the frequencies identified in Figure 5.4 all agree well with those from the theory, as compared with the theoretical results: (i) $\omega_{bl,2}$ = 8.16 Hz and $\omega_{br,2}$ = 8.48 Hz for the second left- and right-shifted bridge frequencies, and (ii) $\omega_{bl,3}$ = 18.48 Hz and $\omega_{br,3}$ = 18.96 Hz for the third left- and right-shifted bridge frequencies. By simply averaging the left- and right-shifted bridge frequencies of the same mode, the desired bridge frequency can be computed as

$$\omega_{b,n} = \frac{\omega_{bl,n} + \omega_{br,n}}{2}. \tag{5.22}$$

As indicated by the spectrum in Figure 5.4, the first bridge frequency can be identified with no difficulty, while frequencies of the higher modes are much less visible, due to their relatively smaller amplitudes compared with the first mode. In a field test, it is likely that the frequencies of the higher modes (i.e., marked as 3L, 3R, and 5) will become invisible, since they are rather small in magnitudes and may be polluted by noises generated by the mechanical components of the vehicle traveling over rough surface. To improve the visibility of the higher modes, the EMD will first be applied to processing the vehicle's response, prior to the Fourier spectral analysis. Three examples will be studied to explore the feasibility of such a technique for treating the vehicle response.

### 5.5.1 Example 1: Single Moving Vehicle

The time–history acceleration response of the vehicle shown in Figure 5.3 will be decomposed using the EMD to yield the IMF components, $c_1$ to $c_4$, and the final residue, $r_5$, as in Figure 5.5. It can be roughly observed that the data are separated according to the contents of frequencies from high to low in sequence.

The Fourier spectra of the first three IMF components, carrying more useful vibrational messages, have been plotted in Figure 5.6a–c. From Figure 5.6a, it can be observed that the dominant frequency set, marked as 2L and 2R, represents the second bridge frequencies with left and right shifts, respectively. In comparison with the Fourier spectrum of the original data shown in Figure 5.4, the visibility of the second bridge-related frequencies has been greatly enhanced via use of the EMD. Even bridge frequencies of the higher modes (e.g., the ones associated with the third mode, marked as 3L and 3R), are made more visible in the spectrum of $c_1$. Although the peaks of the second and third

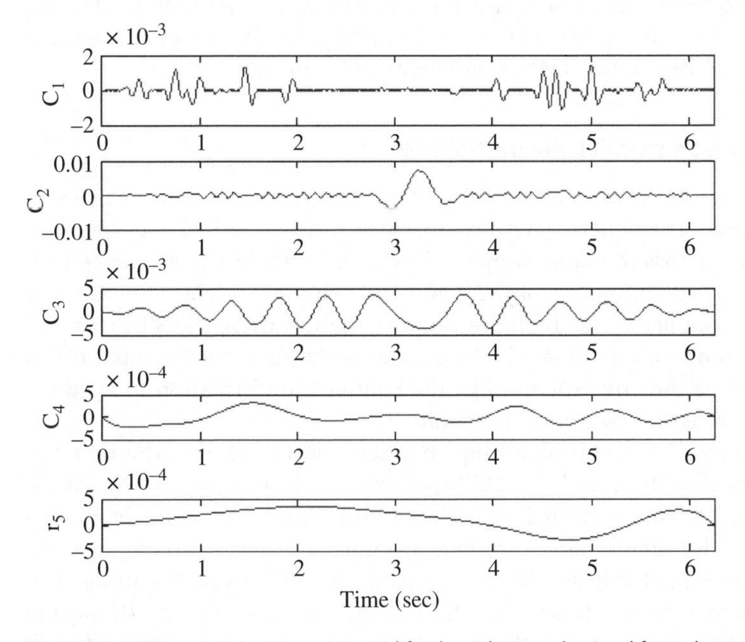


**Figure 5.5** IMF components, $c_1$ to $c_4$, and final residue, $r_5$, obtained from the acceleration response of the vehicle.

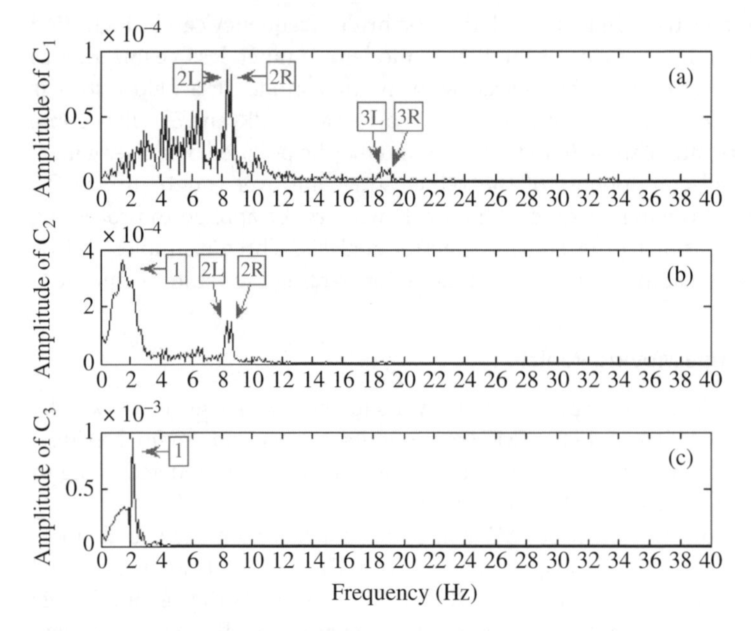


**Figure 5.6** Fourier spectra of IMF components: (a) $c_1$; (b) $c_2$; and (c) $c_3$.

modes are not as distinct as the first peak in the spectrum of $c_3$, they become more visible in terms of the relative magnitudes, as compared with the Fourier spectrum of the original data given in Figure 5.4.

The first bridge frequency set can be clearly identified in the Fourier spectra of $c_2$ and $c_3$, as shown in Figure 5.6b and c, respectively. Undoubtedly, this frequency remains the most dominant one for the last two IMF components in Figure 5.6.

### 5.5.2 Example 2: Five Sequential Moving Vehicles

The purpose of this example is to investigate the effect of sequential moving vehicles on the extraction of bridge frequencies. Assume that five identical vehicles of identical interval $s = 5$ m travel across the same simple beam as in Example 1 in sequence with the speed $v = 10$ m/s. The properties of the beam and vehicles are the same as those previously adopted. Conduct again the finite element analysis using the VBI element. The acceleration response of the third vehicle, i.e., the test vehicle, has been plotted in Figure 5.7, which can be directly processed by the Fourier transformation to yield the spectrum in Figure 5.8, as was conventionally done.

As indicated by Figure 5.8, the bridge frequency sets associated with the first and second modes, marked as No. 1 and No. 2L/2R, are rather clear. A comparison of the spectrum in Figure 5.8 for the present case with that in Figure 5.4 for a single moving vehicle indicates that the simultaneous presence of moving vehicles can enhance the visibility of higher modes of the beam. Nevertheless, the amplitudes for the bridge frequency sets associated with the fourth and fifth modes, marked as No. 4L/4R and 5L/5R, are not as visible compared with the first two modes. In practice, such peaks will likely be polluted by noises of various sources and become generally invisible.

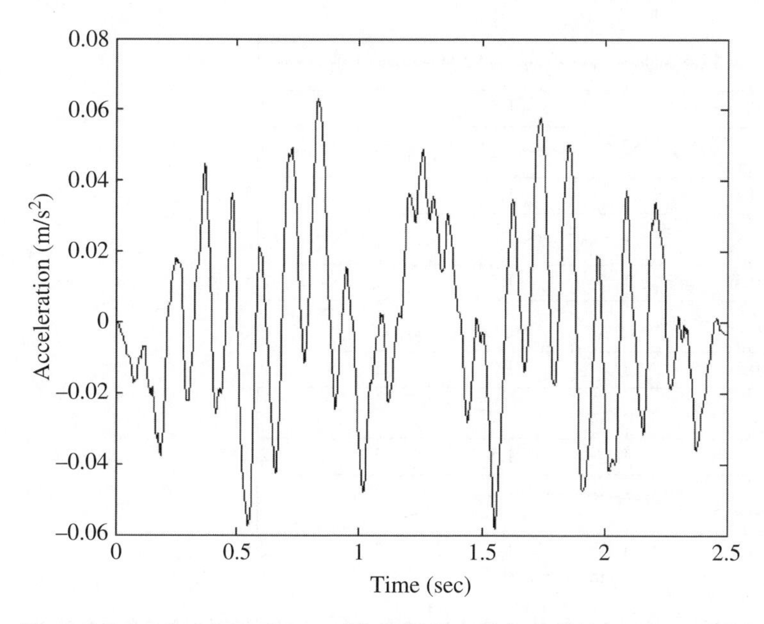


**Figure 5.7** Acceleration response of the middle vehicle.

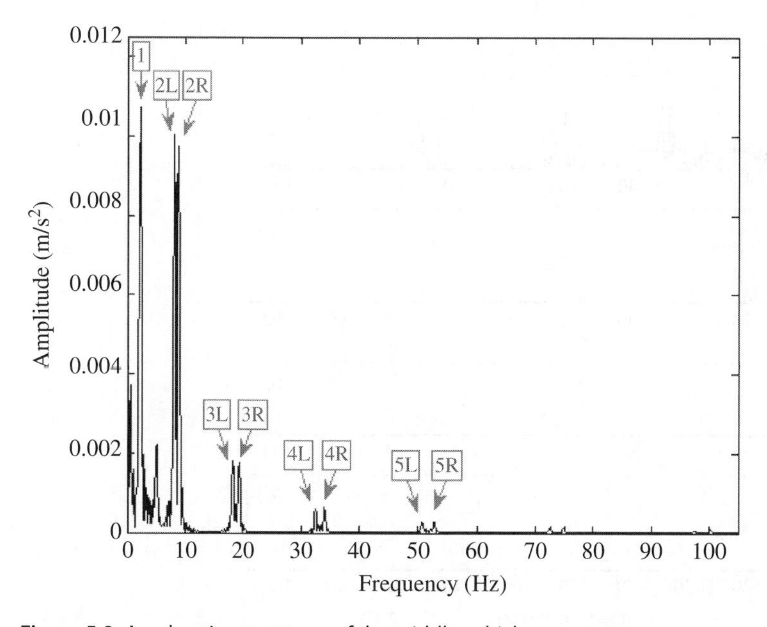


**Figure 5.8** Acceleration spectrum of the middle vehicle.

The five IMF components, $c_1$ to $c_5$, and final residue, $r_6$, computed from the acceleration response of the middle vehicle have been plotted in Figure 5.9, along with the corresponding Fourier spectra in Figure 5.10. Four obvious sets of amplitude peaks can be observed from Figure 5.10a, which represent the frequency sets associated with the fourth mode (4L/4R), fifth mode (5L/5R), sixth mode (6L/6R), and seventh mode (7L/7R)

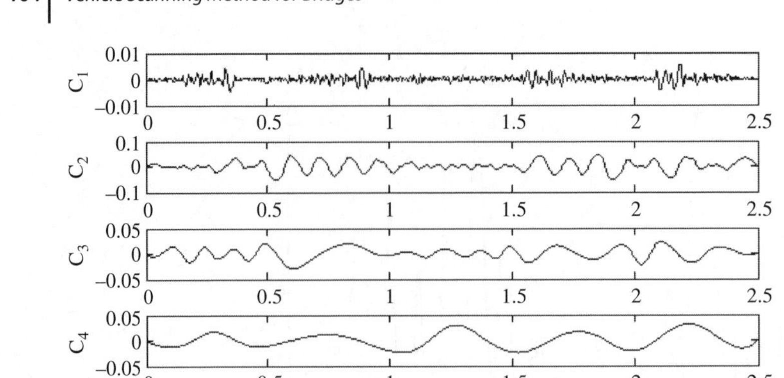


Figure 5.9 IMF components, $c_1$ to $c_5$, and final residue, $r_6$, obtained from the acceleration response of the middle vehicle.

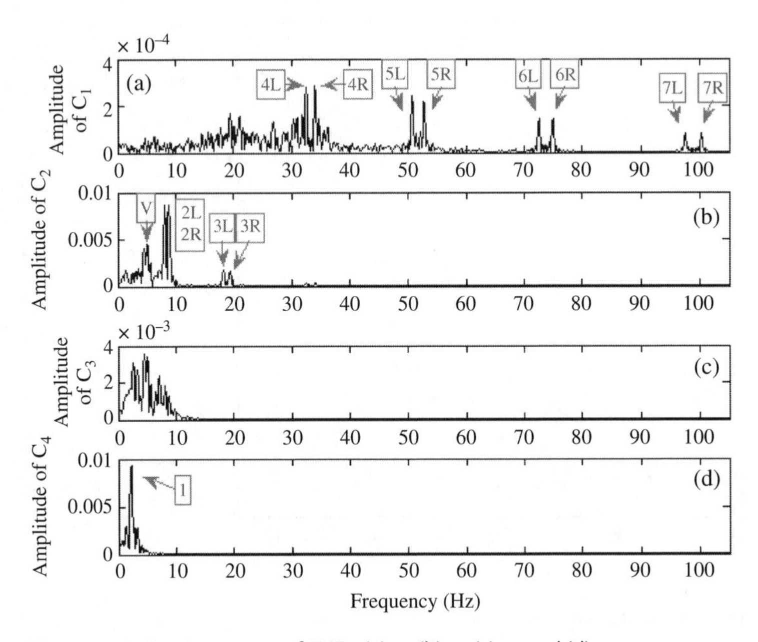


Figure 5.10 Fourier spectra of IMFs: (a) $c_1$; (b) $c_2$; (c) $c_3$; and (d) $c_4$.

in sequence. Obviously, higher modes such as the sixth and seventh modes, which are almost invisible in the original spectrum of Figure 5.8, have become visible due to the preprocessing by EMD. This is clearly an evidence of the capability of the EMD to improve the visibility of bridge frequencies of the higher modes.

The first three identifiable frequency sets in the spectrum of the original data remain identifiable in the spectra of the last few IMFs, e.g., the one associated with the first mode in Figure 5.10d and the ones associated with the second and third modes in Figure 5.10b. It should be mentioned that the peak marked as letter V represents the vehicle frequency, which is not of concern in this chapter.

### 5.5.3 Example 3: Five Random Moving Vehicles

The third example is to simulate the most general case in reality – random traffic. Five different vehicles, as represented by five sprung masses of randomly assigned values, enter the same bridge as in Example 1 with randomly assigned but constant speeds at different times (see Table 5.2). For our purposes, the first vehicle is designated as the test vehicle, of which the dynamic responses will be used to extract the bridge frequencies.

The acceleration response time history and its Fourier spectrum are shown in Figures. 5.11 and 5.12, respectively. As can be observed from Figure 5.12, the first three sets of frequency, marked as 1, 2L/2R, and 3L/3R, are rather clear compared with the results in Example 1 for the case with single moving mass, which indicates that the existence of random traffic may excite the dynamic responses of higher modes to a certain level due to the relatively higher amount of energy input. Nevertheless, the fourth and fifth sets of frequency, 4L/4R and 5L/5R, remain only observable, but not clear enough.

The IMF components obtained by the EMD technique, i.e., $c_1$ to $c_6$, and final residue, $r_7$, have been plotted in Figure 5.13, along with the corresponding spectra in Figure 5.14. It is noteworthy that the third to fifth sets of frequency, 3L/3R, 4L/4R, and 5L/5R, become more identifiable in Figure 5.14a than in Figure 5.12. The dominant frequencies in the second to fourth spectra of Figure 5.14 represent the first few sets of frequency of the bridge.

## 5.6 Experimental Studies

The response history data in the cases studied above are obtained by the finite element simulation, which represent the ideal cases that are free of any noise pollution. Nevertheless, the above analyses did indicate that visibility of higher frequencies can be

**Table 5.2** Randomly assigned values of vehicle mass and speed.

| Vehicle Number | Mass (kg) | Speed (m/s) | Initial Spacing[a] (m) | Remark |
|---|---|---|---|---|
| 1 | 500 | 4 | – | Test vehicle |
| 2 | 800 | 15 | 1 | |
| 3 | 1000 | 5 | 3 | |
| 4 | 400 | 12 | 0 | |
| 5 | 1200 | 8 | 2 | |

[a] Initial spacing is the spacing between each vehicle and the test vehicle when the test vehicle enters the bridge.

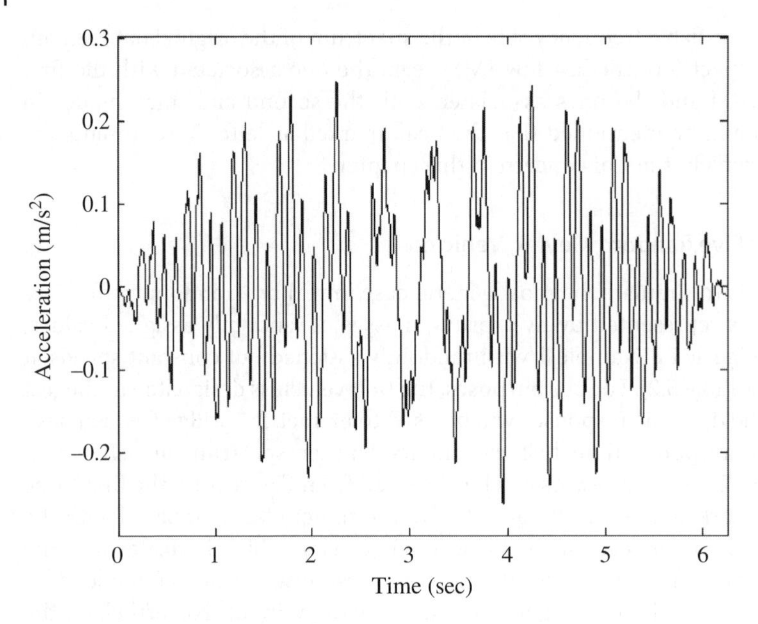


**Figure 5.11** Acceleration response of the test vehicle.

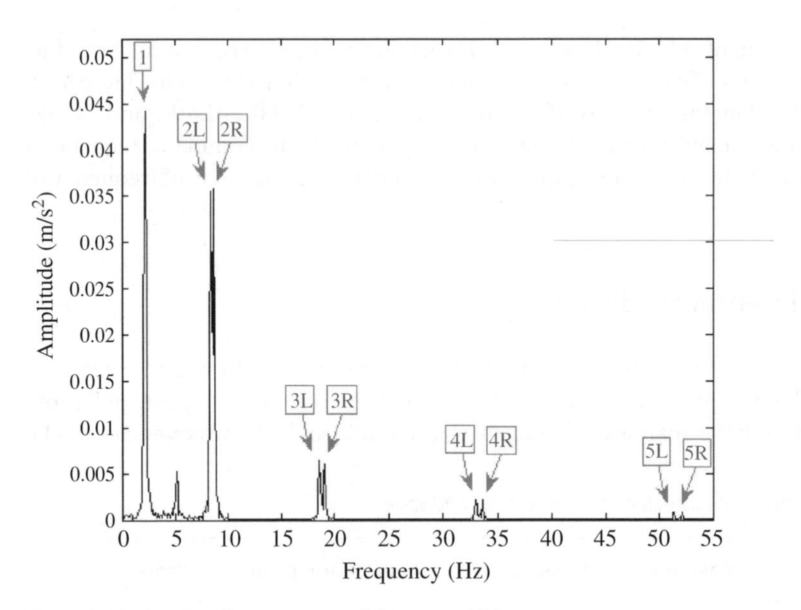


**Figure 5.12** Acceleration spectrum of the test vehicle.

enhanced through preprocessing by the EMD technique. The objective of this section is to explore if the EMD technique remains applicable for enhancing the higher frequencies contained in the data recorded by the moving test vehicle in the field. To this end, a set of field experiments using the tractor-trailer system was conducted for a bridge located in northern Taiwan, as in Chapter 4 and Lin and Yang (2005). For the purpose of comparison, the data recorded will be processed by the FFT with or without the preprocessing by the EMD technique.

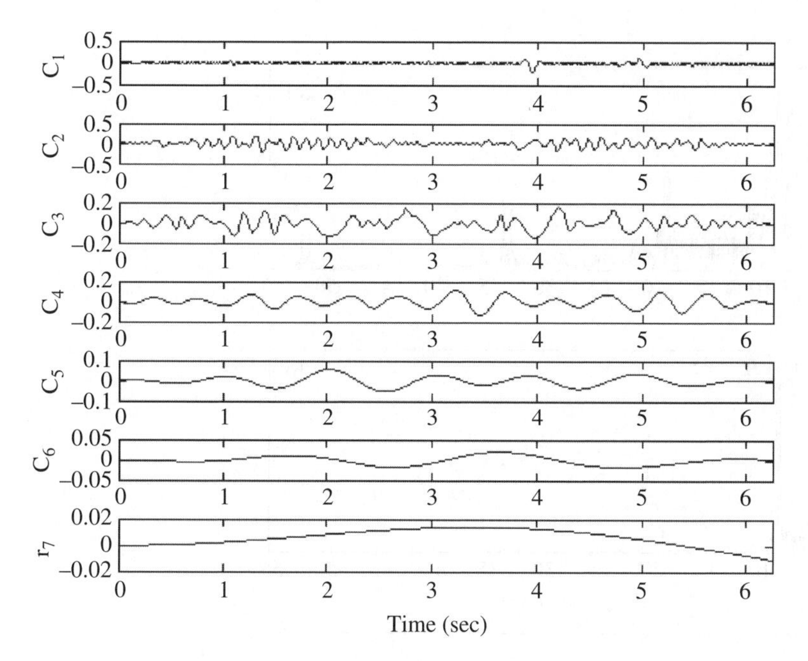


**Figure 5.13** The resulting IMF components, $c_1$ to $c_6$, and final residue, $r_7$, of the signal of acceleration response of the test vehicle.

As was described previously, the tractor is a four-wheel recreation vehicle (RV), which serves to excite the bridge into vibration by its movement over the bridge, thereby playing the role of exciter to the bridge. The trailer, towed by the RV, is a two-wheel cart, which will be excited by the bridge in vibration, thereby playing the role of receiver for the bridge excitation. The linkage between the RV and the trailer is free to rotate in all directions, meaning that there exists no moment transmission between the two vehicles. The vertical response of the trailer is recorded by an accelerometer mounted at a location above the midpoint of its axle, which is connected to the data acquisition system.

Figure 5.15 shows the acceleration response and Fourier spectrum of the trailer moving at 10 km/h over the bridge. In Figure 5.15b, two major frequencies, marked as $V_1$ and $V_2$, are identified, which represent the first and second frequencies of the vehicle. In addition, a group of minor frequencies with a band centering at around 5 Hz, marked as $V_3$, can also be observed, which represent a cluster of the vehicle's dominant frequencies with a magnitude larger than the noise. The existence of such a cluster of frequencies implies that the trailer cannot be idealized as a single sprung mass in reality, which may relate to vibrations of the trailer in other directions, of the suspension, or of the linking action between the tractor and trailer. They are expected to exist in the field experiment for the tractor-trailer. It should be noted that the symbols $V_1$, $V_2$, $V_3$ are used herein only for convenience of interpretation, which do not imply any relative magnitudes.

To increase the reliability of the test results, the field experiment is repeated for three times under the same conditions, namely, by allowing the tractor-trailer to move over the same bridge at the same speed of 10 km/h. The time–history acceleration responses, along with their Fourier spectra, recorded for the trailer during its three passages over

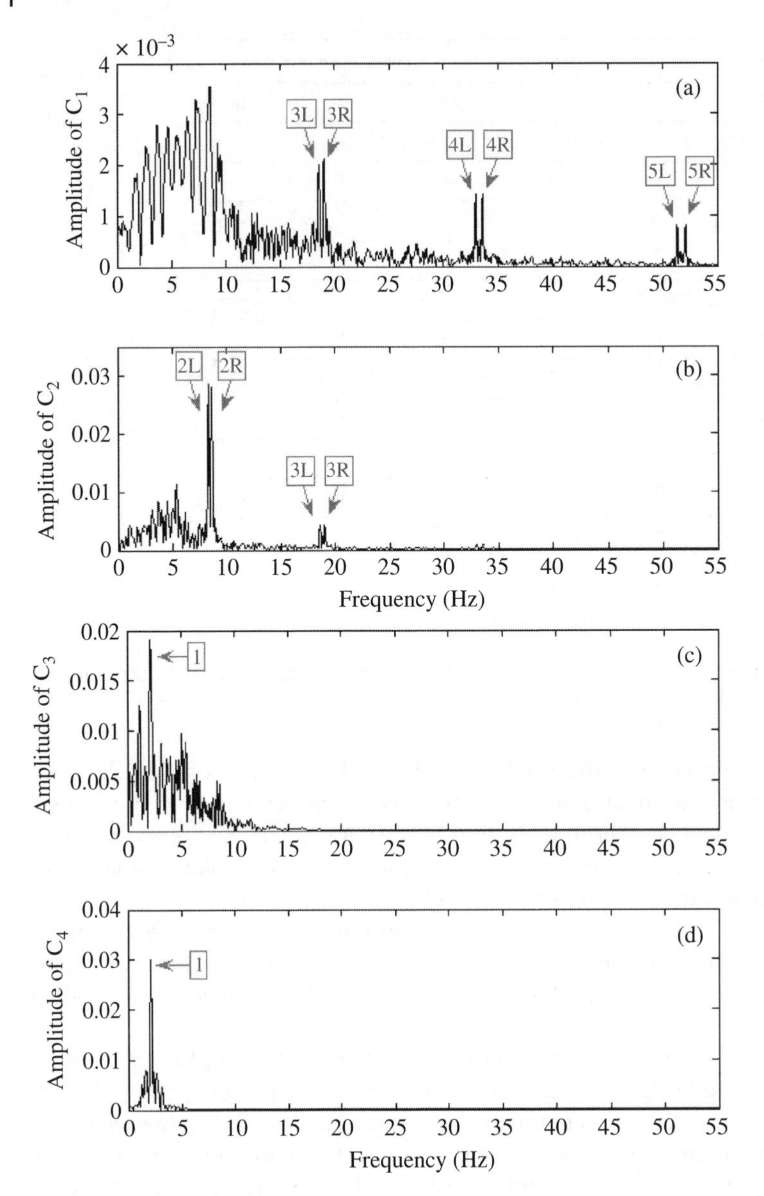


**Figure 5.14** Fourier spectrum of: (a) $c_1$; (b) $c_2$; (c) $c_3$; (d)$c_4$.

the test bridge have been plotted in Figures 5.16–5.18. The principle adopted herein is that only the frequencies that can be repeatedly detected from each test run will be regarded as the ones to be extracted. In other words, only the frequencies simultaneously existing in the three spectra will be adopted as the frequencies identified. Other frequencies appearing in only one of the spectra will simply be discarded, as they may be caused by transient impulses caused by the surface roughness or other factors.

As was indicated by Figures 5.16b, 5.17b, and 5.18b, three frequencies associated with the peaks, marked as $V_1$, $V_2$, and $B_2$, can be identified. In comparison with the spectrum

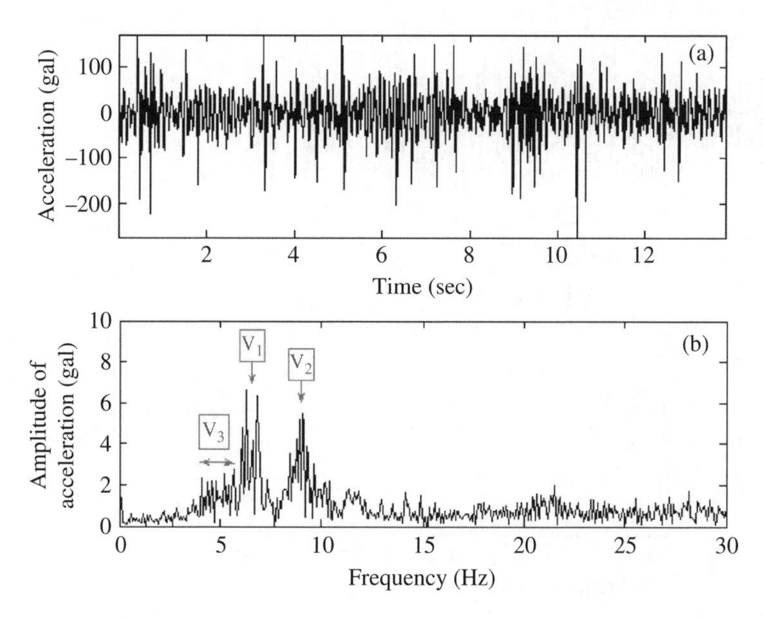


**Figure 5.15** Trailer in surface test: (a) acceleration response; (b) acceleration spectrum.

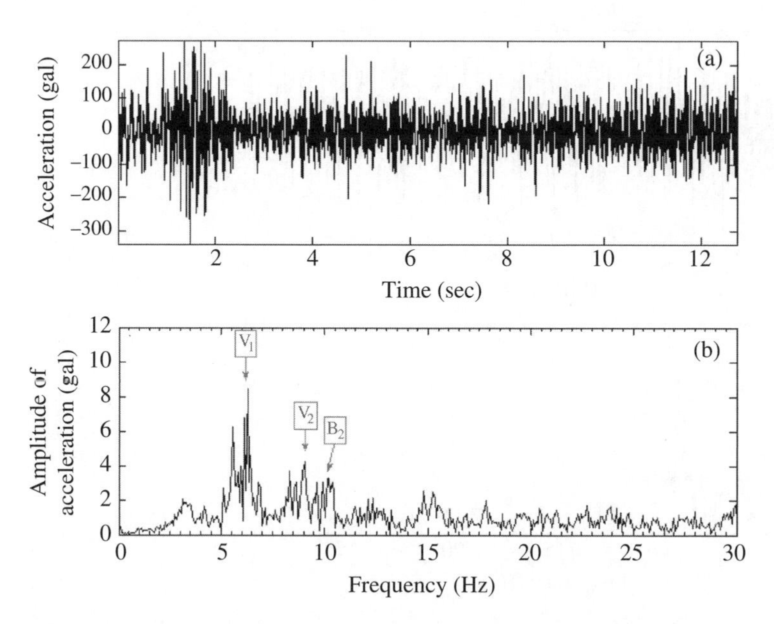


**Figure 5.16** Trailer in field test (first run): (a) acceleration response; (b) acceleration spectrum.

obtained for the surface test, the frequencies $V_1$ and $V_2$ are known as the first two frequencies of the trailer. Due to its repetitiveness, the remaining frequency, $B_2$, is regarded as one of the bridge frequencies of interest. So far, only one bridge frequency and two vehicle frequencies have been identified directly from the Fourier spectra of the vehicle response, with no aid of the EMD technique.

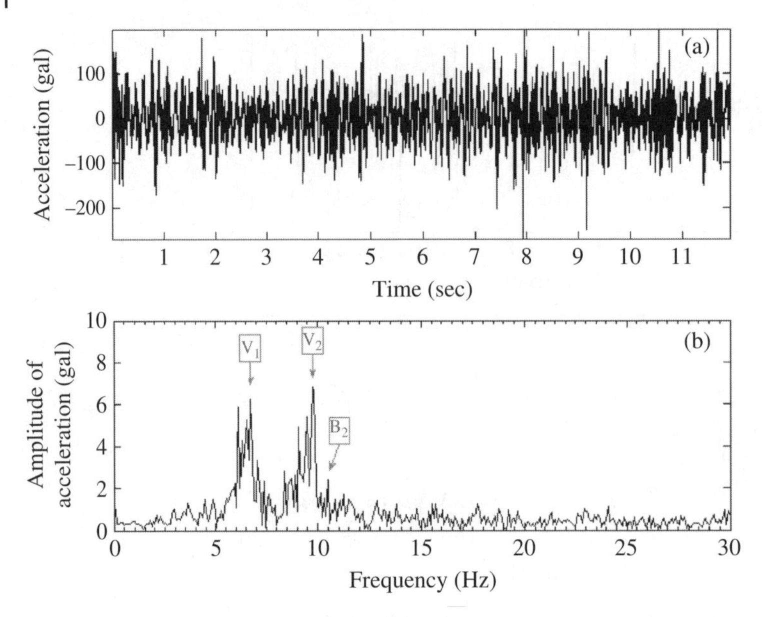


**Figure 5.17** Trailer in field test (second run): (a) acceleration response; (b) acceleration spectrum.

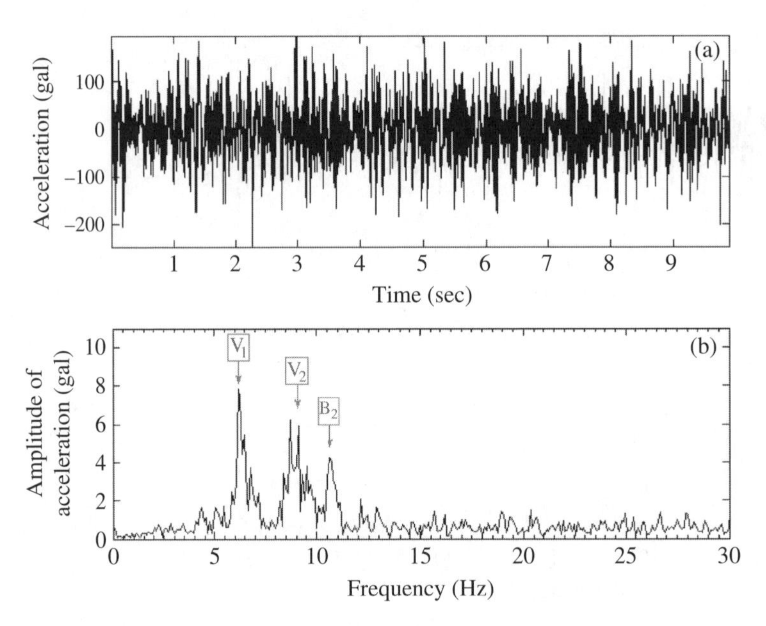


**Figure 5.18** Trailer in field test (third run): (a) acceleration response, (b) acceleration spectrum.

The EMD technique is applied to preprocessing the above three acceleration responses of the trailer, with the resulting Fourier spectra of the first four IMFs plotted in Figures 5.19–5.21. In Figures 5.19a, 5.20a, and 5.21a for the IMF spectra $c_1$, the frequency components are distributed rather uniformly, with generally low average amplitudes, which should be regarded as the noises originating from various sources (e.g., from the roughness in pavement, mechanical parts of the trailer, collision via the linkage between the tractor and trailer, etc.).

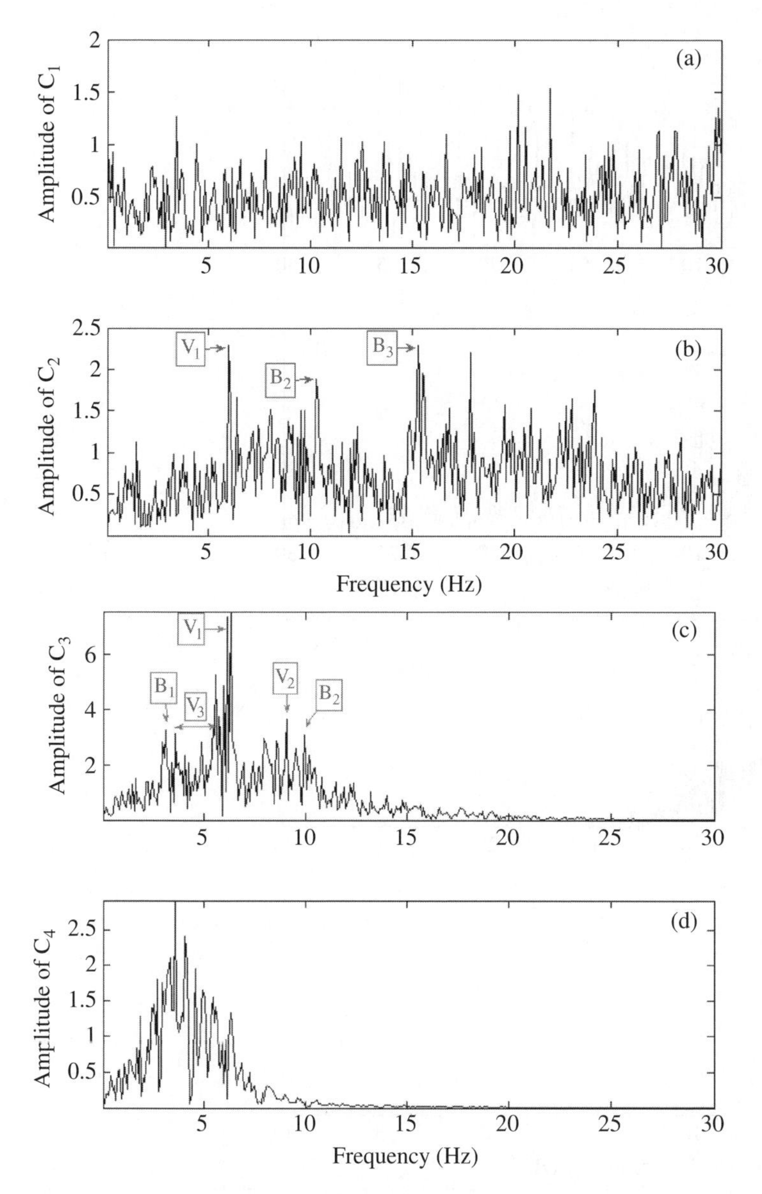


**Figure 5.19** Fourier spectrum of: (a) $c_1$; (b) $c_2$; (c) $c_3$; (d) $c_4$ (first run).

Following the previous principle, only the frequencies appearing repeatedly in each test run will be recognized as the accepted frequencies. In the remaining parts of each figure, the IMF spectra for $c_2$–$c_4$ of the first run in Figure 5.19b–d, of the second run in Figure 5.20b–d, and of the third run in Figure 5.21b–d, five dominant frequencies, marked as $V_1$, $V_2$, $V_3$, $B_1$, and $B_2$, are identified due to their simultaneous presence in each figure. In comparison with the spectrum obtained from the surface test, the frequencies $V_1$, $V_2$, and cluster $V_3$ are known as the first, second, and third (cluster) frequencies of the trailer. The remaining two frequencies, $B_1$ and $B_2$, should therefore be regarded as the frequencies identified of the bridge, among which the lower one ($B_1$) is regarded as the first bridge frequency and the higher one ($B_2$) the second bridge frequency.

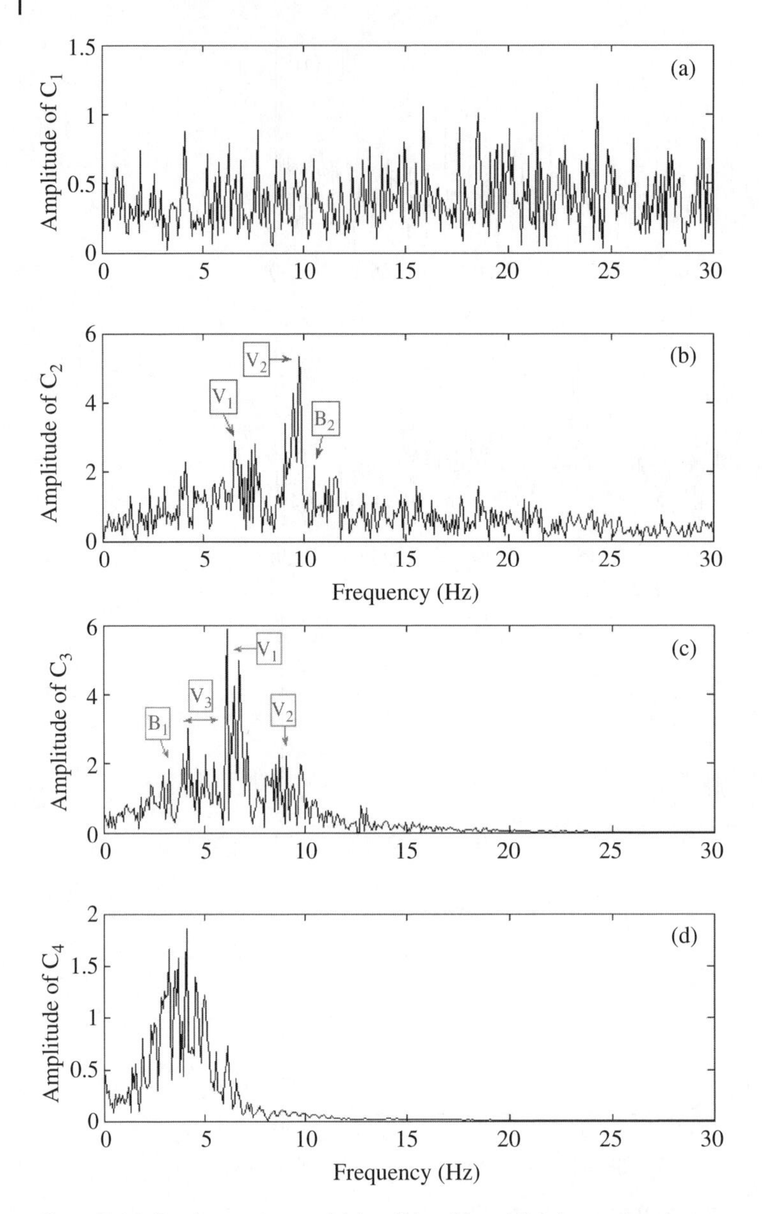


**Figure 5.20** Fourier spectrum of: (a) $c_1$; (b) $c_2$; (c) $c_3$; (d) $c_4$ (second run).

For comparison, an ambient vibration test was conducted directly for the bridge to measure its frequencies, which is known as the direct approach. The acceleration response of the bridge and its Fourier spectrum measured from such a test have been plotted in Figure 5.22. As can be seen, the first bridge frequency, identified from the spectral analysis of the trailer preprocessed by the EMD technique, and the second bridge frequency, identified from the spectral analyses with and without preprocessing by the EMD technique, match very well with the test result. That the second bridge frequency ($B_2$) is more visible than the first one ($B_1$) in the trailer acceleration response

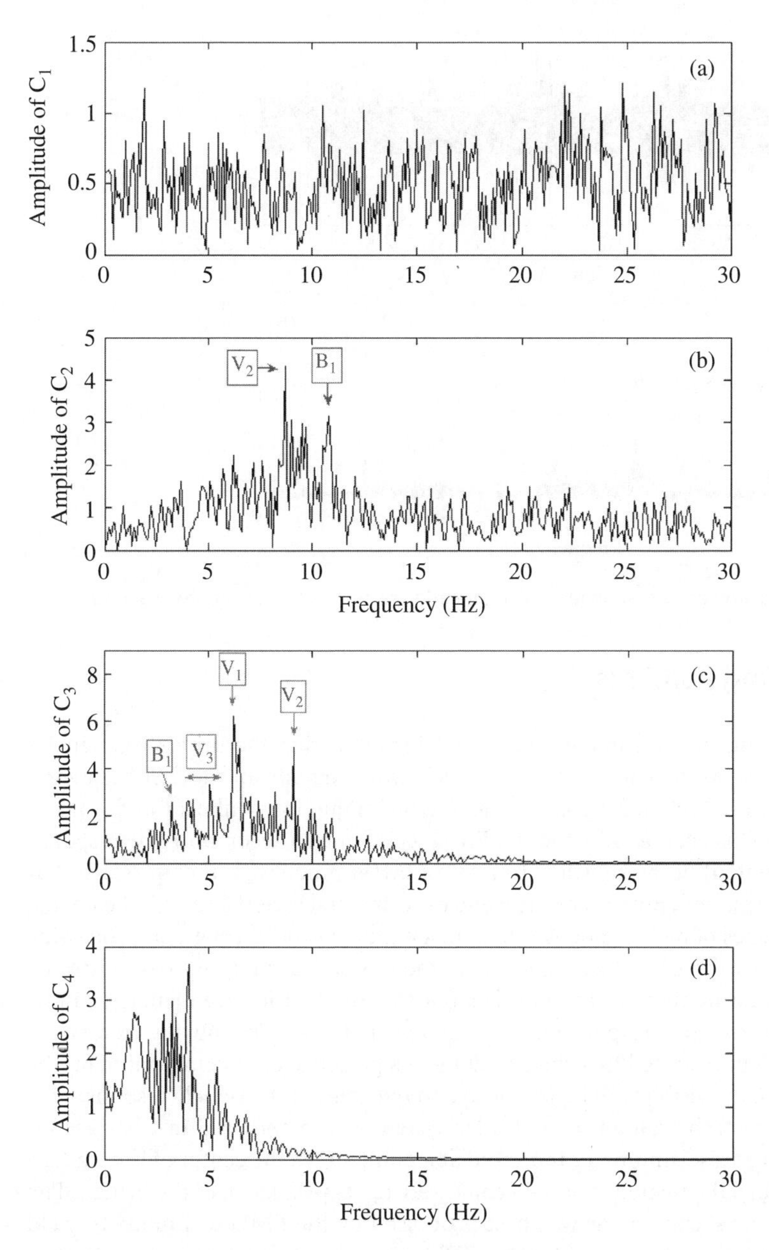


**Figure 5.21** Fourier spectrum of: (a) $c_1$; (b) $c_2$; (c) $c_3$; (d) $c_4$ (third run).

can be attributed to the fact that it is closer to the vehicle frequencies and therefore can be more easily amplified by the trailer. However, the first bridge frequency ($B_1$) becomes identifiable after preprocessing by the EMD technique. This is certainly a contribution of the EMD technique for filtering out the noises. From the viewpoint of field measurement, it is therefore suggested that the data recorded from the trailer always be preprocessed by the EMD technique, so as to enhance the visibility of higher frequencies of the bridge.

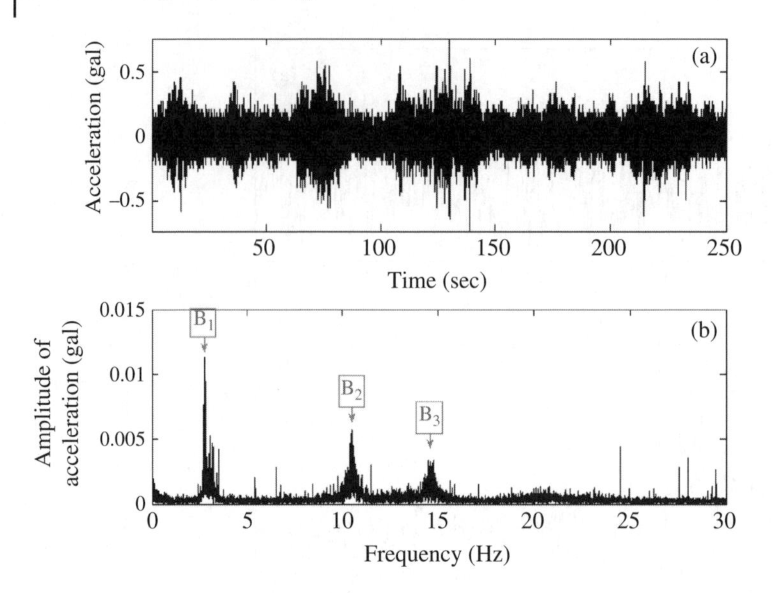


**Figure 5.22** Bridge in ambient vibration test: (a) acceleration response; (b) acceleration spectrum.

## 5.7 Concluding Remarks

The vehicle scanning method initially proposed for extracting the bridge frequencies from the dynamic response of a passing test vehicle works mainly for the first frequency. The EMD is a newly developed signal-processing technique for nonlinear and nonstationary problems. This chapter adopted the EMD technique to decompose the response recorded (or computed) of the vehicle into a set of IMFs, which were then processed by the Fourier transformation. Both theoretical and experimental investigations indicate that the bridge frequencies of higher modes can be made more visible through preprocessing of the vehicle data by the EMD technique. From the theoretical study, we demonstrated that not only the first frequency of the bridge, but also the first few frequencies, can be extracted from the dynamic response of the test vehicle for various moving load cases. The chapter also demonstrated that the simultaneous presence of ongoing traffic on the bridge can enhance the visibility of higher bridge frequencies in the vehicle response.

As for the field test, the same tractor-trailer system as the one used in Chapter 4 is adopted. Specifically, a seismometer mounted at a point above the center of the trailer's axle is used to record the response of the vehicle during its passage over the bridge. The data recorded by the seismometer was first processed by the EMD technique to yield the IMFs, which are then processed by the FFT to yield the spectra for extracting the bridge frequencies. To avoid pollution by various sources, at least three runs are performed for the tractor-trailer under the same conditions (i.e., by allowing it to move over the same bridge at the same speed). The principle adopted herein is that only the frequencies that can be repeatedly detected from each test run are regarded as the ones to be extracted. By comparison with the ambient vibration test on the bridge using the direct approach, it was demonstrated that not only the first frequency of the bridge but also the second frequency can be extracted from the vehicle response through the enhancement by the EMD technique.

# 6

# Effect of Road Roughness on Extraction of Bridge Frequencies

The vehicle scanning method for measuring bridge frequencies is a potentially powerful technique for its mobility, movability, and economy, compared with the conventional technique that requires vibration sensors to be directly installed on the bridge. However, road roughness may pollute the vehicle spectrum obtained and render the bridge frequencies unidentifiable. The objective of this chapter is to study such an effect. First, a numerical simulation is conducted using the vehicle–bridge interaction (VBI) element developed previously to demonstrate how the road roughness affects the vehicle response. Then, an approximate theory in closed form is presented, for physically interpreting the role and range of influence of road roughness on the identification of bridge frequencies. The latter is then expanded to include the action of two connected vehicles of identical frequency moving over the same roughness. It was found that by deducting the response of one vehicle from the other, the identifiability of bridge frequencies will be enhanced in the residue vehicle response. This chapter are based primarily on Yang et al. (2012a,b), with slight modifications on some equations according to Chapter 3.

## 6.1 Introduction

As previously described, the direct approach identifies bridge frequencies through vibration sensors deployed on the bridge. This approach is the one frequently adopted by researchers and engineers. In contrast, the vehicle scanning method is based on the vertical acceleration response recorded as a test vehicle passes over the bridge. This approach is advantageous in that no vibration sensors need to be mounted on the bridge.

Previous research on identification of bridge frequencies by the direct approach has been voluminous. The following is only a partial review. Abdel-Ghaffar and Scanlan (1985) identified the modal parameters of the Golden Gate Bridge using the ambient vibrations tests along with the peak-picking method. Wilson and Liu (1991) performed microvibration tests on the Quincy Bayview Bridge in Illinois, using the accelerators installed on the bridge deck to record the vertical, torsional, and transverse vibrations. The bridge frequencies were also identified by the peak-picking method, which compared well with the finite element analysis results.

Due to the advancement in measurement devices and analysis methods, system identification of bridge modal properties has received increasing attention from researchers

*Vehicle Scanning Method for Bridges*, First Edition. Yeong-Bin Yang, Judy P. Yang, Bin Zhang and Yuntian Wu.

in the past two decades. Ren et al. (2004) used both the peak-picking method and stochastic subspace identification method to identify from the field measurement data the frequencies and modal shapes of the Tennessee River Bridge, a nine-span steel arch bridge, and compared their results with the finite element analyses. He et al. (2009) conducted a system identification study on the Alfred Zampa Memorial Bridge using the dynamic field test data. Brownjohn et al. (2010) conducted an ambient vibration retesting and operational modal analysis of the Humber Bridge.

As was stated in Chapter 1, Yang, et al. first proposed the idea of using an instrumented vehicle to measure the bridge frequencies in 2004 after a long period of research on VBI problems. The idea was experimentally verified to be feasible in the next year (Lin and Yang 2005), and potential applications of VBI properties were identified (Yang and Lin 2005). Extended studies along these lines include those of Bu et al. (2006), McGetrick et al. (2009), Yang and Chang (2009a,b), Chang et al. (2010), Xiang et al. (2010), and Nguyen and Tran (2010). The review compiled in Chapter 1 along these lines was based mainly on work of Yang and Yang (2018).

From the above review, it is realized that the conventional approach for measuring bridge frequencies is a rather mature technique. By installing vibration sensors on a bridge, various sources of excitation can be utilized to identify the dynamic parameters of a bridge. However, deploying vibration sensors on the bridge is generally costly and laborious. For example, the field study of the Quincy Bayview Bridge required a team of four persons to work for five days in the field (Wilson and Liu 1991). Because of these costs, it is not economically feasible to monitor a large number of bridges in a short time to detect damage, such as would be desired after, say, a major earthquake. In comparison, the vehicle scanning method has the potential of becoming an effective tool for periodically monitoring the frequencies of a large number of bridges. It is mobile and economical, although the technique itself should be further enhanced.

This chapter focuses on the effect of road roughness that is critical to the extraction of bridge frequencies from the recorded acceleration response of the passing test vehicle. First, numerical simulations are conducted using the VBI elements to demonstrate the polluting effect of road roughness on the vehicle response. Then, an approximate theory in closed form is presented for physically interpreting and assessing the role and range of influence of road roughness. Finally, a dual vehicle technique is presented for reducing or eliminating the effect of road roughness in the vehicle response spectrum, by reducing the response recorded for one vehicle from the other.

## 6.2 Simulation of Roughness Profiles

The power spectral density (PSD) functions defined by ISO 8608 (1995) for the road profiles will be adopted herein. According to this specification, the road profile is divided into eight classes, with Class A indicating the best surface and Class H the poorest. The PSD function $G_d(n)$ for the road profile is defined as

$$G_d(n) = G_d(n_0)\left(\frac{n}{n_0}\right)^{-w}, \tag{6.1}$$

where $n$ denotes the spatial frequency per meter, $w$ is a constant equal to 2, $n_0 = 0.1$ cycle/m, and the functional value $G_d(n_0)$ is determined by the roughness class, as given in ISO 8608 (1995). The amplitude $d$ for each class of roughness selected is determined by

$$d = \sqrt{2G_d(n)\Delta n} \tag{6.2}$$

where $\Delta n$ is the sampling interval of the spatial frequency. However, the amplitude of roughness so obtained is too large to reflect the bridge surface roughness observed in the field. The reason is that most roughness profiles previously measured were primarily good for general roads but not for bridge surfaces; the latter are known to be in better condition. For this reason, in the study of moving vehicles over bridge surfaces, a square root is arbitrarily taken of the geometric mean of the functional value provided by ISO 8608 (1995) herein. Besides, to simulate the perfect surface condition (i.e., roughness level A) for bridges, a very small value of $0.001 \times 10^{-6}$ is assigned for the functional value (geometric mean) of the class without taking the square root. Consequently, the functional values adopted for the three classes of roughness in the study of bridges are: Class A: $G_d^*(n_0) = 0.001 \times 10^{-6} m^3$; Class B: $G_d^*(n_0) = \sqrt{64} \times 10^{-6} m^3$; and Class C: $G_d^*(n_0) = \sqrt{256} \times 10^{-6} m^3$.

Then, the road roughness can be superimposed as follows:

$$r(x) = \sum_i d_i \cos(n_{s,i}x + \theta_i) \tag{6.3}$$

where $n_{s,\,i}$ is the $i$th spatial frequency considered, and $d_i$ and $\theta_i$ denote the amplitude, as given in Eq. (6.2), and the random phase angle, respectively, of the $i$th cosine function. In this study, the sampling interval $\Delta n_s$ for the spatial frequency is taken as 0.04 cycle/m, and the range of spatial frequency $n_s$ is taken as 1–100 cycle/m.

## 6.3 Simulation of Bridges with Rough Surface

This section summarizes the procedure for simulating the VBI system and for considering the road roughness by the finite element method. For illustration, only simply supported beams are considered, and the vehicle is modeled as a single concentrated mass supported by a spring-dashpot unit. In the numerical simulation, a beam is divided into a number of finite elements, each of six degrees of freedom (DOFs) for the present two-dimensional case.

For an element of length $l$ directly under the action of the vehicle, it is modeled as a VBI element with the effects of vehicle action and surface roughness included (see Figure 6.1). The equation of motion for the VBI element is (see Appendix or Chang et al. 2010):

$$\begin{bmatrix} m_v & 0 \\ 0 & [\mathbf{m_b}] \end{bmatrix} \begin{Bmatrix} \ddot{q}_v(t) \\ \{\ddot{\mathbf{q}}_\mathbf{b}(t)\} \end{Bmatrix} + \begin{bmatrix} c_v & -c_v\{\mathbf{N}(x_c)\}^T \\ -c_v\{\mathbf{N}(x_c)\} & [\mathbf{c_b}] + c_v\{\mathbf{N}(x_c)\}\{\mathbf{N}(x_c)\}^T \end{bmatrix} \begin{Bmatrix} \dot{q}_v(t) \\ \{\dot{\mathbf{q}}_\mathbf{b}(t)\} \end{Bmatrix}$$
$$+ \begin{bmatrix} k_v & -k_v\{\mathbf{N}(x_c)\}^T - c_v v\left[\{\mathbf{N}'(x)\}^T\right]\Big|_{x=x_c} \\ -k_v\{\mathbf{N}(x_c)\} & [\mathbf{k_b}] + c_v v\{\mathbf{N}(x_c)\}\left[\{\mathbf{N}'(x)\}^T\right]\Big|_{x=x_c} + k_v\{\mathbf{N}(x_c)\}\{\mathbf{N}(x_c)\}^T \end{bmatrix} \begin{Bmatrix} q_v(t) \\ \{\mathbf{q_b}(t)\} \end{Bmatrix}$$
$$= \begin{Bmatrix} c_v v r'(x)\big|_{x=x_c} + k_v r(x_c) \\ -c_v\left[r'(x)\right]\big|_{x=x_c}\{\mathbf{N}(x_c)\} - k_v r(x_c)\{\mathbf{N}(x_c)\} - m_v g\{\mathbf{N}(x_c)\} \end{Bmatrix} \tag{6.4}$$

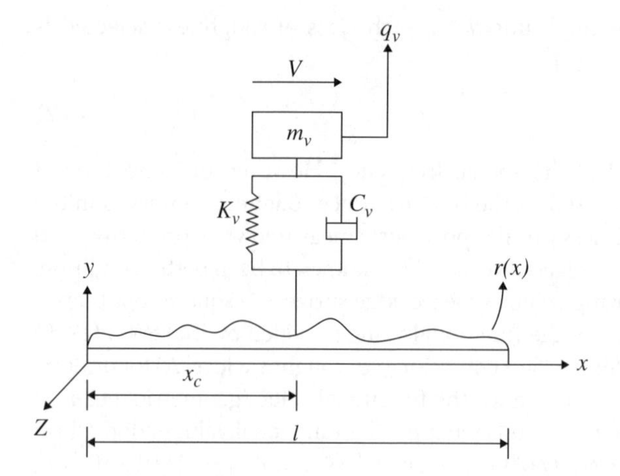


Figure 6.1 Vehicle-bridge interaction (VBI) element with surface roughness.

where the parameters related to the vehicle are: $q_v$ = vertical displacement, $m_v$ = lumped mass, $c_v$, $k_v$ = damping and stiffness coefficients of the suspension system, and $v$ = speed; the parameters related to the beam element are: $\{\mathbf{q_b}\}$ = vector of the nodal DOFs, $[\mathbf{m_b}]$, $[\mathbf{c_b}]$, $[\mathbf{k_b}]$ = mass, damping, and stiffness matrices, and $\{\mathbf{N}\}$ = cubic Hermitian interpolation functions; $g$ = acceleration of gravity; $r(x)$ = road profile; $x_c$ = contact position of the vehicle on the element; and a prime and overdot denote, respectively, differentiation of the quantity with respect to coordinate and time.

The term on the right-hand side of Eq. (6.4) represents the contact or interaction forces between the vehicle and beam caused by the vehicle weight, $-m_v g\{N(x_c)\}$, and road roughness $r(x)$. In particular, the road roughness $r(x)$ affects only the contact forces, but not the system matrices appearing on the left-hand side of Eq. (6.4). Evidently, the presence of road roughness does not alter the frequency contents of the whole VBI system.

The remaining parts of the beam not directly acted on by the vehicle are modeled by the conventional 6-DOF beam elements, of which the equation of motion is

$$[\mathbf{m_b}]\{\ddot{\mathbf{q}}_\mathbf{b}\}+[\mathbf{c_b}]\{\dot{\mathbf{q}}_\mathbf{b}\}+[\mathbf{k_b}]\{\mathbf{q_b}\}=\{\mathbf{0}\} \tag{6.5}$$

where each of the terms has already been defined following Eq. (6.4). By the finite element procedure of assembly, the equation of motion can be established for the whole VBI system as follows:

$$[\mathbf{M}]\{\ddot{\mathbf{u}}\}+[\mathbf{C}]\{\dot{\mathbf{u}}\}+[\mathbf{K}]\{\mathbf{u}\}=\{\mathbf{P}\} \tag{6.6}$$

where $[\mathbf{M}]$, $[\mathbf{C}]$, and $[\mathbf{K}]$ denote, respectively, the mass, damping, and stiffness matrices of the system, and $[\mathbf{P}]$ denotes the forces acting on the system. In this study, the preceding equation will be solved by the Newmark-$\beta$ method with constant average acceleration (i.e., with $\beta = 1/4$ and $\gamma = 1/2$) for its unconditional stability.

## 6.4 Effect of Road Roughness on Vehicle Response

Three classes of surface roughness, Classes A–C, will be considered for the bridge that is to be traveled by the test vehicle. The properties adopted for the vehicle lumped as a single-DOF sprung mass shown in Figure 6.1 are: $m_v$ = 1000 kg, $k_v$ = 170 kN/m, and

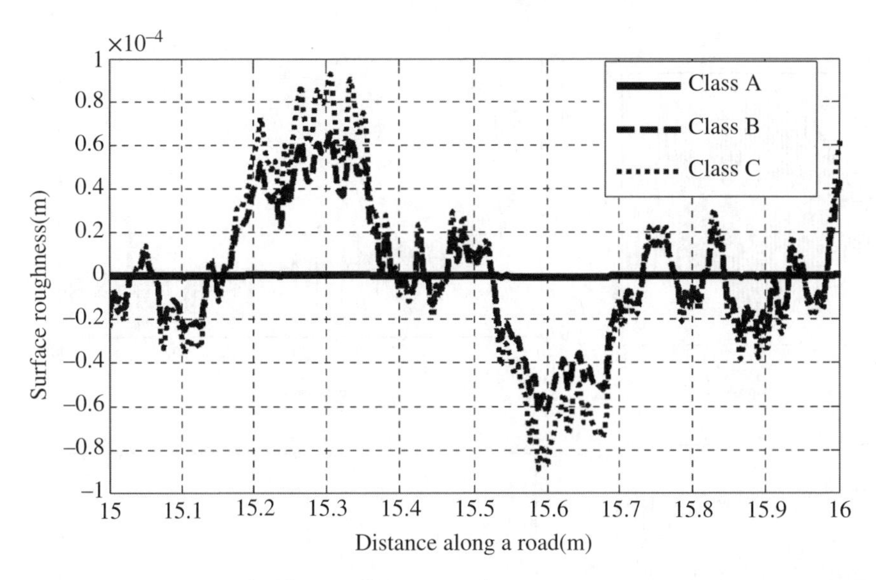


**Figure 6.2** A segment of surface profile generated.

$v = 2\,\text{m/s}$. The beam simply supported at both ends is divided into 20 elements with the following properties: elastic modulus $E = 27.5\,\text{GPa}$, moment of inertia $I_b = 0.175\,\text{m}^4$, mass per unit length $\bar{m} = 1000\,\text{kg/m}$, and cross-sectional area $A = 2\,\text{m}^2$.

As was mentioned previously, the functional values $G_d(n_0)$ to be used for the PSD function $G_d(n)$ in Eq. (6.1) are modified from the geometric means provided by ISO 8608 (1995) as $0.001 \times 10^{-6}$, $8 \times 10^{-6}$, and $16 \times 10^{-6}$, respectively, for roughness Classes A, B, and C. A segment of the surface profile of roughness generated using Eq. (6.3), along with Eqs. (6.2) and (6.1), for each of the three classes has been plotted in Figure 6.2.

### 6.4.1 Case 1: Vehicle Frequency Less than Any Bridge Frequencies

With the previously assumed data, the vehicle frequency computed is less than any of the bridge frequencies. Figures 6.3a–c show the time–history acceleration response and frequency spectrum of the test vehicle during its passage over the bridge for the three classes of road roughness considered. The observations are as follows: For road profile of the best quality, i.e., Class A, the first three bridge frequencies, 3.867, 15.27, and 34.3 Hz, can be clearly identified, along with the vehicle frequency of 2.067 Hz. For road profile of the second best quality, i.e., Class B, only the first bridge frequency can be identified, along with the vehicle frequency. And for the poorest road profile, i.e., Class C, the only frequency that can be identified is the vehicle frequency, while all the bridge frequencies have been completely hidden. From this analysis, it is clear that road roughness is vitally importance in identifying the bridge frequencies from the vehicle response spectrum. It is concluded that the rougher the pavement surface, the less number of bridge frequencies can be identified, or the higher the difficulty exists for identifying the bridge frequencies from the vehicle response.

### 6.4.2 Case 2: Vehicle Frequency Greater than the First Bridge Frequency

The previous analysis was conducted for the case when the vehicle frequency is smaller than any of the bridge frequencies. To evaluate the effect of the vehicle/bridge frequency

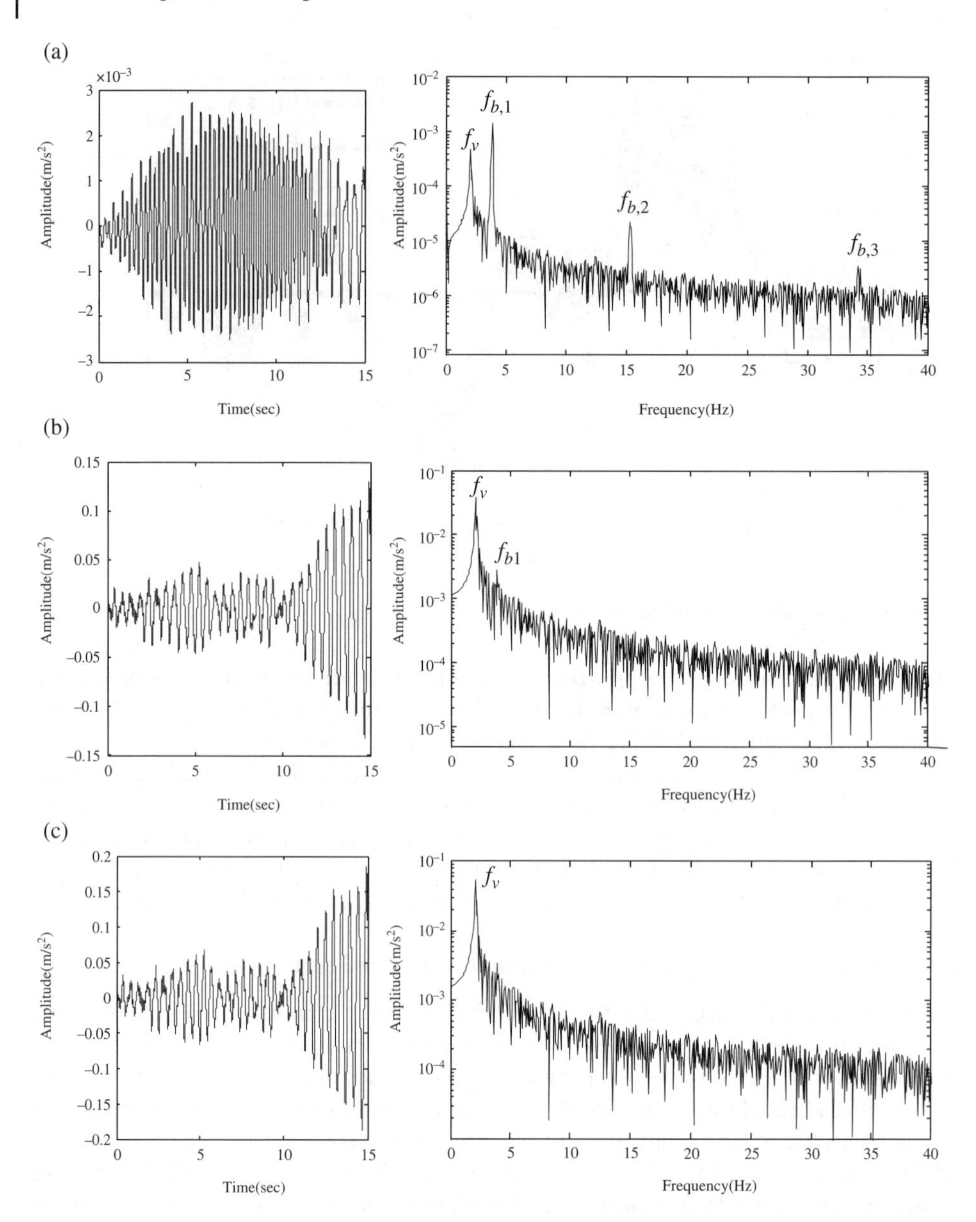


**Figure 6.3** Case 1 – Acceleration and frequency responses of test vehicle for Class: (a) A; (b) B; (c) C.

ratio, one considers herein another case with the vehicle frequency significantly higher than the first frequency of the bridge. In other words, the vehicle frequency is raised from 2.067 to 10 Hz by increasing the vehicle stiffness from 170 to 3.947 MN/m, while the other properties remain unchanged. For this case, the time–history acceleration response and frequency spectrum solved of the test vehicle during its passage over the bridge for the three classes of roughness considered have been plotted in Figures 6.4a–c.

An observation from Figures 6.4a–c is that the amplitude of the vehicle frequency and roughness frequencies are amplified as the roughness level increases. Moreover, the

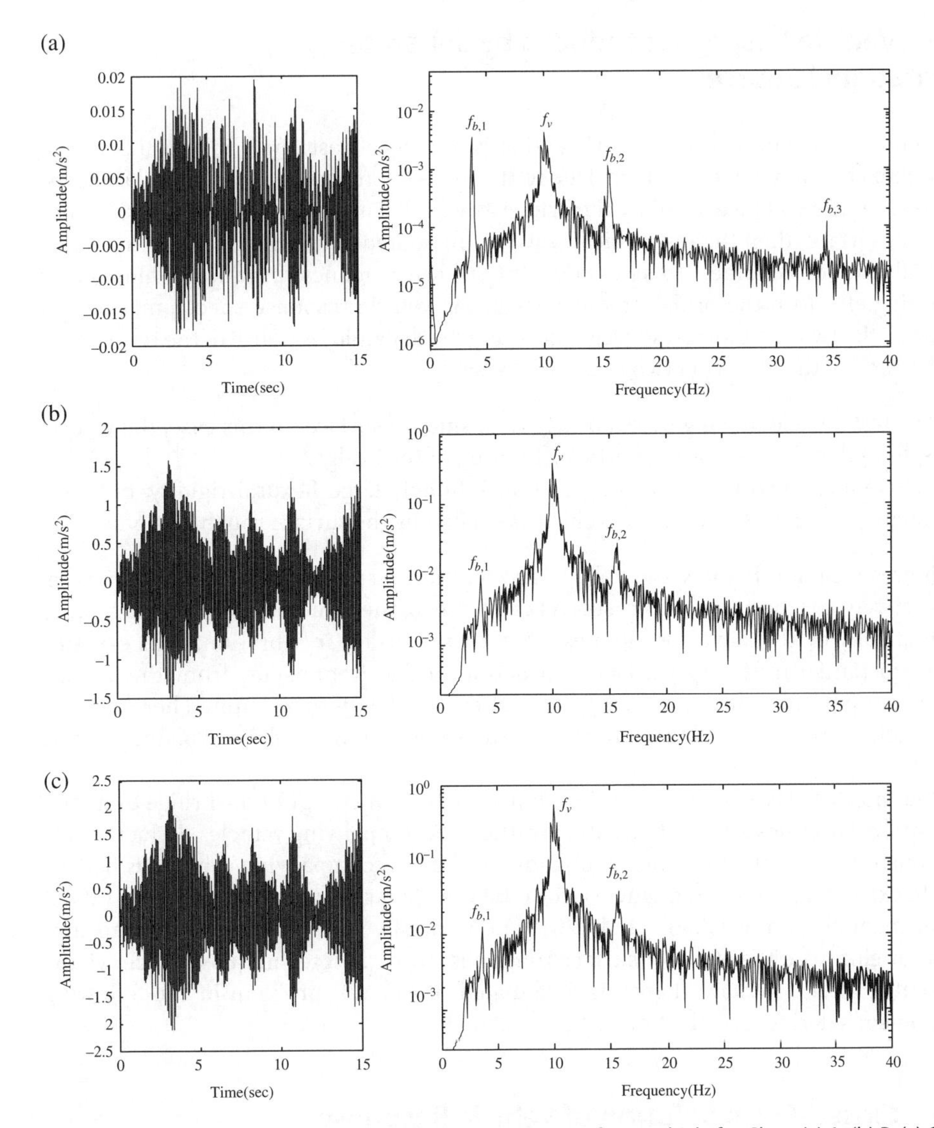


**Figure 6.4** Case 2 – Acceleration and frequency responses of test vehicle for Class: (a) A; (b) B; (c) C.

third bridge frequency is hidden for Class B and C, while the amplitudes of the first and second bridge frequency appear to be generally visible. Further, by comparing Figures 6.4 with 6.3, one observes that in Case 1 only the first bridge frequency remains visible for all three classes of roughness, while in Case 2, both the first and second bridge frequencies are visible for all classes of roughness considered. Thus, from the point of practice, it is advantageous to use a test vehicle with a frequency "greater than" the first frequency of the bridge, in order to increase the visibility of bridge frequencies in the vehicle response. In general, the above analyses also indicate that bridge frequencies of higher modes may be blurred or hidden by the presence of roughness, which therefore is crucial to the successful identification of bridge frequencies using the test vehicle.

## 6.5 Vehicle Responses Induced by Separate Excitational Sources

For a test vehicle moving over a bridge that is initially at rest, the bridge will be set in motion by the moving test vehicle. Due to the interaction between the two subsystems, the test vehicle will be excited by the bridge as well during its passage. If the bridge has a rough surface, then the test vehicle will also be excited by the surface roughness in a spatially random manner. To assess how the surface roughness produces a blurring or polluting effect on the bridge frequencies in the vehicle response spectrum, one can separate the two excitational sources and study the dynamic response of the test vehicle under each of the following two extreme cases:

1) The test vehicle moving over a bridge with smooth surface: In this case, the vehicle is excited exclusively by the vertical vibration of the bridge.
2) The vehicle moving over a bridge of an infinitely large flexural rigidity, but with rough surface: In this case, the vehicle is excited by the surface roughness only.

Figures 6.5a and b show the vehicle acceleration response and Fourier spectrum, respectively, for the above two cases. As can be seen, the vehicle is excited much more dramatically by the surface roughness than by the bridge in vibration. This explains why substantial difficulty exists in extracting bridge frequencies from the vehicle response spectrum, once the surface roughness is taken into account. There are two approaches that can be adopted to resolve such a problem, as will be explained in this chapter.

One approach is to increase the vibration amplitude or energy of the bridge by allowing the bridge to be exposed to existing traffic or accompanying vehicles. In fact, it was demonstrated that the existence of ongoing traffic or accompanying vehicles is beneficial to extracting bridge frequencies from the vehicle response by Lin and Yang (2005) experimentally and by Chang et al. (2010) numerically. The other approach is to suppress or eliminate the effect of surface roughness by certain techniques, which will be presented after a general theory is formulated of the VBI problem in the following section considering the effect of surface roughness.

## 6.6 Closed-Form Solution of Vehicle Response Considering Road Roughness

It has been illustrated in preceding section that the vehicle passing a bridge will be excited more seriously by road roughness than by the bridge vibration. Such a phenomenon can be physically interpreted by the analytical formulation below.

Figure 6.6 shows the two-dimensional VBI model of concern herein, where the vehicle is modeled as a moving sprung mass $m_v$ supported by a spring of stiffness $k_v$, and the simple beam is modeled as a Bernoulli-Euler beam of length $L$, mass density $\bar{m}$ per unit length, and bending rigidity $EI$. The road roughness profile is denoted by $r(x)$, a function of the beam axis $x$. The damping effects of both the vehicle and beam are neglected, since the responses considered herein are mainly of the transient nature, for which the damping effects can be ignored.

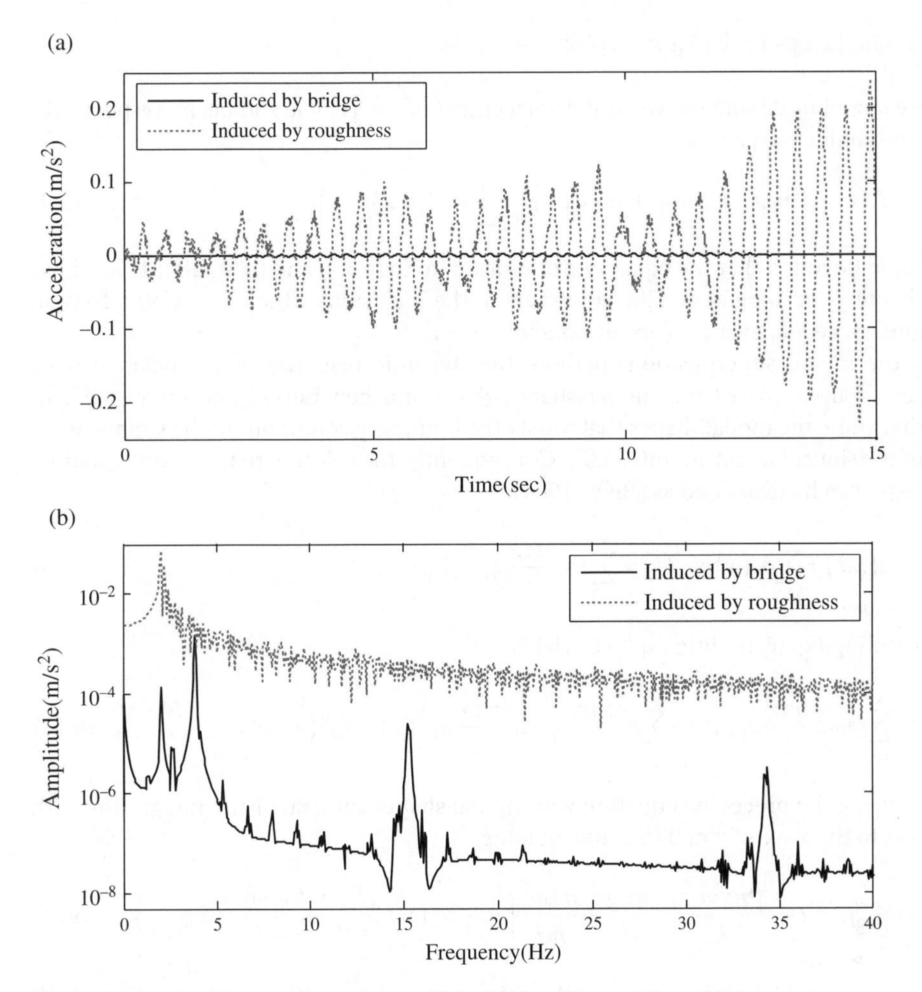


**Figure 6.5** Acceleration responses of the vehicle: (a) time history; (b) amplitude spectrum.

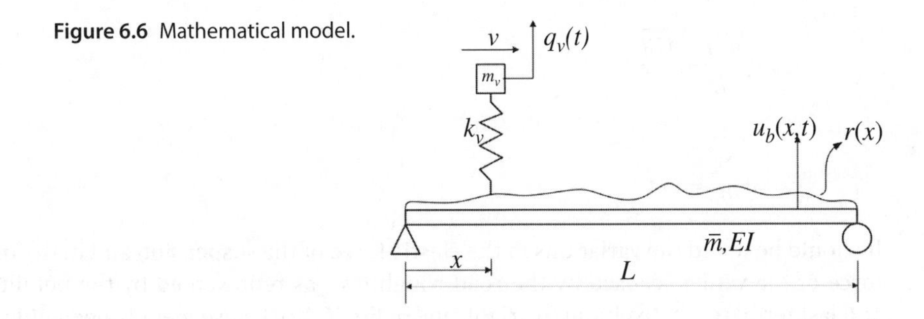


**Figure 6.6** Mathematical model.

Let the vehicle move over the beam with speed $v$. The equations of motion for both the vehicle and beam at time $t$ can be expressed as follows:

$$m_v \ddot{q}_v(t) + k_v \left[ q_v(t) - u_b(x,t)\big|_{x=vt} - r(x)\big|_{x=vt} \right] = 0 \tag{6.7}$$

$$\bar{m}\ddot{u}_b(x,t)+EIu_b''''(x,t)=f_c(t)\delta(x-vt) \tag{6.8}$$

where $q_v$ and $u_b$ denote the vertical displacement of the vehicle and beam, respectively, and the contact force $f_c$ is

$$f_c(t)=-m_v g+k_v\left[q_v(t)-u_b(x,t)\big|_{x=vt}\right]-k_v\left[r(x)\big|_{x=vt}\right] \tag{6.9}$$

It should be noted that the equations of motion in Eqs. (6.7) and (6.8) differ from those of Chapter 2 or Yang et al. (2004a) merely in the inclusion of the term related to road roughness, i.e. $-k_v[r(x)|_{x=vt}]$ in the contact force $f_c$.

By the modal superposition method, the dynamic response of the beam can be expressed in terms of the modal shapes $\phi_n(x)$ and generalized coordinates $q_{b,n}(t)$. Furthermore, the modal shapes that satisfy the boundary conditions of the simple beam are of the sinusoidal form, $\sin(n\pi x/L)$. Consequently, the solution to the beam equation, Eq. (6.8), can be expressed as (Biggs 1964):

$$u(x,t)=\sum_{n=1}^{\infty}\phi_n(x)q_{b,n}(t)=\sum_{n=1}^{\infty}\sin\frac{n\pi x}{L}q_{b,n}(t) \tag{6.10}$$

Substituting Eq. (6.10) into Eq. (6.8) yields

$$\sum_{n=1}^{\infty}\bar{m}\sin\frac{n\pi x}{L}\ddot{q}_{b,n}(t)+\sum_{n=1}^{\infty}EI\left(\frac{n\pi}{L}\right)^4\sin\frac{n\pi x}{L}q_{b,n}(t)=f_c\delta(x-vt) \tag{6.11}$$

Multiplying the preceding equation with modal shapes $\sin(n\pi x/L)$ and integrating with respect to the $x$-axis from 0 to $L$, one obtains

$$\ddot{q}_{b,n}+\omega_{b,n}^2 q_b=2\sin\left(\frac{n\pi vt}{L}\right)\left\{\frac{-m_v g}{\bar{m}L}+\frac{m_v\omega_v^2}{\bar{m}L}\left[q_v-u_b(x,t)\big|_{x=vt}\right]-\frac{m_v\omega_v^2}{\bar{m}L}r(x)\big|_{x=vt}\right\} \tag{6.12}$$

where $\omega_{b,n}$ is the beam frequency of the $n$th mode and $\omega_v$ the vehicle frequency, which are respectively defined as

$$\omega_{b,n}=\frac{n^2\pi^2}{L^2}\sqrt{\frac{EI}{\bar{m}}} \tag{6.13}$$

$$\omega_v=\sqrt{\frac{k_v}{m_v}} \tag{6.14}$$

It should be noted the variations in the elastic force of the suspension and in the inertial force of the vehicle caused by the road roughness, as represented by the penultimate and last term, respectively, on the right side of Eq. (6.12), have generally negligible effect on the response of the beam (Yau et al. 1999). By neglecting these two terms, the equation of motion for the beam in Eq. (6.12) reduces to

$$\ddot{q}_{b,n}+\omega_{b,n}^2 q_b=2\sin\frac{n\pi vt}{L}\left\{\frac{-m_v g}{\bar{m}L}\right\} \tag{6.15}$$

Assuming zero initial conditions for the beam, one can solve Eq. (6.15) to obtain the generalized coordinate $q_{b,n}$ of the $n$th mode as

$$q_{b,n}(t)=\frac{\Delta_{st,n}}{1-S_n^2}\left[\sin\frac{n\pi vt}{L}-S_n\sin\omega_{b,n}t\right] \tag{6.16}$$

where $\Delta_{st,n}$ is the static deflection caused by the weight of the vehicle,

$$\Delta_{st,n}=\frac{-2m_v gL^3}{n^4\pi^4 EI} \tag{6.17}$$

and $S_n$ is a nondimensional speed parameter,

$$S_n=\frac{n\pi v}{L\omega_{b,n}} \tag{6.18}$$

Substituting Eq. (6.16) back into Eq. (6.10) yields the general solution of the vertical displacement of the beam as

$$u(x,t)=\sum_n\frac{\Delta_{st,n}}{1-S_n^2}\left\{\sin\frac{n\pi x}{L}\left[\sin\frac{n\pi vt}{L}-S_n\sin\omega_{b,n}t\right]\right\} \tag{6.19}$$

It should be noted that, even with the neglect of the effect of road roughness, Eq. (6.19) is generally accurate for predicting the dynamic response of the beam (Yau et al. 1999).

At this stage, one can proceed to solve the vehicle response. First, the equation of motion for the vehicle in Eq. (6.7) may be rewritten as

$$\ddot{q}_v(t)+\omega_v^2 q_v(t)=\omega_v^2 r(x)\big|_{x=vt}+\omega_v^2 u_b(x,t)\big|_{x=vt} \tag{6.20}$$

Clearly, the vehicle is subjected to two excitational sources: the beam's vibration and road roughness, as represented by the terms on the right-hand side of Eq. (6.20). Substituting Eq. (6.19) for $u(x, t)$and Eq. (6.3) for $r(x)$ into Eq. (6.20), one can obtain by Duhamel's integral the displacement response of the vehicle as

$$\begin{aligned} q_v(t)=\sum_{n=1}^{\infty}\bigg[&A_{1n}+A_{2n}\cos\left(\frac{2n\pi v}{L}\right)t+A_{3n}\cos(\omega_v t)\\ &+A_{4n}\cos\left(\omega_{b,n}-\frac{n\pi v}{L}\right)t+A_{5n}\cos\left(\omega_{b,n}+\frac{n\pi v}{L}\right)t\bigg]\\ +\sum_{i=1}\frac{\omega_v^2 d_i}{\omega_v^2-(n_{s,i}v)^2}&\left[\cos(n_{s,i}vt+\theta_i)-\cos(\theta_i)\cos(\omega_v t)+\frac{n_{s,i}v}{\omega_v}\sin(\theta_i)\sin(\omega_v t)\right] \end{aligned} \tag{6.21}$$

where the coefficients are

$$A_{1n}=\frac{\Delta_{stn}}{2\left(1-S_n^2\right)}, \tag{6.22a}$$

$$A_{2n}=\frac{\Delta_{stn}}{2\left(1-S_n^2\right)\left(4\mu_n^2 S_n^2-1\right)}, \tag{6.22b}$$

$$A_{3n} = \frac{2\Delta_{stn} S_n^2 \mu_n^4 \left(2 + \mu_n^2 S_n^2 - \mu_n^2\right)}{\left(4\mu_n^2 S_n^2 - 1\right)\left[1 - 2\mu_n^2\left(1 + S_n^2\right) + \mu_n^4\left(1 - S_n^2\right)^2\right]}, \tag{6.22c}$$

$$A_{4n} = -\frac{\Delta_{stn} S_n}{2\left(1 - S_n^2\right)\left[1 - \mu_n^2 (1 - S_n)^2\right]}, \tag{6.22d}$$

$$A_{5n} = \frac{\Delta_{stn} S_n}{2\left(1 - S_n^2\right)\left[1 - \mu_n^2 (1 + S_n)^2\right]}, \tag{6.22e}$$

Differentiating Eq. (6.21) with respect to $t$ twice, the acceleration response of the vehicle can be obtained as

$$\begin{aligned} \ddot{q}_v(t) = \sum_{n=1}^{\infty} \Bigg\{ & \bar{\bar{A}}_{2,n} \cos\left(\frac{2n\pi v}{L}\right)t + \bar{\bar{A}}_{3,n} \cos(\omega_v t) \\ & + \bar{\bar{A}}_{4,n} \cos\left(\omega_{b,n} - \frac{n\pi v}{L}\right)t + \bar{\bar{A}}_{5,n} \cos\left(\omega_{b,n} + \frac{n\pi v}{L}\right)t \Bigg\} \\ & + \sum_{i=1} \frac{\omega_v^2 d_i}{\omega_v^2 - (n_{s,i} v)^2} \Big[ -(n_{s,i} v)^2 \cos(n_{s,i} v t + \theta_i) \\ & + \omega_v^2 \cos(\theta_i)\cos(\omega_v t) - (n_{s,i} v \omega_v)\sin(\theta_i)\sin(\omega_v t) \Big] \end{aligned} \tag{6.23}$$

where the coefficients are

$$\begin{aligned} & \bar{\bar{A}}_{2,n} = -\left(\frac{2n\pi v}{L}\right)^2 \times A_{2,n}, \quad \bar{\bar{A}}_{3,n} = -\omega_v^2 \times A_{3,n}, \\ & \bar{\bar{A}}_{4,n} = -\left(\omega_{b,n} - \frac{n\pi v}{L}\right)^2 \times A_{4,n}, \quad \bar{\bar{A}}_{5,n} = -\left(\omega_{b,n} + \frac{n\pi v}{L}\right)^2 \times A_{5,n} \end{aligned} \tag{6.24a–d}$$

As can be seen from Eq. (6.23), besides the following three groups of frequency that have been previously identified in the vehicle response (see Chapter 3 or Yang and Lin 2005): driving frequencies $(n \pm 1)\pi v/L$, bridge-related frequencies $\omega_{b,n} \pm n\pi v/L$, and vehicle frequency $\omega_v$, the acceleration response of the vehicle is also affected by the roughness-related frequencies $n_{s,i}v$, defined as the product of the spatial frequency $n_{s,i}$ of road roughness and the vehicle speed $v$. The solution for the acceleration response of the vehicle in Eq. (6.23) differs from the one in Chapter 3 or Yang and Lin (2005) in the appearance of the terms due to road roughness, which are denoted as $\ddot{q}_{v,r}$, namely,

$$\begin{aligned} \ddot{q}_{v,r} = \sum_{i=1} \frac{\omega_v^2 d_i}{\omega_v^2 - (n_{s,i} v)^2} \Big[ & -(n_{s,i} v)^2 \cos(n_{s,i} v t + \theta_i) \\ & + \omega_v^2 \cos(\theta_i)\cos(\omega_v t) - (n_{s,i} v \omega_v)\sin(\theta_i)\sin(\omega_v t) \Big] \end{aligned} \tag{6.25}$$

The preceding equation indicates that the roughness term $\ddot{q}_{v,r}$ is dominated by the roughness frequencies $n_{s,i}v$ and vehicle frequency $\omega_v$. This term has the function of introducing the roughness frequencies to the vehicle response, while amplifying the amplitudes of the vehicle frequency, especially when any of the roughness frequencies $n_{s,i}v$ is close to the vehicle frequency $\omega_v$, as revealed by the denominator in Eq. (6.25). Both effects are unfavorable to identification of bridge frequencies from the vehicle response. Evidently, the present closed-form solution offers a theoretical basis for interpreting the influence of road roughness on the vehicle response, aside from the numerical analyses presented in previous sections.

## 6.7 Reducing the Impact of Road Roughness by Using Two Connected Vehicles

From Eq. (6.20), it is clear that the test vehicle is subjected to two excitational sources: the beam's vibration (as represented by the term $m_v \times \omega_v^2 u_b(x,t)|_{x=vt}$) and road roughness (as represented by the term $m_v \times \omega_v^2 r(x)|_{x=vt}$). As was stated previously, to enhance the visibility of bridge frequencies from the vehicle spectrum, one approach is to amplify the vibration of the bridge by exposing it to existing traffic or accompanying vehicles (Lin and Yang 2005; Chang et al. 2010), which is not the focus herein.

The other approach is to reduce the impact of road roughness by recording the responses of two connected vehicles of identical frequency during their passage over the same bridge, and then by deducting the response recorded for one vehicle from the other to eliminate the effect of road roughness (Yang et al. 2012b). The latter approach has been verified to be feasible in the numerical simulation using the VBI element. In this section, a theoretical formulation along with physical interpretation will be presented for such an approach. The mathematical model considered for the problem is shown in Figure 6.7, in which two connected vehicles with fixed spacing $s$ are allowed to pass through a simple beam at constant speed $v$. The rear vehicle is labeled as 1 and the front one as 2. The symbols used for the beam and vehicle are identical to those previously introduced.

The equations of motion for both vehicles can be expressed as follows.

$$m_{v,1}\ddot{q}_{v,1}(t)+k_{v,1}\left[q_{v,1}(t)-u_b(x,t)\Big|_{x=vt}-r(x)\Big|_{x=vt}\right]=0 \tag{6.26}$$

$$m_{v,2}\ddot{q}_{v,2}(t)+k_{v,2}\left[q_{v,2}(t)-u_b(x,t)\Big|_{x=vt_a}-r(x)\Big|_{x=vt_a}\right]=0 \tag{6.27}$$

where the time variable $t_a$ for the front vehicle is defined as the sum of time $t$ and the delay $t_s$:

$$t_a \equiv t+\frac{s}{v}=t+t_s \tag{6.28}$$

The equation of motion for the beam is

$$\bar{m}\ddot{u}_b(x,t)+EIu_b''''(x,t)=f_{c,1}(t)\delta(x-vt)+f_{c,2}(t)\delta(x-vt_a) \tag{6.29}$$

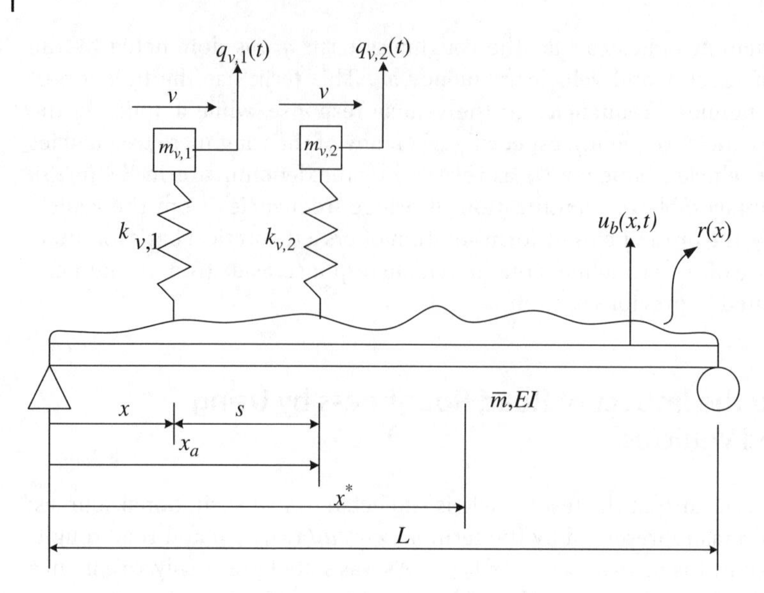


**Figure 6.7** Model of two connected vehicles passing through a simple beam.

where $f_{c,1}$ and $f_{c,2}$ are the contact forces for the rear and front vehicles, respectively,

$$f_{c,1} = -m_{v,1}g + k_{v,1}\left[ q_{v,1}(t) - u_b(x,t)\big|_{x=vt} - r(x)\big|_{x=vt} \right] \tag{6.30}$$

$$f_{c,2} = -m_{v,2}g + k_{v,2}\left[ q_{v,2}(t) - u_b(x,t)\big|_{x=vt_a} - r(x)\big|_{x=vt_a} \right] \tag{6.31}$$

By the modal superposition method, one can substitute the beam displacement in Eq. (6.10) into Eq. (6.29) to obtain

$$\sum_{n=1}^{\infty} \bar{m} \sin\frac{n\pi x}{L} \ddot{q}_{b,n}(t) + \sum_{n=1}^{\infty} EI\left(\frac{n\pi}{L}\right)^4 \sin\frac{n\pi x}{L} q_{b,n}(t) = f_{c,1}\delta(x - vt) + f_{c,2}\delta(x - vt_a) \tag{6.32}$$

Multiplying the preceding equation with $\sin(n\pi x/L)$ and integrating with respect to $x$ from 0 to $L$ yields the following equation:

$$\begin{aligned} \ddot{q}_{b,n} + \omega_{b,n}^2 q_b = {} & 2\sin\frac{n\pi vt}{L}\left\{ \frac{-m_{v,1}g}{\bar{m}L} + \frac{m_{v,1}\omega_{v,1}^2}{\bar{m}L}\left[ q_{v,1}(t) - u_b(x,t)\big|_{x=vt} - r(x)\big|_{x=vt} \right] \right\} \\ & + 2\sin\frac{n\pi vt_a}{L}\left\{ \frac{-m_{v,2}g}{\bar{m}L} + \frac{m_{v,2}\omega_{v,2}^2}{\bar{m}L}\left[ q_{v,2}(t) - u_b(x,t)\big|_{x=vt_a} - r(x)\big|_{x=vt_a} \right] \right\} \end{aligned} \tag{6.33}$$

where $\omega_{b,n}$ is the beam frequency of $n$th mode, as defined in Eq. (6.13), and $\omega_{v,j}$ is the frequency of the rear ($j$ = 1) or front vehicle ($j$ = 2), defined as

$$\omega_{v,j} = \sqrt{\frac{k_{v,j}}{m_{v,j}}} \tag{6.34}$$

Similarly, as far as the response of the beam is concerned, the terms inside the brackets of Eq. (6.33) are negligibly small compared with their preceding terms. Accordingly, Eq. (6.33) can be reduced to

$$\ddot{q}_{b,n} + \omega_{b,n}^2 q_{b,n} = 2\sin\frac{n\pi vt}{L}\left\{\frac{-m_{v,1}g}{\bar{m}L}\right\} + 2\sin\frac{n\pi vt_a}{L}\left\{\frac{-m_{v,2}g}{\bar{m}L}\right\} \tag{6.35}$$

Solving the above equation yields the generalized coordinate of the $n$th mode as

$$\begin{aligned} q_{b,n}(t) = &\frac{-\Delta_{st,n}^2}{1-S_n^2}\sin\frac{n\pi vt_s}{L}\cos\omega_{b,n}t - S_n\left[\frac{\Delta_{st,n}^1}{1-S_n^2} + \frac{\Delta_{st,n}^2}{1-S_n^2}\cos\frac{n\pi vt_s}{L}\right]\sin\omega_{b,n}t \\ &+\left[\frac{\Delta_{st,n}^1}{1-S_n^2} + \frac{\Delta_{st,n}^2}{1-S_n^2}\cos\frac{n\pi vt_s}{L}\right]\sin\frac{n\pi vt}{L} + \left[\frac{\Delta_{st,n}^2}{1-S_n^2}\sin\frac{n\pi vt_s}{L}\right]\cos\frac{n\pi vt}{L} \end{aligned} \tag{6.36}$$

where $S_n$ is the speed parameter as defined in Eq. (6.18), and $\Delta_{st,n}^j$ is the static displacement induced by the rear ($j$ = 1) or front vehicle ($j$ = 2),

$$\Delta_{st,n}^j = \frac{-2m_{v,j}gL^3}{n^4\pi^4 EI} \tag{6.37}$$

Substituting Eq. (6.36) back to Eq. (6.10) yields the general displacement of the beam:

$$\begin{aligned} u_b(x,t) = &\sum_{n=1}^{\infty}\left\{\frac{\Delta_{st,n}^1}{1-S_n^2}\sin\frac{n\pi x}{L}\left[\sin\frac{n\pi vt}{L} - S_n\sin\omega_{b,n}t\right]\right\} \\ &+\sum_{n=1}^{\infty}\left\{\frac{\Delta_{st,n}^2}{1-S_n^2}\sin\frac{n\pi x}{L}\left[\sin\frac{n\pi vt_a}{L} - \sin\frac{n\pi vt_s}{L}\cos\omega_{b,n}t - S_n\cos\frac{n\pi vt_s}{L}\sin\omega_{b,n}t\right]\right\} \end{aligned} \tag{6.38}$$

The preceding equation is an approximate, but quite accurate, expression of the displacement of the beam subjected to two passing connected vehicles with constant speed and spacing. Clearly, the first braced term is contributed by the rear vehicle and the second by the front vehicle.

With the response of the beam made available, the responses of the two vehicles can be solved as follows. First, the equation of motion for the rear vehicle can be rewritten as

$$\ddot{q}_{v,1} + \omega_{v,1}^2 q_{v,1} = \omega_{v,1}^2\, r(x)\big|_{x=vt} + \omega_{v,1}^2\, u_b(x,t)\big|_{x=vt} \tag{6.39}$$

Substituting Eq. (6.3) for $r(x)$ and Eq. (6.38) for $u_b(x, t)$ into the preceding equation, the displacement of the rear vehicle can be solved. Herein, only the part of the response induced by road roughness is of our concern, which can be given as follows:

$$q_{v1,r}(t) = \sum_{i=1}^{\infty} d_i \frac{\omega_{v,1}^2}{\omega_{v,1}^2 - (n_{s,i}v)^2}\left[\cos(n_{s,i}vt + \theta_i) - \cos\theta_i\cos\omega_{v,1}t + \frac{n_{s,i}v}{\omega_{v,1}}\sin\theta_i\sin\omega_{v,1}t\right] \tag{6.40}$$

Differentiating twice with respect to $t$ yields the acceleration of the rear vehicle as

$$\begin{aligned} \ddot{q}_{v1,r}(t) = &\sum_{i=1}^{\infty} d_i \frac{\omega_{v,1}^2}{\omega_{v,1}^2 - (n_i v)^2}\Big[-(n_i v)^2\cos(n_i vt + \theta_i) \\ &+\omega_{v,1}^2\cos\theta_i\cos\omega_{v,1}t - (n_i v\omega_{v,1})\sin\theta_i\sin\omega_{v,1}t\Big] \end{aligned} \tag{6.41}$$

Similar to Eq. (6.25), the rear vehicle response $\ddot{q}_{v1,r}$ is governed by the roughness frequencies $n_{s,i}v$ and vehicle frequency $\omega_v$. Let us denote the component related to the roughness frequencies as $\ddot{R}_{1,r}$:

$$\ddot{R}_{1,r}(t) = \sum_{i=1}^{\infty} d_i \frac{\omega_{v,1}^2}{\omega_{v,1}^2 - (n_i v)^2} \left[ -(n_i v)^2 \cos(n_i vt + \theta_i) \right] \tag{6.42}$$

It is known that $\ddot{R}_{1,r}$ may overshadow the bridge-frequency-related components in the vehicle spectrum, thereby making them difficult to identify.

Similarly, the acceleration of the front vehicle due to road roughness is

$$\begin{aligned} \ddot{q}_{v2,r}(t) = \sum_{i=1}^{\infty} d_i \frac{\omega_{v,2}^2}{\omega_{v,2}^2 - (n_i v)^2} \Big[ & -(n_i v)^2 \cos(n_i vt + \theta_a) \\ & + \omega_{v,2}^2 \cos\theta_a \cos\omega_{v,2} t - (n_i v \omega_{v,2}) \sin\theta_a \sin\omega_{v,2} t \Big] \end{aligned} \tag{6.43}$$

and the part of response $\ddot{R}_{2,r}$ directly related to roughness frequencies is

$$\ddot{R}_{2,r}(t) = \sum_{i=1} d_i \frac{\omega_{v,2}^2}{\omega_{v,2}^2 - (n_i v)^2} \left[ -(n_i v)^2 \cos(n_i vt + \theta_a) \right] \tag{6.44}$$

where the phase angle $\theta_a$ is defined as $\theta_a = n_i(vt_s) + \theta_i = n_i s + \theta_i$.

A comparison of the roughness responses for $\ddot{R}_{1,r}$ and $\ddot{R}_{2,r}$ in Eqs. (6.42) and (6.44) indicates that their amplitudes $A_{1r,i}$ and $A_{2r,i}$ have similar expressions:

$$A_{1r,i} = \left| d_i \frac{\omega_{v,1}^2 (n_i v)^2}{\omega_{v,1}^2 - (n_i v)^2} \right| \tag{6.45}$$

$$A_{2r,i} = \left| d_i \frac{\omega_{v,2}^2 (n_i v)^2}{\omega_{v,2}^2 - (n_i v)^2} \right| \tag{6.46}$$

Specifically, they are equal to each other when the two vehicles have the same frequency, i.e,. $\omega_{v,1} = \omega_{v,2} = \omega_v$:

$$A_{1r,i} = A_{2r,i} = \left| d_i \frac{\omega_v^2 (n_i v)^2}{\omega_v^2 - (n_i v)^2} \right| \tag{6.47}$$

Theoretically, the above equality offers an attractive clue for reducing the effect of road roughness. Namely, for two vehicles with identical frequency, i.e., $\omega_{v,1} = \omega_{v,2} = \omega_v$, passing through the beam with the same roughness profile, their response amplitudes associated with roughness frequencies are same and therefore can be eliminated by subtracting one response spectrum from the other, which offers the theoretical basis for the idea proposed by Yang et al. (2012b).

Moreover, for two vehicles with identical frequency, the responses $\ddot{R}_{1,r}$ and $\ddot{R}_{2,r}$ in Eqs. (6.42) and (6.44) differ only in the phase angle term, i.e., with $\theta_i$ for $\ddot{R}_{1,r}$ and $\theta_a$ for $\ddot{R}_{2,r}$, meaning that the roughness frequency-related components for the two vehicles are identical when evaluated at the same position $x^*$, but with a time delay $s/v$. This can be proved by letting $t_1{}^* = x^*/v$ and $t_2{}^* = t_1{}^* - s/v$ for the rear and front vehicle, respectively, which shows that the results obtained from Eqs. (6.42) and (6.44) for $\ddot{R}_{1,r}$ and $\ddot{R}_{2,r}$, respectively, are identical. As a result, one may also work in the time domain to eliminate the effect of surface roughness, by first synchronizing the accelerations of the two vehicles for the same contact points, and then subtracting the synchronized responses from each other.

## 6.8 Numerical Studies

It has been theoretically proved in the preceding section that the effect of road roughness can be eliminated by adopting two vehicles of identical frequency and by subtracting the response of one vehicle from the other. The feasibility of such an idea will be illustrated through several examples in this section. The properties adopted of the simple beam remain the same as those previously used, i.e., $L = 30\,\text{m}$, $E = 27.5\,\text{GPa}$, $I_b = 0.175\,\text{m}^4$, $\bar{m} = 1000\,\text{kg/m}$, and $A = 2\,\text{m}^2$.

### 6.8.1 Example 1. Two Identical Vehicles Moving over the Bridge of Class A Roughness

To ensure the frequencies are the same, two vehicles with identical properties are adopted: $m_{v,1} = m_{v,2} = 1000\,\text{kg}$, $k_{v,1} = k_{v,2} = 170\,\text{kN/m}$. They move with speed $v = 2\,\text{m/s}$ and spacing $s = 3\,\text{m}$ over the bridge of roughness class A, i.e., the best quality. Figure 6.8a shows the amplitude spectrum of the acceleration response calculated of the rear vehicle, in which only the vehicle frequency $f_v$, but no bridge-related frequencies, can be identified. As was stated previously, the difficulty in bridge frequency identification from the vehicle spectrum is mainly due to the fact that the roughness-induced responses are so large that the bridge-induced responses are overshadowed.

To reduce the effect of surface roughness in time domain, one first synchronizes the acceleration responses of the two vehicles with respect to the same contact points, and then subtracts the synchronized response for one vehicle from the other. Figure 6.8b shows the amplitude spectrum of the subtracted response. As can be seen, the effect of roughness has been largely reduced, making it easier to identify the bridge frequencies of the first and second modes, $f_{b,1}$ and $f_{b,2}$, for this example.

### 6.8.2 Example 2. Two Identical Vehicles Moving over the Bridge of Class C Roughness

In this example, all the properties adopted for the bridge and vehicle are identical to those of Example 1, except that roughness class C (normal quality) is adopted. Figures 6.9a and b, respectively, show the original and subtracted amplitude spectra of the acceleration response of the rear vehicle. As can be seen, the effect of road surface roughness has been largely reduced by subtracting the response obtained for one vehicle from the other. For this example, the first three bridge frequencies, $f_{b,1}$ to $f_{b,3}$, can be clearly identified, even with a poorer surface condition.

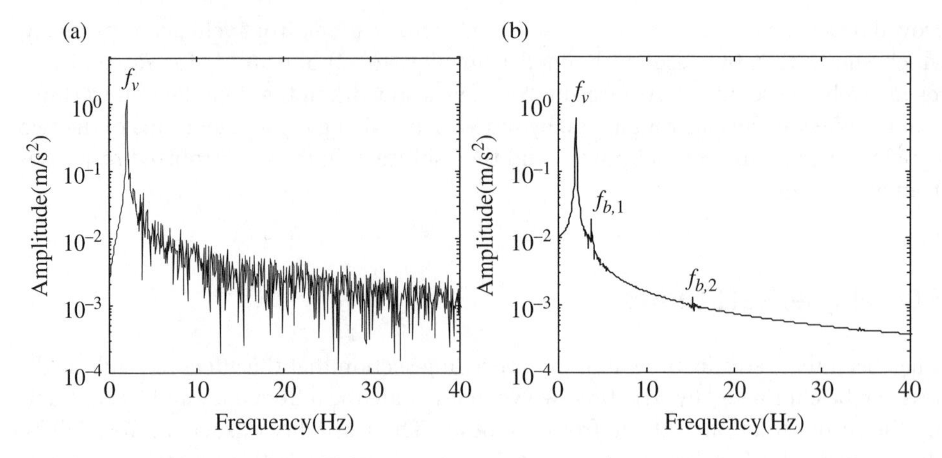


**Figure 6.8** Example 1: Acceleration response spectrum of rear vehicle: (a) original; (b) subtracted.

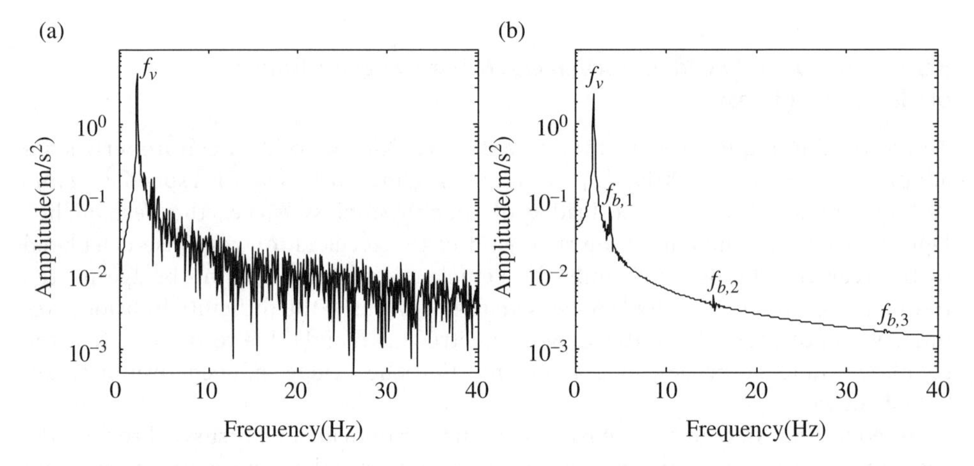


**Figure 6.9** Example 2: Acceleration response spectrum of rear vehicle: (a) original; (b) subtracted.

### 6.8.3 Example 3. Two Vehicles of Identical Frequency but Different Properties

It was shown in Eqs. (6.42) and (6.44) that the vehicle responses induced by road roughness of the same profile are identical, if the two vehicles adopted are of identical frequency. Certainly, for two vehicles to have the same frequency, one may just adopt two vehicles of the same physical properties, which are exactly the cases shown in Examples 1 and 2. In this example, two vehicles of identical frequency, but of different physical properties, will be adopted. To this end, the same physical properties are adopted for the rear vehicle: $m_{v,1}$ = 1000 kg and $k_{v,1}$ = 170 kN/m, but the front vehicle is assumed to be much lighter and softer, $m_{v,2}$ = 100 kg and $k_{v,2}$ = 17 kN/m. It is easy to see that both vehicles have an identical frequency of 2.08 Hz. We shall let the two vehicles move over the bridge with constant speed $v$ = 2 m/s and spacing $s$ = 3 m.

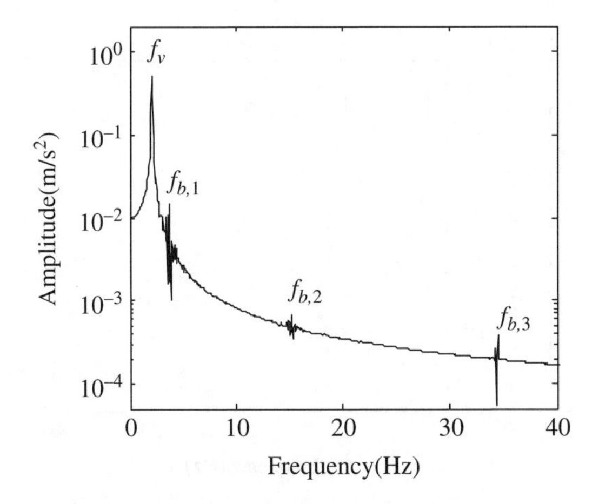


**Figure 6.10** Example 3: Subtracted amplitude spectrum of the response.

**Table 6.1** Vehicle spacing's adopted in Example 4.

| Case | *s* (m) | *s*/*L* (%) |
|---|---|---|
| 1 | 3 | 10 |
| 2 | 6 | 20 |
| 3 | 9 | 30 |
| 4 | 12 | 40 |
| 5 | 15 | 50 |

Figure 6.10 shows the amplitude spectrum of the response obtained by subtracting the synchronized acceleration response of the front vehicle from that of the rear vehicle. This figure indicates that as long as two vehicles have identical frequency, the effect of road surface roughness can be largely reduced through subtraction, so as to enhance the visibility of the bridge frequencies, even though the two vehicles may be physically different.

### 6.8.4 Effect of Vehicle Spacing on Identification of Bridge Frequencies

In this example, the effect of vehicle spacing on the accuracy of the bridge frequencies identified will be studied. Two identical vehicles are adopted, with the same properties as those adopted in Example 1. They are allowed to pass at constant speed $v = 2\,\text{m/s}$ through the simple beam with road roughness of Class A. The spacing $s$ between the two vehicles varies from 3 to 15 m, as listed in Table 6.1.

Figure 6.11 shows the amplitude spectrum of the subtracted response for the five cases considered, and Table 6.2 summarizes the bridge and vehicle frequencies identified. It is observed in each case that the effect of road surface roughness can be reduced, thereby making it easier to identify the bridge frequencies. No obvious discrepancy in the amplitude spectra and bridge frequencies identified is observed as far as the

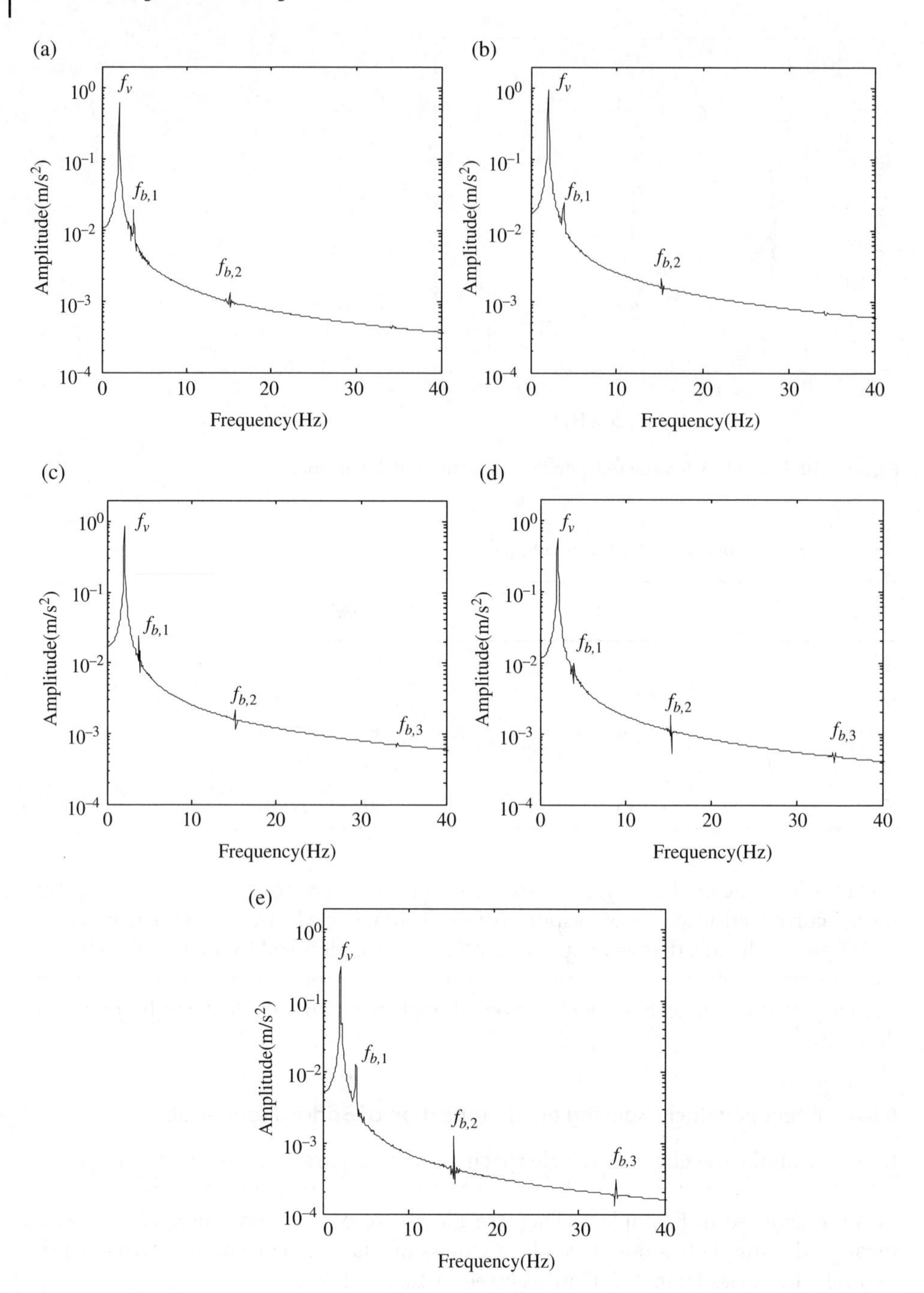


**Figure 6.11** Amplitude spectrum of the subtracted response: (a) Case 1, (b) Case 2, (c) Case 3, (d) Case 4, and (e) Case 5.

**Table 6.2** Bridge and vehicle frequencies identified.

| Case | $s$(m) | $f_{b,1}$(Hz)[a] | $f_{b,2}$(Hz)[a] | $f_{b,3}$(Hz)[a] | $f_v$(Hz)[b] |
|---|---|---|---|---|---|
| 1 | 3 | 3.83 | 15.37 | 34.53 | 2.07 |
| 2 | 6 | 3.80 | 15.37 | 34.37 | 2.07 |
| 3 | 9 | 3.77 | 15.37 | 34.33 | 2.07 |
| 4 | 12 | 3.90 | 15.30 | 34.33 | 2.07 |
| 5 | 15 | 3.90 | 15.30 | 34.33 | 2.07 |

[a] Analytical bridge frequencies of the first three modes: 3.83, 15.32, and 34.46 Hz.
[b] Analytical vehicle frequency: 2.08 Hz.

visibility of bridge frequencies is concerned. This result indicates that vehicle spacing is not a key parameter concerning the visibility of bridge frequencies in the vehicle response. However, the spacing between two connected vehicles is a mechanical issue that should be considered from the point of maneuverability or dynamic stability, to ensure that the two connected vehicles can move smoothly together at the desired speeds in practice.

## 6.9 Concluding Remarks

The effect of road roughness on the response of a moving vehicle aimed at the visibility of bridge frequencies is studied herein by an analytical approach. First, from the numerical simulations using the VBI element, it is illustrated that road roughness is an effect of crucial importance, which may excite the vehicle to a level higher than that by the bridge in vibration, such that the roughness-related frequencies may overshadow the bridge frequencies in the vehicle spectrum.

The effect of road roughness is also investigated by an approximate, but quite accurate, theory in closed form, from which the effects of two excitational sources to the vehicle, i.e., road roughness and bridge's vibration, can be clearly interpreted. The theory is then expanded to include the action of two connected vehicles moving over the bridge. From the analysis, it is concluded that, if two connected vehicles of identical frequency pass at constant speed through the same road roughness, the roughness-related responses for the two vehicles are identical when traveling to the same contact points, but differ by a phase angle. Consequently, one can work in the time domain to eliminate the effect of road roughness by first synchronizing the responses of the two connected vehicles with respect to the same contact points, and then by subtracting the synchronized response of one vehicle from the other. The two vehicles adopted in this regard are required to be identical in frequency but not in physical properties. The feasibility of such an approach is verified in the numerical studies. It was demonstrated that the spacing of the two connected vehicles is not a key parameter for identification of bridge frequencies; it should be determined according to the maneuverability or dynamic stability of the two connected vehicles, such that they can move smoothly together at the desired speeds in practice.

## 6.9 Concluding Remarks

# 7

# Filtering Technique for Eliminating the Effect of Road Roughness

By letting a test vehicle move over a bridge, one can identify the frequencies of the bridge by picking the peaks in the Fourier response spectra of the test vehicle. One problem with this approach is that the vehicle frequency (undesired) may appear as a dominant peak in the spectrum, making it difficult to identify the bridge frequencies (desired). To enhance the visibility of bridge frequencies, an effective filter is needed to filter out the vehicle frequency. As a preliminary attempt in this study, three filters are adopted, the *band-pass filter* (BPF), *singular spectrum analysis* (SSA), and the *singular spectrum analysis with band-pass filter* (SSA-BPF) proposed by Yang et al. (2013a) as a combination of the above two. Through numerical study on two cases with the vehicle frequency smaller or larger than the first bridge frequency, the SSA-BPF technique has been shown to be most effective for extracting the bridge frequencies, due to its salient feature that there is no need to select the number of singular values, as required by SSA, while the unexpected peaks encountered by the BPF are avoided. The materials presented in this chapter are based primarily on Yang et al. (2013a).

## 7.1 Introduction

By letting a properly instrumented test vehicle travel over a bridge, one may extract the vibration frequencies of the supporting bridge from the dynamic response recorded for the test vehicle during its passage over the bridge. Yang et al. (2004a) proposed this *indirect approach* for measuring the bridge frequencies, as it works solely on the vehicle, rather than the bridge response. This technique was later renamed the vehicle scanning method for bridges to make it self-explanatory. This technique requires no instrumentation on the bridge. Instead, the bridge frequencies are identified directly from the bridge response, which requires the deployment of vibration sensors and transmission devices on the bridge. Though the vehicle scanning method is still in the development stage, it has the potential of becoming an economic, convenient, and portable tool for extracting the bridge frequencies. For this reason, an increasing number of researches have been conducted along these lines, following the advent of the vehicle-based approach in 2004, covering not only bridge frequency extractions (Lin and Yang 2005; Chen and Xia 2009; Yang and Chang 2009a,b; Gomez et al. 2011), but also bridge

*Vehicle Scanning Method for Bridges*, First Edition. Yeong-Bin Yang, Judy P. Yang, Bin Zhang and Yuntian Wu.

damage detections (Bu et al. 2006; Wang et al. 2008; McGetrick et al. 2009; Nguyen and Tran 2010; Xiang et al. 2010) and other related issues (Yin and Tang 2011).

When using the moving test vehicle as the tool of measurement, one can identify the bridge frequencies by picking the major peaks in the Fourier response spectra of the test vehicle. Other than the bridge frequencies, the vehicle frequency also appears as a dominant peak in the vehicle spectrum, in some cases with an amplitude much higher than those of the bridge frequencies. If the amplitude of the vehicle's peak happens to be too high compared with the remaining peaks, the bridge frequencies may be shadowed and even made invisible. One cause for the appearance of extremely high peak for the vehicle frequency is the presence of road surface roughness (Chang et al. 2010), a factor that cannot be circumvented in practice. To render the vehicle scanning method for measurement of bridge frequencies feasible, it is essential that some data processing or filtering techniques be developed to remove, or at least to reduce, the dominant peak associated with the vehicle frequency. In this regard, it is assumed that the vehicle frequency has been made available prior to the field tests for bridge frequencies, by means of a dynamic test (Lin and Yang 2005) on the test vehicle to be used.

For the reasons already stated, the objective of this chapter is to search for a workable *filter technique* for use with the test vehicle in identifying the bridge frequencies, through comparison of the effects of selected filters on the test vehicle responses. Such a theoretical study is necessary before the filtering techniques can be successfully implemented for field tests, for which the environmental noises can be another concern. In this theoretical study, the response of the vehicle moving over the bridge will be artificially generated by the finite element simulation, as to be outlined first. Next, a brief introduction of the filters will be given in terms of their operating algorithms. As a preliminary attempt, three filters are studied herein, i.e., the band-pass filter (BPF), singular spectrum analysis (SSA), and the singular spectrum analysis with band-pass filter (SSA-BPF) proposed by Yang et al. (2013a) as a combination of the above two. Then, the filters will be performed on the vehicle responses for the two cases either with the vehicle frequency higher or lower than the first bridge frequency. From the two cases studied, the advantages and disadvantages of each filter will be discussed. Unsurprisingly, it is concluded that the SSA-BPF outperforms the other two in that it does not require the number of singular values as required by the SSA to be determined beforehand, while the unexpected peaks generated by BPF are totally circumvented.

## 7.2 Numerical Simulations for Vehicle Responses

Using the vehicle scanning method, the bridge frequencies are extracted from the dynamic responses of the test vehicle moving over the bridge. In order to concentrate on the evaluation of various filters, the dynamic responses of the test vehicle will be generated by the finite element simulations, rather than by the field measurement. The mathematical model and procedure for calculating the dynamic responses of the test vehicle are briefly described in the following.

The two-dimensional vehicle-bridge interaction (VBI) model is shown in Figure 7.1, where the test vehicle is modeled as a sprung mass $m_v$ of single degree of freedom (DOF), supported by a spring of elastic constant $k_v$, and the bridge is modeled as a simple beam of length $L$, flexural rigidity $EI$, and mass per unit length $m^*$. The surface

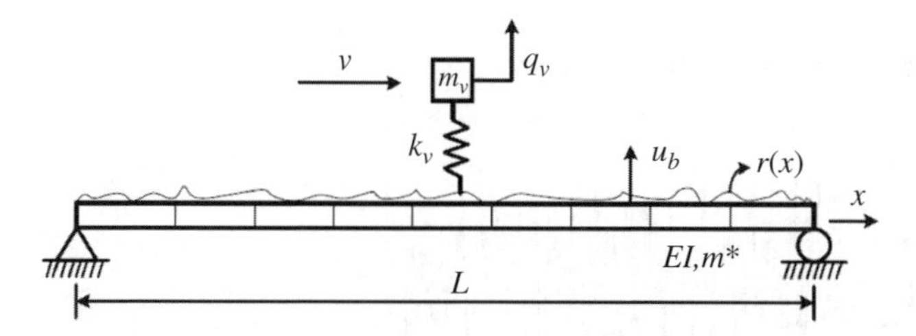


**Figure 7.1** Vehicle-bridge interaction (VBI) model.

roughness $r(x)$ of the bridge is a function of the axis $x$. In practice, a two-wheel trailer (towed by a tractor) that can be modeled as a single DOF system of zero damping is used as the test vehicle (Lin and Yang 2005). Normally, the test vehicle is assumed to travel at constant speed $v$, but it was demonstrated that the variation in vehicle speed causes only minor effects on the bridge frequency extraction (Chen and Xia 2009).

As shown in Figure 7.1, the beam is discretized into a number of finite elements. The beam element directly acted upon by the sprung mass is modeled as a VBI element, of which the element matrices based on Chang et al. (2010) are summarized in Appendix, as a modification from the ones by Yang and Yau (1997) and Yang et al. (2004b). The remaining beam elements not directly acted upon by the sprung mass are modeled as the conventional six-DOF beam elements, of which the element matrices are available in most finite element textbooks, such as Cook et al. (2002) and Zienkiewicz and Taylor (2005). By assembling the VBI element with the other conventional beam elements, one can establish the equations of motion for the entire VBI system. The equations of motion for the system are solved by Newmark's $\beta$ method step-by-step to yield the dynamic responses of the vehicle and the beam at any instant. For the sake of unconditional stability in numerical integration, the parameters $\beta$ and $\gamma$ are selected as 1/4 and 1/2, respectively (Clough and Penzien 1995).

Let us take a typical example to illustrate the need of using the filter techniques to enhance the visibility of bridge frequencies in the vehicle response. In this case, the test vehicle is set to travel with a constant seed of $v$ = 2 m/s through a simple bridge with rough surface. The properties selected for the vehicle are: $m_v$ = 1000 kg, $c_v$ = 0, and $k_v$ = 170 kN/m, and those for the bridge are: $m^*$ = 1000 kg/m, $E$ = 27.5 GPa, $I$ = 0.175 m$^4$, and $L$ = 30 m. The time step size is set as 0.001 second. The surface roughness profile is generated using the power spectrum density curve of class A given by ISO 8608 (1995), as shown in Figure 7.2.

Figure 7.3a shows the simulated acceleration response of the test vehicle during its passage over the bridge, as generated by the finite element analysis program. By performing the fast Fourier transform (FFT) to the vehicle acceleration, one obtains the Fourier spectrum as in Figure 7.3b. For the sake of brevity, the term *response* will be used to refer to the acceleration response and *spectrum* the Fourier spectrum hereinafter.

From Figure 7.3b, one observes that, except for the dominant peak at around 2 Hz, no other remarkable peaks exist. The dominant peak of around 2 Hz corresponds exactly to the natural frequency of the test vehicle, calculated as $(k_v/m_v)^{1/2}$ = 13.04 rad/sec = 2.08 Hz. This figure indicates that the vehicle frequency has such a high peak that all the other bridge frequencies are made invisible, making it difficult for the latter to be

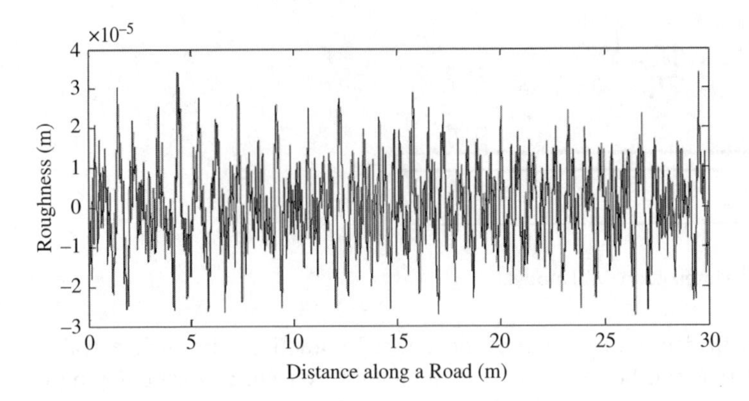


**Figure 7.2** Roughness profile.

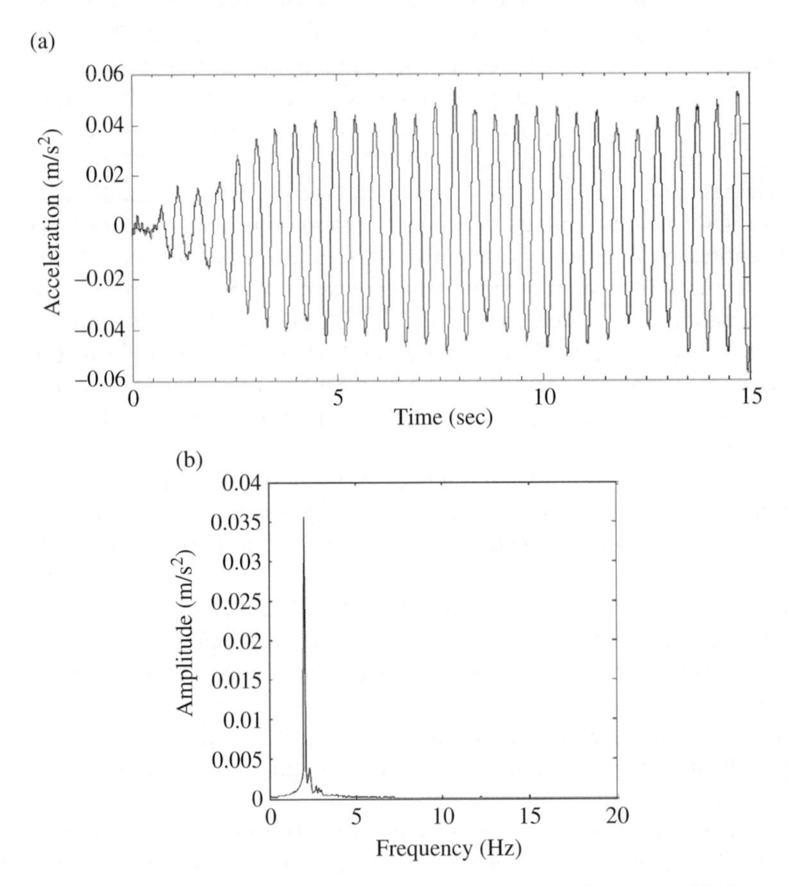


**Figure 7.3** Acceleration response of the test vehicle: (a) time history, and (b) Fourier spectrum.

identified. To enhance the visibility of the bridge frequencies in the vehicle's spectrum, there is a need to remove or reduce the influence of the vehicle frequency by letting the test vehicle's spectrum be pre-processed by some appropriate filters, as will be discussed in the following sections.

## 7.3 Filtering Techniques

As a preliminary attempt, three filters are adopted in this study, the BPF or band-stop filter (BSF), SSA, and the SSA-BPF, proposed by Yang et al. (2013a) as a combination of the above two. Among the three filters, the BPF is a popular data-processing technique used in frequency domain (Moschas and Stiros 2011; Wallin et al. 2011; Magalhães et al. 2012), and the SSA is a relatively new technique performed in time domain. The BPF and SSA filters have their own advantages and disadvantages. The third filter, the SSA-BPF, combines the advantages of both the BPF and SSA, while removing their disadvantages, by introducing a band-pass filtering criterion through revision of the conventional SSA algorithm. The respective advantages and disadvantages of each of the three filters will be highlighted in the case studies later on, after a summary of their operating algorithms in this section.

### 7.3.1 Band-Pass Filter (BPF)/Band-Stop Filter (BSF)

Consider a set of data in time series $f(t) = [f_0, f_1, f_2, \ldots, f_{N\text{-}1}]$ of length $N$, where the elements $f_0, \ldots, f_{N\text{-}1}$ are values of certain physical quantity sampled at time steps $t_0, \ldots, t_{N\text{-}1}$. In this study, the series of data may represent a response history of the test vehicle. The algorithm for processing the series of data with the BPF or BSF is as follows.

*Step 1: Fast Fourier transform (FFT)*
Performing the FFT to the time series data $\boldsymbol{f}(t)$, one can obtain an amplitude representation function $\boldsymbol{F}(f)$ in frequency domain as $\boldsymbol{F}(f) = \mathbf{FFT}\{\boldsymbol{f}(t)\} = [F_0, F_1, F_2, \ldots, F_{N\text{-}1}]$, where the elements $F_0, \ldots, F_{N\text{-}1}$ denote the amplitudes corresponding to the discrete frequencies $f_1, \ldots, f_{N\text{-}1}$.

*Step 2: Window function multiplication*
Select a *window function* $\boldsymbol{W}(f)$, i.e., a function by which a weight of unity is assigned for the frequency band to be retained and a weight of zero for the remaining to be eliminated in frequency domain. Usually, the interface between the retained and eliminated bands is designed as a transitional zone by which the weight may decrease from unity to zero or increase from zero to unity in a gradual manner, so as to avoid a drastic jump; see Figure 7.4 for illustration. Then, we can multiply $\boldsymbol{W}(f)$ with $\boldsymbol{F}(f)$ to obtain the filtered frequency function $\boldsymbol{F}_w(f)$, which allows the original frequency function to pass a certain assigned band and to stop outside the band. For this reason, this filter has been referred to as the BPF, or complementarily the BSF, with the weights of the retained and eliminated bands reversed.

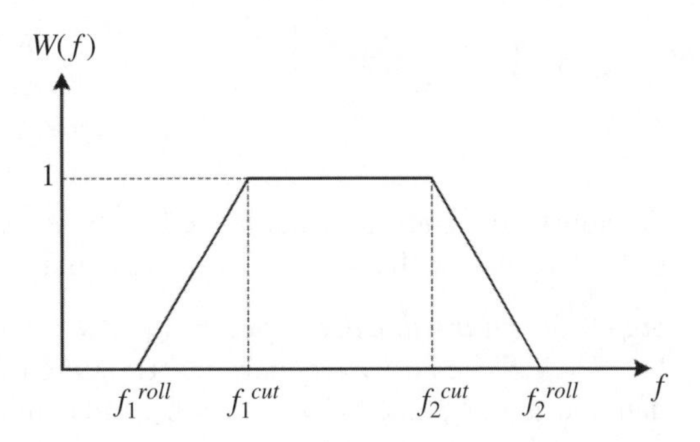


**Figure 7.4** Trapezoidal window function for band-pass filter (BPF).

From the spectrum plotted for $\boldsymbol{F}_w(f)$ with respect to the frequency $f$, the dominant frequencies can be identified by peak-picking. In addition, one may perform the inverse FFT to $\boldsymbol{F}_w(f)$ to obtain the filtered time series if necessary. The *window function* adopted in this study is of the trapezoidal form as shown in Figure 7.4, with $f_1^{\mathrm{cut}}$ and $f_2^{\mathrm{cut}}$ denoting the lower and upper cutoff frequencies, respectively, and $f_1^{\mathrm{roll}}$ and $f_2^{\mathrm{roll}}$ the lower and upper roll-off frequencies. The range between $f_1^{\mathrm{cut}}$ and $f_2^{\mathrm{cut}}$ is exactly the passband. It should be noted that the filtered data in time domain is not of concern in this study, since all the filtering and identification tasks are performed in frequency domain. As such, the simplest window function of trapezoidal type is adopted in this study, even though it does not work as well as the usual cosine or bell ones in energy compensation.

### 7.3.2 Singular Spectrum Analysis (SSA)

The SSA is a time-domain data analysis technique, by which a given set of data in time series is decomposed into a finite number of interpretable components according to their respective singular values. These components represent the *trends, oscillatory components, noises,* or other physical phenomena. In the measurement of bridge frequencies using the moving test vehicle, the vehicle spectrum contains a dominant component related to the vehicle frequency, as was illustrated in the previous example. The vehicle-related frequency component can be regarded as the "trend" of the vehicle response, since it is the principal or dominant component that will occur in the vehicle response, and it is the one (undesired) that will make the bridge-related frequencies (desired) not so visible or even invisible in the vehicle spectrum. Accordingly, the SSA is to filter out the vehicle-frequency component, leaving the bridge-frequency components in the remaining response so that the bridge frequencies can be easily identified.

There are four major steps for performing the conventional SSA, i.e., *embedding, singular value decomposition* (SVD), *grouping,* and *skew diagonal averaging.* A brief description is given as follows, whereas more details can be found in the book by Golyandina et al. (2001).

*Step 1: Embedding*

Given the same set of time series data $f(t)$, one may first select a window length $L$ (an integer in the range $1 < L < N$) and embed the elements of the data $f(t)$ into $K$ *lagged vectors* $\boldsymbol{L}_i$ of length $L$ as $\boldsymbol{L}_i = [f_{i-1}, f_i, \ldots, f_{i+L-2}]^T$, $i = 1, 2, \ldots, K$, where $K$ is calculated as $K = N - L + 1$. Then, the $K$ lagged vectors can further be embedded into a *trajectory matrix* $\boldsymbol{X}$ as

$$\boldsymbol{X} = \left[\boldsymbol{L}_1, \boldsymbol{L}_2, \ldots, \boldsymbol{L}_K\right] = \begin{bmatrix} f_0 & f_1 & \cdots & f_{K-1} \\ f_1 & f_2 & \cdots & f_K \\ \vdots & \vdots & \ddots & \vdots \\ f_{L-1} & f_L & \cdots & f_{N-1} \end{bmatrix}, \tag{7.1}$$

Obviously, the trajectory matrix $\boldsymbol{X}$ is a $L \times K$ *Hankel matrix,* since all elements in the skew diagonals with $j + i$ = constant are equal.

*Step 2: Singular value decomposition (SVD)*

Let $\boldsymbol{S} = \boldsymbol{X}\boldsymbol{X}^{\mathrm{T}}$ be an $L \times L$ square matrix. One may find $L$ eigenvalues for the matrix $\boldsymbol{S}$, denoted as $\lambda_1, \lambda_2, \ldots, \lambda_L$ ($\lambda_1 > \lambda_2 > \ldots > \lambda_L$), and $L$ unit eigenvectors, denoted as $\boldsymbol{U}_1, \boldsymbol{U}_2, \ldots, \boldsymbol{U}_L$.

Supposing that $d$ ($d \leq L$) is the number of positive eigenvalues, the squared roots of the first $d$ eigenvalues, $\sqrt{\lambda_1}, \sqrt{\lambda_2}, \ldots, \sqrt{\lambda_d}$, are referred to as the *singular values* of the trajectory matrix $\boldsymbol{X}$. Accordingly, the number of singular values of the trajectory matrix $\boldsymbol{X}$ is synonymous to the number $d$ of positive eigenvalues for $\boldsymbol{S}$. In practice, some singular values may be extremely small. Consequently, only singular values that are of practical contributions will be included in the decomposition process. Thus, the number $d$ used will be less than the theoretical number of singular values, without losing its accuracy.

Each singular value $\sqrt{\lambda_i}$ corresponds to an elementary matrix $\boldsymbol{X}_i$ given as

$$\boldsymbol{X}_i = \sqrt{\lambda_i}\boldsymbol{U}_i\boldsymbol{V}_i^T \tag{7.2}$$

where $\boldsymbol{V}_i$ is another unit singular vector corresponding to the singular value of concern, calculated as: $\boldsymbol{V}_i = \boldsymbol{X}^{\mathrm{T}}\ \boldsymbol{U}_i\ /\ \sqrt{\lambda_i}$. The two vectors $\boldsymbol{U}_i$ and $\boldsymbol{V}_i$ are referred to as the *left* and *right singular vectors,* respectively. Finally, the trajectory matrix $\boldsymbol{X}$ can be expressed as the summation of the $d$ elementary matrices as

$$\boldsymbol{X} = \boldsymbol{X}_1 + \boldsymbol{X}_2 + \ldots + \boldsymbol{X}_d \tag{7.3}$$

In this way, the trajectory matrix $\boldsymbol{X}$ is decomposed into $d$ elementary matrices of rank 1 with a norm equal to the singular value. The above expression is also called the SVD of the trajectory matrix $\boldsymbol{X}$.

*Step 3: Grouping*

In this step, all the elementary matrices obtained in the last step are collected into a finite number of groups, say $m$ groups, according to pre-set criteria. The selection of the grouping criteria depends on the expected function of the SSA, e.g. denoising, smoothing, oscillatory component extracting, etc. some of which have been discussed in Golyandina et al. (2001) and Alonso et al. (2005). Summing all the elementary matrices in the same group for each group and denoting the $m$ resultant matrices as $\boldsymbol{X}_{G1}, \boldsymbol{X}_{G2}, \ldots, \boldsymbol{X}_{Gm}$, the original trajectory matrix $\boldsymbol{X}$ can be expressed as follows:

$$\boldsymbol{X} - \boldsymbol{X}_{G1} + \boldsymbol{X}_{G2} + \ldots + \boldsymbol{X}_{Gm} \tag{7.4}$$

which is referred to as the *grouped decomposition.*

The grouping criterion in this study is rather simple. It can be proved that the norm of an elementary matrix $\boldsymbol{X}_i$ equals the singular value $\sqrt{\lambda_i}$, representing its contribution to the trajectory matrix $\boldsymbol{X}$. Since the singular values are arranged in a descending order, the first few elementary matrices contribute more than the other ones to the trajectory matrix. In this study, the major elementary matrices will be collected in a group related to the *vehicle frequency,* since they contribute most to the dynamic response of the test vehicle. The remaining elementary matrices will be collected in another group associated with the *nonvehicle frequencies,* which represents exactly the part of concern for further identification of the bridge frequencies.

*Step 4: Skew diagonal averaging*

This step serves to convert each resultant matrix into a new set of time-series data of length $N$. If the resultant matrix happens to be a Hankel matrix, the conversion can be accomplished in a simple manner that is exactly the reverse of the embedding procedure. However, this is not generally true in practice. To recover the time series from a non-Hankel matrix, a skew diagonal averaging procedure should be adopted. Let $\boldsymbol{Y}$ be any of the resultant matrices $\boldsymbol{X}_{Gi}$, with the elements denoted as $y_{ij}$, $i = 1 \sim L$, $j = 1 \sim K$. For $L < K$, the recovered time-series data $\boldsymbol{g} = [g_0, g_1, \ldots, g_{N-1}]$ is given by

$$g_k = \begin{cases} \dfrac{1}{k+1}\displaystyle\sum_{m=1}^{K+1} y_{m,k-m+2} & \text{for } 0 \le k \le L-1 \\ \dfrac{1}{L}\displaystyle\sum_{m=1}^{L} y_{m,k-m+2} & \text{for } L-1 \le k \le K \\ \dfrac{1}{N-k}\displaystyle\sum_{m=k-K+2}^{N-K+1} y_{m,k-m+2} & \text{for } K \le k \le N \end{cases} \tag{7.5}$$

For $L > K$, the preceding expressions remain valid except that the length $L$ should be switched with $K$.

### 7.3.3 Singular Spectrum Analysis with Band-Pass Filter (SSA-BPF)

The SSA-BPF proposed by Yang et al. (2013a) combines the advantages of both the SSA and BPF by introducing a band-pass filtering criterion to the grouping step of the conventional SSA with slight revisions. Figure 7.5 shows the flowchart of the SSA-BPF, which

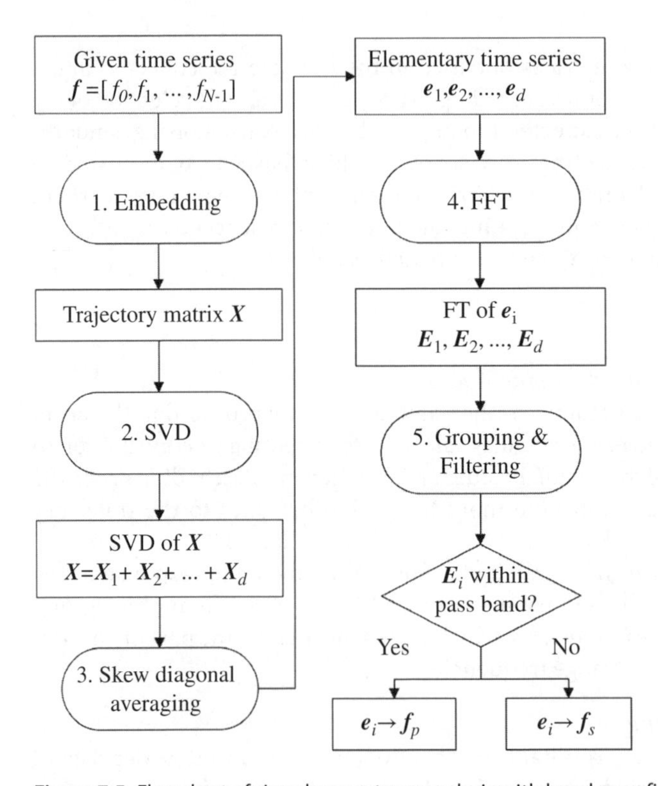


**Figure 7.5** Flowchart of singular spectrum analysis with band-pass filter (SSA-BPF).

consists of five major steps: (i) embedding; (ii) SVD; (iii) skew diagonal averaging; (iv) FFT; and (v) grouping and filtering.

Consider the same set of data in time series $f(t)$. The first two steps, embedding and SVD, are identical to those for the conventional SSA. As such, an identical trajectory matrix $\boldsymbol{X}$ along with SVD, i.e., $\boldsymbol{X} = \boldsymbol{X}_1 + \boldsymbol{X}_2 + \ldots + \boldsymbol{X}_d$ as in Eq. (7.3), is available at the end of the two steps. The third step is to perform skew diagonal averaging to each elementary matrix $\boldsymbol{X}_i$, $i = 1 \sim d$, to yield a set of elementary time-series data $\boldsymbol{e}_i$ of length $N$ identical to that of the original data. In the fourth step, the FFT is performed on each data set $\boldsymbol{e}_i$ to yield the Fourier spectrum $\boldsymbol{E}_i$. The fifth step is called the grouping and filtering step, in that all the data sets $\boldsymbol{e}_i$'s are grouped according to the following filtering criterion: those with their dominant frequency falling within a preset frequency band are collected into the "pass group," and those outside the band into the "stop group." Finally, by summing the elementary data sets for both groups, one obtains the time series $\boldsymbol{f}_p$ and $\boldsymbol{f}_s$ corresponding to the pass and stop groups, respectively.

## 7.4 Case Studies

Two numerical cases are studied, either with the vehicle frequency smaller or larger than the first frequency of the bridge. For both cases, the dynamic responses of the test vehicle moving over the bridge are generated by the finite element analysis program and then processed by the three filters BPF, SSA, and SSA-BPF. By comparing the resulting amplitude spectra preprocessed by the three filters, the advantages and disadvantages of each filter for use in the indirect measurement for identifying the bridge frequencies will be highlighted.

### 7.4.1 Case 1: Vehicle Frequency Smaller than First Bridge Frequency

The properties adopted of the test vehicle and bridge are the same as those of the previous example, where the time-history response and Fourier spectrum of the test vehicle are also presented. As was mentioned, the known vehicle frequency of 2.08 Hz should be filtered out in order for the bridge frequencies to become visible.

First, the vehicle response history is processed by the BPF with a passband of 3~10 Hz and roll-off frequencies of 2 and 11 Hz, as shown in Figure 7.6. Clearly, the vehicle frequency has been filtered out and a new dominant frequency appears at 3.87 Hz. This dominant frequency matches well with the bridge frequency of the first mode calculated analytically, indicating that the filter works well in suppressing the major peak associated with the vehicle frequency, while enhancing the visibility of the bridge frequencies. However, an artificial, unexpected peak also appears at around 3 Hz, which is seemingly induced by the transition zone between the pass- and stop-band. Such a peak may confuse engineers and give them an erroneous interpretation of the bridge frequencies.

Second, the vehicle response history is processed by the SSA. The total length $N$ of the original response history is 15 000. The window length $L$ is selected as 600 for the embedding step, and the first $d = 200$ singular values are considered sufficient for decomposing the original time history for the SVD step. Since no information is available on the number of singular values to be used for the vehicle-related frequency

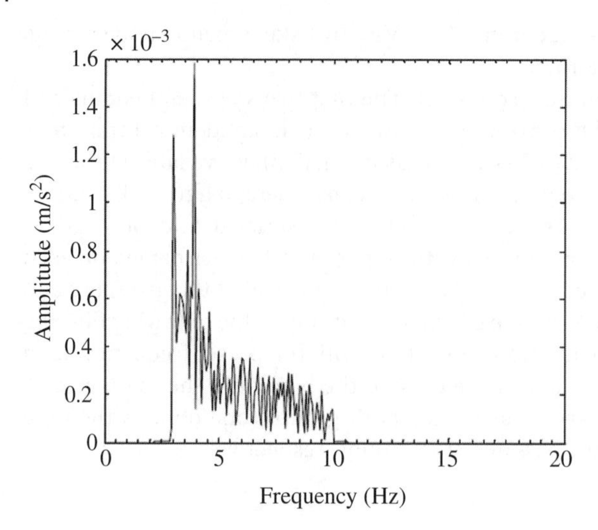


**Figure 7.6** Resultant spectrum processed by band-pass filter (BPF) (Case 1).

component prior to the analysis, and neither is that for the non-vehicle-related frequency component, some tests should be done to get an optimal result. In the comparison study, the vehicle-related frequency group $G_v$ is assumed to be composed of either the first two, four, six, eight, or ten singular values or elementary matrices, and thus be equal to the sum of all the elementary matrices. Thus, the non-vehicle-related frequency group $G_n$ is computed as the sum of all the remaining elementary matrices in the range of the sequential numbers of 3~200, 5~200, 7~200, 9~200, or 11~200. Since only the elements in the non-vehicle-related frequency group $G_n$ is of interest, the resultant matrices for $G_n$ are skew diagonal averaged and then processed by the FFT to yield the Fourier spectra as in Figures 7.7a–e.

From Figure 7.7a, one observes that the vehicle frequency has been clearly filtered out and that the 1st bridge frequency of 3.87 Hz becomes visible. No unexpected peak occurs in this case. However, in Figures 7.7b–e for the analyses with less singular values (and associated elementary matrices) grouped in the non-vehicle group $G_n$, the first bridge frequency disappears, while other higher, but minor, frequencies become visible in the spectra. These figures indicate that the first bridge frequency has been filtered out along with the vehicle frequency, and that the components induced by road surface roughness and other noises have been amplified in the spectra, which is unfavorable for bridge frequency identification. The comparison study reveals that the selection of the number of singular values to be grouped in the non-vehicle-related frequency group $G_n$ requires some engineering judgments, and should be performed with extreme care. Further, since the bridge frequency is not known prior to filtering, it is difficult to determine the number of singular values to be included in the non-vehicle-related frequency group.

Third, the vehicle response history is processed by the SSA-BPF. The window length $L$ is selected to be 600, same as that used by the SSA, and the passband is 3~10 Hz, also the same as that used by the BPF. For this case, the number $d$ of singular values to be included for the SVD step is 20 or 100. Compared with the use of $d = 200$ for the

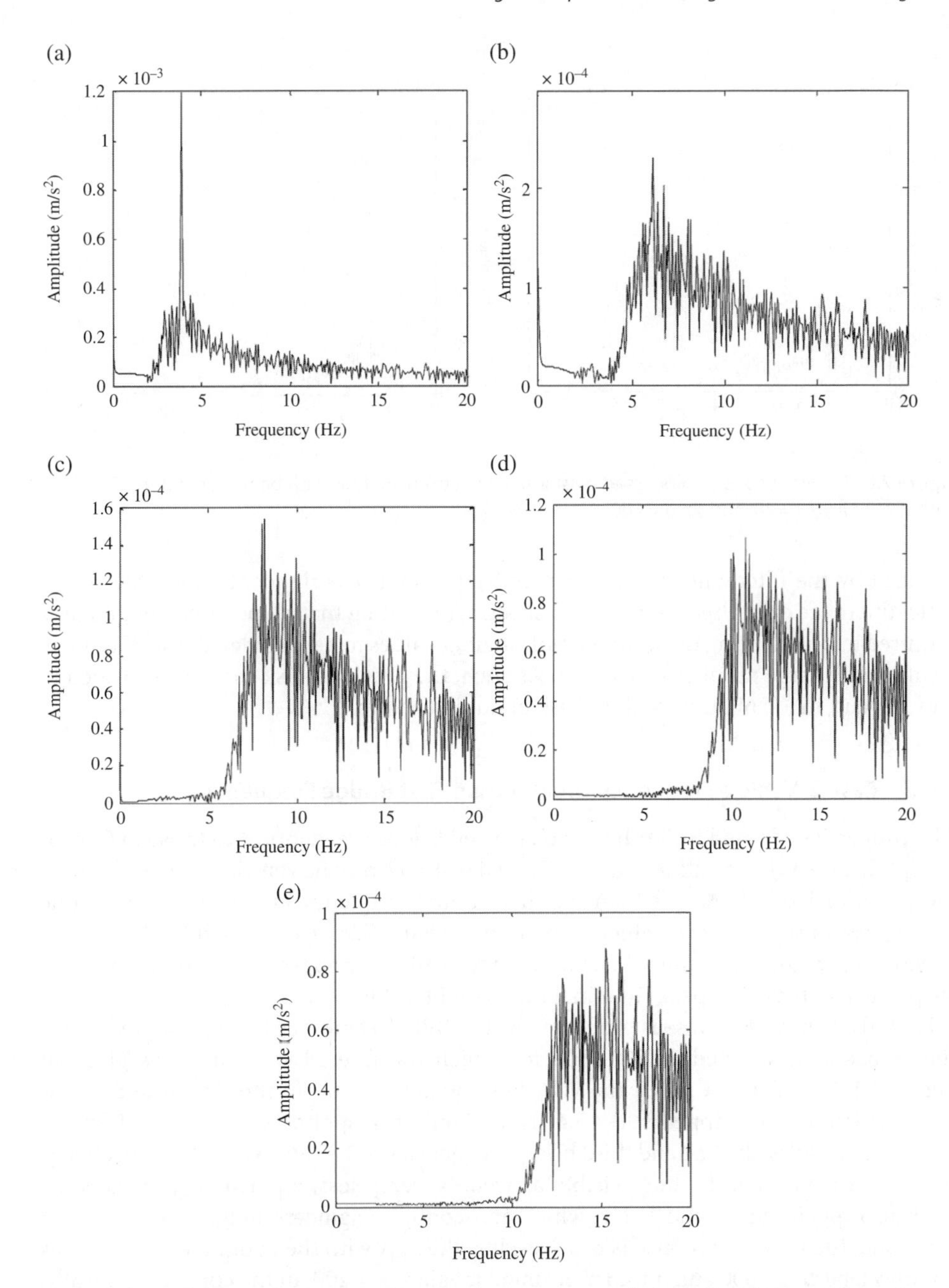


**Figure 7.7** Resultant spectra processed by singular spectrum analysis (SSA) with the numbers of singular values selected as (Case 1): (a) 3~200; (b) 5~200; (c) 7~200; (d) 9~200; and (e) 11~200.

previous cases, the use of two smaller $d$ values is merely to save computation time without losing the accuracy of solution. Figures 7.8a and b show the Fourier spectra of the resultant time series $f_p$ for the pass group of $d = 20$ and 100, respectively. In both spectra, the vehicle frequency has been filtered out completely, with no unexpected peaks

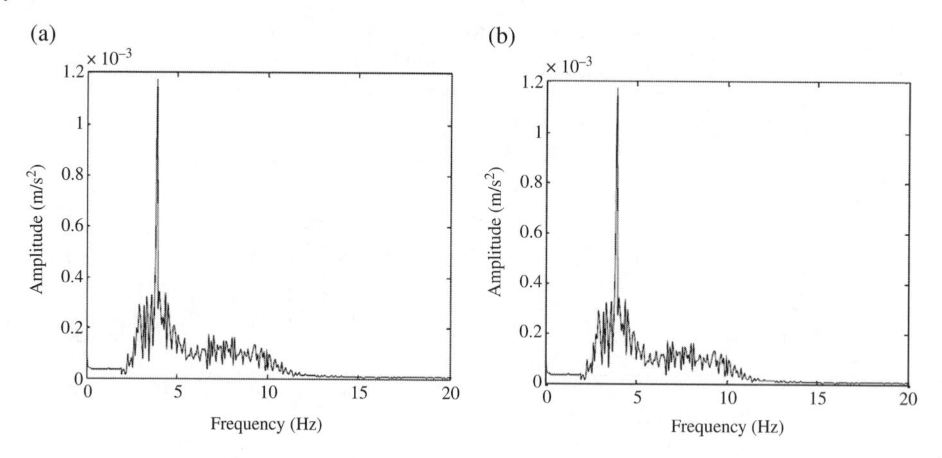


**Figure 7.8** Resultant spectra processed by singular spectrum analysis with band-pass filter (SSA-BPF) with (Case 1): (a) $d = 20$; and (b) $d = 100$.

induced by the filter, and the first bridge frequency is clearly identifiable. Moreover, little difference exists between the two spectra, indicating that no personal judgment is required in determining the number $d$ of singular values to be included in the SSA, once a sufficiently large number is selected. As such, the grouping task can be performed in a rather automatic manner, with no risk of human errors.

### 7.4.2 Case 2: Vehicle Frequency Greater than First Bridge Frequency

The properties adopted for both the vehicle and bridge are identical to those in Case 1, except that the vehicle stiffness $k_v$ is increased to 4 MN/m. The vehicle frequency is thus increased to 10.07 Hz, larger than the first natural frequency of the bridge. The time history response of the test vehicle is plotted in Figure 7.9a. As revealed by the Fourier spectrum in Figure 7.9b, the vehicle frequency of 10.07 Hz appears as a dominant peak, surpassing all the other peaks and making them invisible.

First, the vehicle response is processed by the BPF. To embrace the vehicle frequency, the stopband is selected as 5~11 Hz, for which the filtered spectrum was given in Figure 7.10. Similar to Case 1, the vehicle frequency has been filtered out and a new dominant frequency appears at 15.46 Hz, which corresponds to the second bridge frequency. Besides, the first and third bridge frequencies of 3.87 and 34.79 Hz, respectively, have been made slightly more visible as well. Nevertheless, an artificial, unexpected peak also appears at around 11 Hz, which may confuse engineers' judgment.

Second, the vehicle response is processed by the SSA with the parameters selected as window length $L = 600$ and number of singular values $d = 200$. In the comparison study, the following sequential ranges of singular values or elementary matrices: 3~200, 5~200, 7~200, and 9~200, are respectively grouped and summed for the non-vehicle-related frequency group $G_n$. By the same procedure as that for Case 1, the resultant Fourier spectra were obtained for the above four analyses in Figures 7.11a–d. In Figure 7.11a, the second bridge frequency (15.46 Hz) becomes the dominant frequency, although the vehicle frequency has not been filtered out completely. In Figure 7.11b, which indicates the result obtained with two more singular values being removed from the non-vehicle

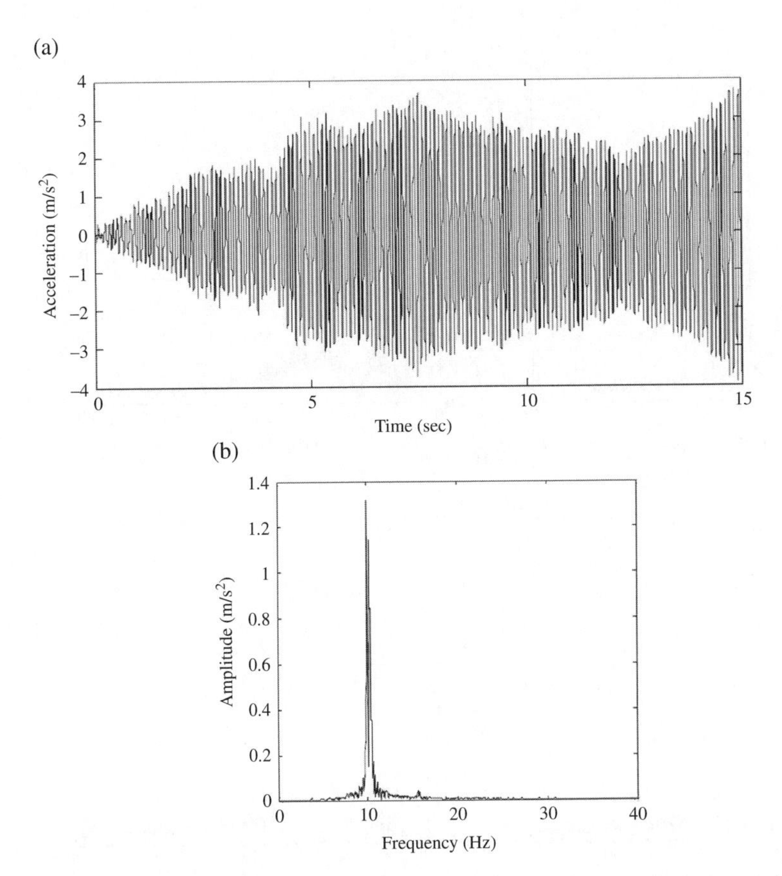


**Figure 7.9** Acceleration response of the test vehicle (Case 2): (a) time history; and (b) Fourier amplitude spectrum.

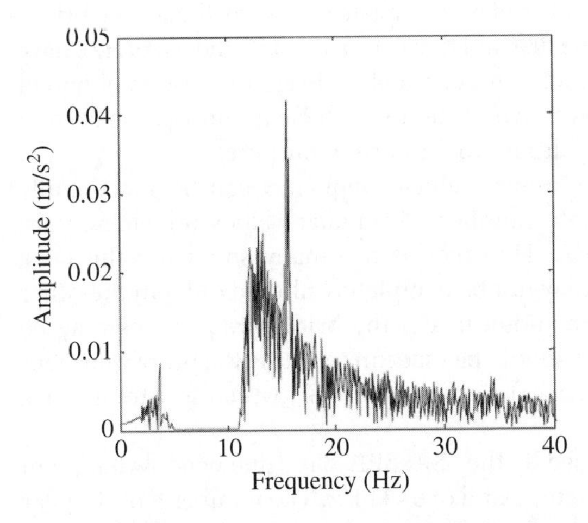


**Figure 7.10** Resultant spectrum processed by band-pass filter (BPF) (Case 2).

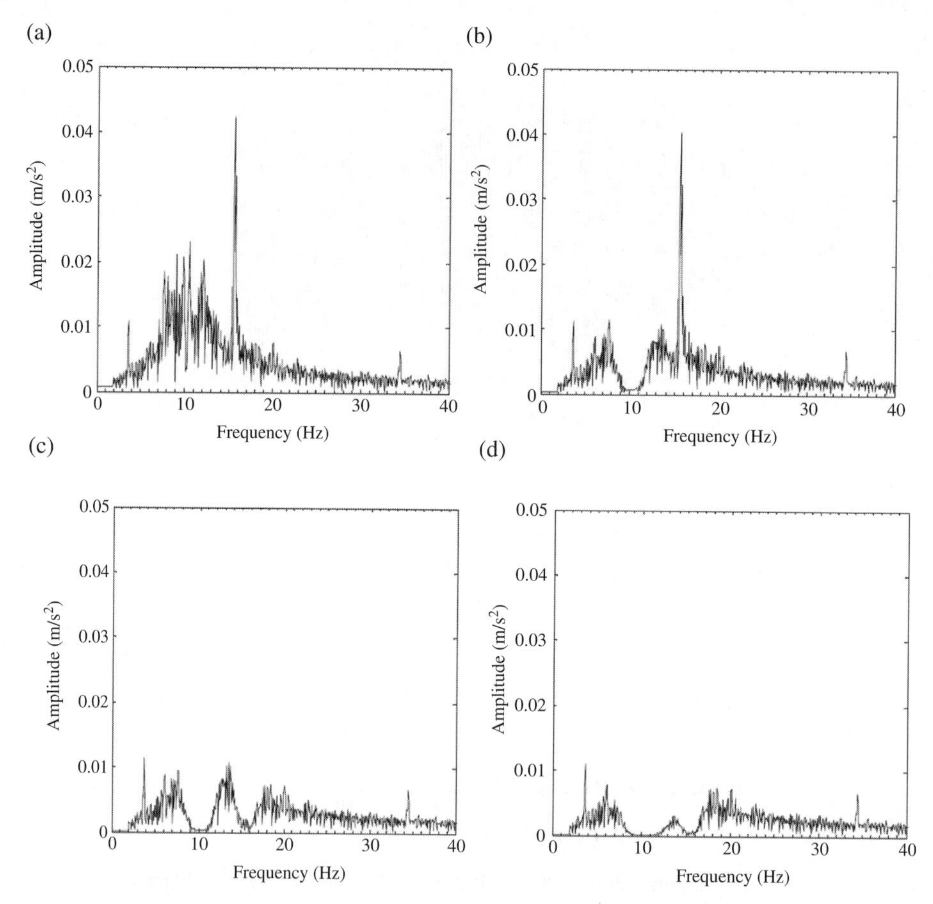


**Figure 7.11** Resultant spectra processed by singular spectrum analysis (SSA) with the numbers of singular values selected as (Case 2): (a) 3~200; (b) 5~200; (c) 7~200; and (d) 9~200.

group $G_n$, the vehicle frequency has been filtered out completely. As such, the second bridge frequency can be clearly identified and the first and third ones (i.e., 3.87 and 34.79 Hz), have also become more visible. In Figures 7.11c and d, which indicate the results obtained with more singular values being removed from the non-vehicle group $G_n$, the second bridge frequency was also filtered out, contrary to the current purpose.

The above results indicate that the known vehicle frequency can be successfully filtered out by selecting an appropriate number of singular values (or elementary matrices) for the nonvehicle group $G_n$. However, if too many singular values are included in $G_n$, the vehicle frequency may not be completely filtered out. On the other hand, if too few singular values are included in $G_n$, the bridge frequencies may be overfiltered. Clearly, personal judgment should be made in selecting the proper number of singular values (or elementary matrices) for the group $G_n$, which is a tough task similar to that for Case 1.

Third, the vehicle response is processed by the SSA-BPF. The same window length of $L = 600$ is selected, along with the same stop band of 5~11 Hz. The number $d$ of singular values to be included for the SVD step is selected as 100 or 600, for which the resultant

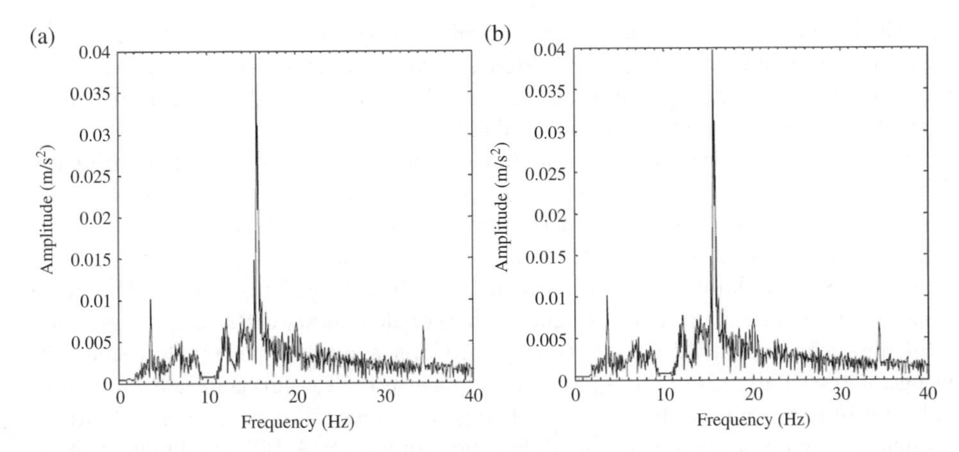


**Figure 7.12** Resultant spectra processed by singular spectrum analysis with band-pass filter (SSA-BPF) with (Case 2): (a) $d = 100$; and (b) $d = 600$.

Fourier spectra are given in Figures 7.12a and b, respectively. Compared with the previous case, one smaller $d$ is selected in order to save the computation time, whereas one larger $d$ is selected to offer a reference for the accuracy of the solution. As can be seen, the vehicle frequency has been filtered out completely in both spectra, with no unexpected peak induced by the filter. The bridge frequencies of the first three modes become eminently visible in comparison with the original spectrum. In addition, little difference can be observed between the two spectra. Therefore, the conclusion drawn from Case 1 for the proposed SSA-BPF still applies herein: no engineering judgment is required of the number of singular values except that it should be sufficiently large, which allows the grouping step to be conducted in a rather automated manner.

The previous two cases illustrate that the SSA-BPF performs much better than the SSA and BPF in filtering out the known vehicle-frequency components for identification of the bridge frequencies from the vehicle response. The reason can be given as follows. It is known that any set of data can be decomposed with SSA into a finite number of components in time domain, but the main issue is how to physically interpret those components and then to group them accordingly. As far as the physical meanings are concerned, the present band-pass filtering criterion included in the grouping step of the conventional SSA offers a more meaningful way. By designating a frequency band that covers the known (usually dominant) vehicle frequency, the decomposed components with frequencies falling within such a band are physically dominated by the vehicle frequency and should be filtered out accordingly. Consequently, the bridge frequencies are made more visible and can be identified from the remaining components (which are less dominant than the vehicle frequency), as was demonstrated previously.

## 7.5 Concluding Remarks

Three filters have been studied for their effectiveness in identifying the bridge frequencies from the test vehicle response: the BPF or BSF, SSA, and the SSA-BPF proposed by Yang et al. (2013a). Each filter was applied to the acceleration response history of the test vehicle generated by the finite element analysis program, considering the existence of pavement roughness. Through the numerical studies on the two cases with the vehicle frequency either larger or smaller than the first bridge frequency, the key features of the three filters were evaluated, as summarized below.

For the BPF, the known vehicle frequency can be completely filtered out with ease. However, an artificial, unexpected peak caused by the interface between the pass- and stop-band may appear in the filtered Fourier spectrum, which tends to confuse engineers and even mislead them into erroneous interpretations.

For the SSA, the known vehicle frequency can be filtered out successfully with a proper selection of the number of singular values or elementary matrices to be grouped in the non-vehicle-related frequency group. Too large a number may result in underfiltering and too small the number in overfiltering. Nevertheless, it is generally difficult to make a proper estimate of the number of singular values to be used, since the bridge frequencies are not known beforehand. For the proposed SSA-BPF, the known vehicle frequency can be filtered out completely, while basically no judgment needs to be made regarding the number of singular values to be included in the SVD step, thereby allowing the grouping task to be conducted in a rather automatic way.

For the reasons already stated, the proposed SSA-BPF is considered the most effective filter for use, along with the vehicle scanning method for measurement of bridge frequencies. The conclusions drawn herein have been based mainly on numerical simulations and the structural data adopted. Further work should be conducted to evaluate the effectiveness of the techniques developed via the field tests.

# 8

# Hand-Drawn Cart Used to Measure Bridge Frequencies

This chapter presents an experimental study of a hand-drawn cart used for measuring bridge frequencies. The idea of developing a hand-drawn test cart is that it allows us to measure the bridge frequencies in a handy and mobile way, while examining each of the parameters of the device in a close way. The dynamic characteristics of the test cart are crucial to the successful identification of the bridge frequencies of the bridge. This chapter primarily focuses on two issues: the elastic properties of the cart wheels/tires and the reliability of bridge frequencies extracted from the test cart. Tests are conducted under various operating conditions, including the ambient vibration test, free vibration test, ground dynamic test, and field test. It is demonstrated that the hand-drawn cart presented herein can be reliably used in the field for measuring bridge frequencies. Furthermore, some qualitative guidelines essential to the development of feasible test vehicles are drawn from the field test. The materials presented in this chapter are based primarily on Yang et al. (2013b).

## 8.1 Introduction

The frequencies of a bridge are among the most important information for health monitoring of the bridge, as they relate closely to its overall stiffness or loading capacity. A drop in any frequency of the bridge implies some deterioration in the stiffness of the structure in the global sense, which may be caused by damages in certain components, failures in support conditions, or other defects encountered in practice. The approaches for measuring the bridge frequencies can be classified into two major categories – the direct approach and indirect approach, or vehicle scanning method.

The direct approach, measuring vibrations of the bridge itself, is most common. As such, it requires installation of quite a number of detection sensors on various points of the bridge in order to grasp the modal characteristics of vibration of the bridge under the excitation of external sources, such as ambient vibrations, forced vibrations, traffic loads, and so on (see, e.g., Huang et al. 1999b). By analyzing the bridge responses recorded by the sensors, the dynamic and particularly the modal properties of the bridge, such as natural frequencies, modal shapes, and damping ratios, can be obtained. However, this approach suffers from the drawback that the acquisition of the vibration data usually requires a huge number of vibration sensors and equipment, for which the

*Vehicle Scanning Method for Bridges*, First Edition. Yeong-Bin Yang, Judy P. Yang, Bin Zhang and Yuntian Wu.

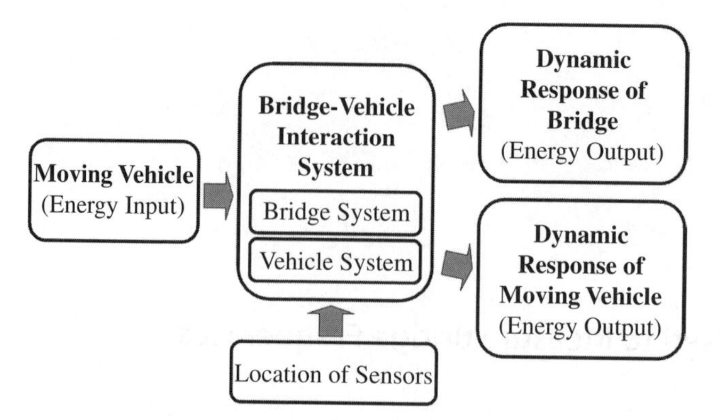


**Figure 8.1** Vehicle–bridge interaction system.

labor cost involved in the deployment stage is generally high, while to ensure the long-term functioning of the system, the maintenance and surveillance cost is also high. Since the monitoring system established for a particular bridge can hardly be transported to and work for another bridge, the direct approach of bridge measurement is limited in scope of application, for instance, it does not allow us to monitor all the bridges in an area after a major natural hazard, such as earthquake or flooding, a problem commonly encountered these days.

In comparison, the indirect approach or vehicle scanning method proposed by Yang et al. (2004a) is much more versatile. Such an idea was proposed after their decade-long research on vehicle–bridge interaction problems (Figure 8.1). By this approach, the dynamic response of a test vehicle moving over a bridge is recorded during its travel and processed to yield the bridge frequencies. The original idea of the indirect approach is as follows. Due to the interaction of the moving vehicle with the bridge, the moving vehicle plays the dual role of an exciter (by its horizontal movement) to the bridge and a receiver of the (vertical) response transmitted from the bridge. Only a small number of vibration sensors need to be installed on the moving vehicle to record the vehicle's vertical response. As far as bridge frequencies are concerned, the vehicle scanning method of measurement is superior to the conventional direct approaches in terms of portability, convenience, and economy, since the vibration sensors need only to be installed on the test vehicle, rather than on the bridge.

The feasibility of the concept of the vehicle scanning method for measuring the bridge frequencies was first verified theoretically by Yang et al. (2004a) and Yang and Lin (2005). By deriving the closed-form solution to the vehicle-bridge equations of motion, along with finite element simulations, it was confirmed that bridge frequencies dominate the dynamic response of a vehicle moving over the bridge, in addition to the self-frequency of the vehicle, and therefore can be extracted from the vertical response recorded of the moving vehicle during its travel. Such a concept was experimentally verified to be feasible by Lin and Yang (2005) by letting a test vehicle move over a highway bridge in northern Taiwan.

Various factors affecting the effectiveness of the vehicle scanning method for measuring bridge frequencies have been studied. Chen and Xia (2009) demonstrated that lower vehicle speeds and smaller mass ratios of the vehicle to the bridge result in better accuracy for identifying the bridge frequencies, while bridge damping and varying vehicle speeds are rather minor effects. McGetrick et al. (2009) indicated that better accuracy can be achieved for the bridge frequencies at lower speeds and for smooth road profiles. Yang and Chang (2009b) have shown in their parametric study that the initial vehicle/bridge acceleration amplitude ratio plays an important role in identifying the bridge frequency, namely, a smaller vehicle/bridge acceleration amplitude ratio is beneficial to the successful identification of bridge frequencies. In general, road surface roughness has an adverse effect on the extraction of bridge frequencies from the test vehicle. In this regard, the resolution of bridge frequencies can be enhanced by introducing multiple test vehicles or by raising the moving speed of accompanying vehicles (Chang et al. 2010; Yang et al. 2012b). As for processing the recorded data of the test vehicle, the resolution of bridge frequencies can be enhanced by pre-processing the recorded vehicle response with the empirical mode decomposition (EMD) and/or ensemble EMD (Yang and Chang 2009a).

Recently, the vehicle scanning method has been adopted not only in frequency detection but also in damage detection. For instance, Bu et al. (2006) used the measured response of a vehicle moving over a bridge to assess the damage of a bridge deck. Oshima and Yamamoto (2009) processed the dynamic response of the moving vehicle by independent component analysis and auto-regressive model to assess the bridge frequencies. Xiang et al. (2010) used the acceleration response of a passing vehicle, with the enhancement of a tapping device, to obtain bridge damage information. Using the on-vehicle vibration signals, a method based on the wavelet analysis was proposed by Nguyen and Tran (2010) for detecting the multiple cracks of a bridge structure.

Yamamoto et al. (2009) used a train-car-mounted accelerometer to identify the dynamic parameters of a viaduct over which trains run, with the results compared with those directly measured from the bridge. They also showed that the correlation between the signals obtained from the train and bridge were not as perfect as predicted, and little bridge information was gathered from the sensors mounted on the train. Gomez et al. (2011) used six accelerometers mounted on a tanker to extract the bridge frequency of a curved viaduct, and showed that the frequency peaks can be found indistinctly because of the complex dynamic behavior of the tanker. Miyamoto and Yabe (2011) demonstrated the feasibility of the vehicle scanning method using an accelerometer mounted on the chassis of a moving bus to extract the bridge information. Recently, vehicle-based measurement has been extended to construct the mode shapes of bridges (Yang et al. 2014).

Lin and Yang (2005) used a tractor-trailer system, with the tractor serving as the exciter, and the trailer-mounted accelerometer to record its dynamic response. Yang and Chang (2009b) demonstrated that the simpler the dynamic property of the test vehicle, the better the result can be achieved. In addition, Bu et al. (2006), in their numerical study of damage detection for bridges using passing vehicles, showed that different vehicle models lead to slightly different results, but a simple model is better than a complex one.

The idea of extracting bridge frequencies from a passing vehicle is simple in concept. But its successful implementation may not be as straightforward. The real problem is how to design a test cart with the desired dynamic properties such that it can be effectively used in practice. This chapter describes an experimental study of using a hand-drawn cart for measuring the bridge frequencies and its relative advantages of mobility, efficiency, and cost-effectiveness compared with the conventional direct approach. Focus of this chapter is placed on the improvement of the dynamic properties of the test cart using different types of wheels/tires. Tests include the ambient vibration test, free vibration test, and surface dynamic test. Finally, field tests are discussed with regard to the applicability of the self-designed test cart, with emphasis placed on the reliability of the associated devices and technique developed.

## 8.2 Dynamic Properties of the Hand-Drawn Test Cart

To verify the reliability of the self-designed hand-drawn test cart in field applications, two stages of experiments are undertaken:

1) Select the type of wheels/tires that can yield the most favorable dynamic properties for the cart.
2) Investigate the range of application for the cart in terms of self weight, traveling speed, and existing traffic flows.

The test cart studied herein can also be considered as a model of the real-size cart to be used in practice. In this section, the design concept of the test cart will be introduced, with the purpose of enhancing the accuracy and reliability of the test cart for use in the field measurement of bridge frequencies.

The structure of the hand-drawn cart used in this study is made by thin-walled stainless steel bars, and designed with replaceable wheels/tires to offer versatility to meet different needs. The key design concept for the cart is that it will behave like a single-degree-of-freedom (SDOF) system similar to the one theoretically studied in Yang et al. (2004a). As shown in Figure 8.2, the cart is composed of a steel frame with the following properties:

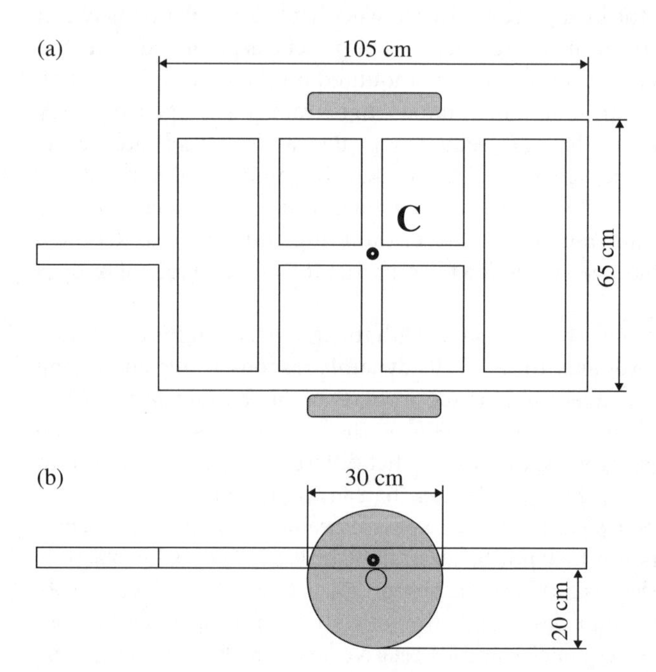


**Figure 8.2** Hand-drawn cart: (a) top view; (b) side view.

**Figure 8.3** Three types of wheels: (a) inflatable wheel; (b) rubber wheel; (c) PU wheel.

weight = 42 kg, width = 65 cm, height = 20 cm. An accelerometer is mounted at the center point C of the axle to record the vertical acceleration of the cart during its movement. Besides, steel plates may be used to add the cart weight, which can be placed in a basket under the frame of the cart below the center point C. The mechanical behavior of the cart (i.e., the dynamic properties and stability), during its movement is crucial to the successful application of the vehicle scanning method for extracting bridge frequencies.

To investigate the mechanism of transmission of energy from the bridge to the cart, three deferent types of wheels/tires that can be easily acquired are used in this study: inflatable wheels, rubber wheels, and polyurethane (PU) wheels, shown in Figure 8.3: The inflatable wheels shown are common commercial wheels with inflated tires, with a diameter of 300 mm a width of 85 mm, as shown in Figure 8.3a. The rubber wheels/tires are stronger due to the fact that the tires are composed of solid rubbers, which has the same diameter of 300 mm, but a width of 50 mm, as shown in Figure 8.3b. And the so-called PU wheels are solid metals wheels surrounded by a thin layer of about 10 mm polyurethane, as shown in Figure 8.3c. The diameter of the PU wheels is about 300 mm and the width is 70 mm.

## 8.3 Basic Dynamic Tests for the Test Cart

The dynamic properties of the test cart used are crucial to identification of the modal properties of a bridge, which depend strongly on the type of wheels/tires used. The natural frequency of the cart is the key parameter that affects the transmission of energy from the bridge to the cart. For better visibility of the bridge frequencies in the cart recorded response, it is preferable that the cart frequency be made larger than the first frequency of the bridge (Yang et al. 2013a). In this study, three kinds of dynamic tests – the ambient vibration test, free vibration test, and surface dynamic test – will be conducted using the three types of wheels/tires shown in Figure 8.3. Based on these results, the most suitable type of wheels/tires for practical applications will be selected.

*The ambient vibration test* has frequently been used to test the dynamic properties of a structure under environmental vibrations. In this test, the cart is allowed to stay on the ground free of external excitations. The results recorded for the accelerometer for three types of wheels/tires have been presented in time domain in Figure 8.4 and in frequency domain via the fast Fourier transform (FFT) in Figure 8.5.

As indicated by the time history responses in Figure 8.4, the acceleration responses for all the three types of wheels/tires under the ambient test are generally less than 0.3 gal. In the frequency domain, no particular peaks can be observed from the responses for the cart installed with rubber wheels and PU wheels. However, a clear resonant peak exists for the cart installed with inflatable wheels in the 16–17 Hz range, which

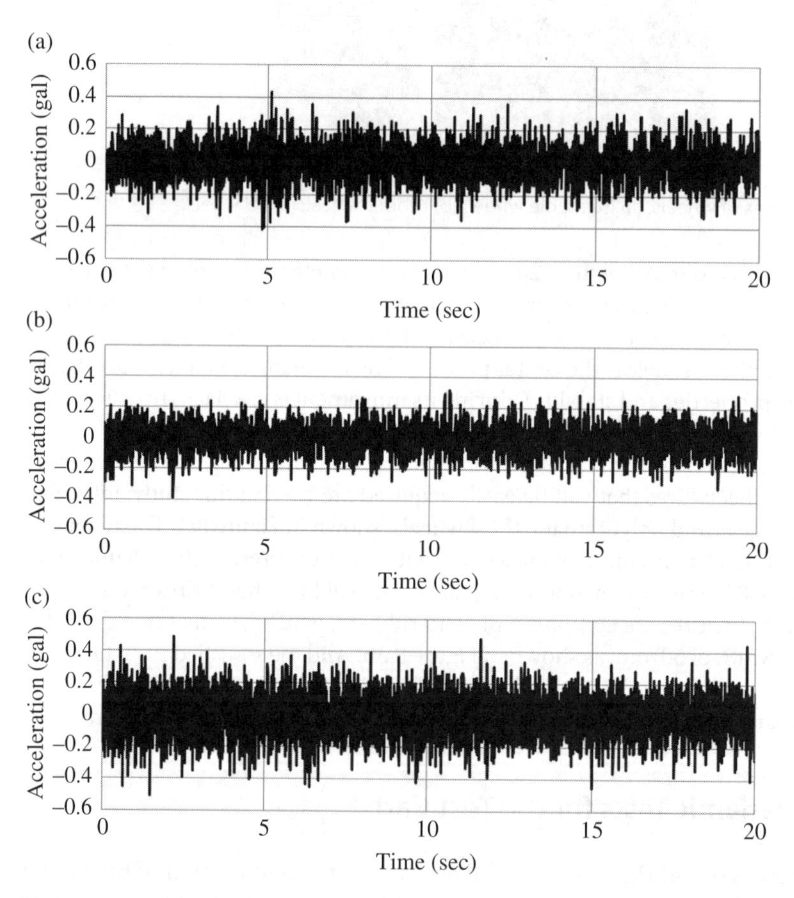


**Figure 8.4** Cart response in time domain under the ambient test: (a) inflatable wheels; (b) rubber wheels, (c) PU wheels.

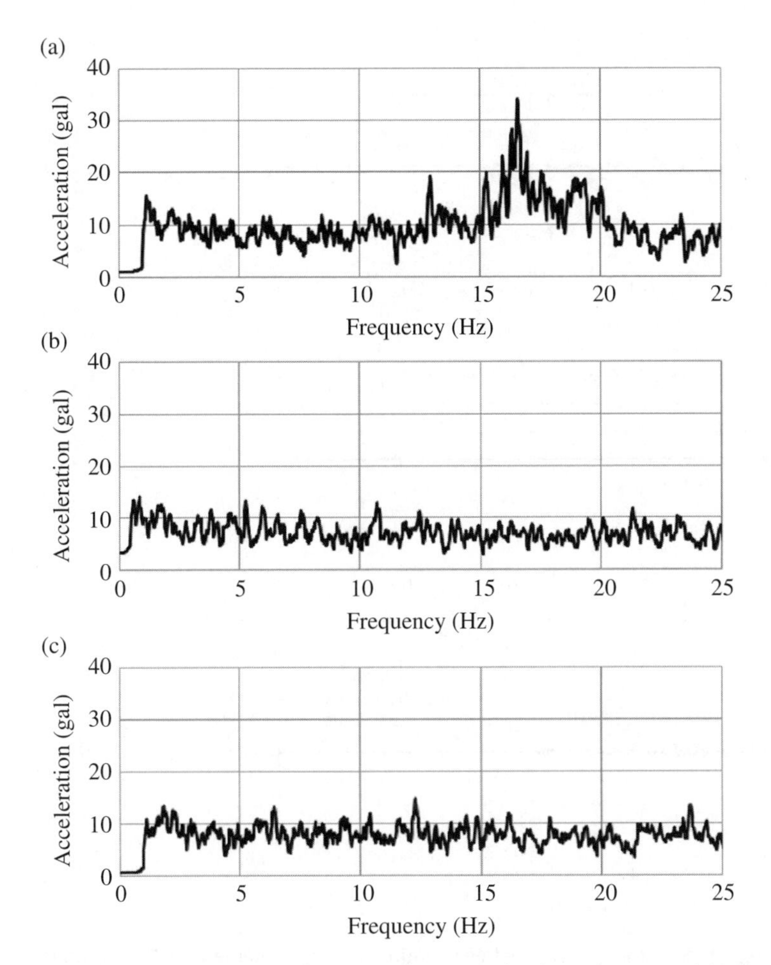


**Figure 8.5** Cart response in frequency domain under the ambient test: (a) inflatable wheels; (b) rubber wheels; (c) PU wheels.

is interpreted as the natural frequency of the test cart. This implies that the cart with inflatable wheels can be more easily excited than the other two types of wheels under the ambient vibration. Therefore, inflatable wheels are not considered a favorable choice compared with the other two types of wheels/tires, as the vibration transmitted from the bridge may be polluted by the self vibration of the test cart.

In the *free vibration test*, the vertical acceleration response of the test cart without motion is recorded when it is subjected to an impulse force, generated by the sudden jump of a person from the test cart. The time-history acceleration responses recorded for the test cart with three different wheels/tires in free vibration have been plotted in Figure 8.6, along with their FFT in Figure 8.7.

As far as the time-history response is concerned, one observes that the cart response in Figure 8.6a shows a complete waveform for the inflatable wheels. However, the same is not equally true for the rubber wheels and PU wheels in Figures 8.6b and c. This observation is generally consistent with the one for the FFT response. For instance,

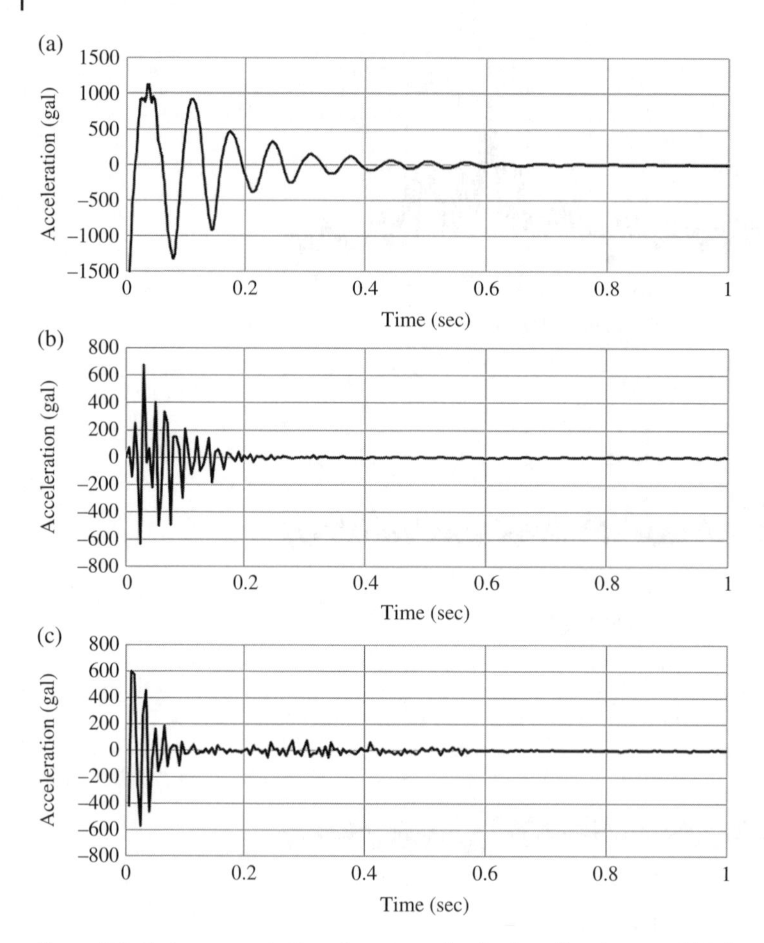


**Figure 8.6** Cart response in time domain under the free vibration test: (a) inflatable wheels; (b) rubber wheels; (c) PU wheels.

there exists a single obvious peak at about 15 Hz for the test cart with inflatable wheels in Figure 8.7a. However, no such concentrated peaks exist for the other two types of wheels/tires in Figures 8.7b and c. In particular, the cart with PU wheels shows a most favorable dynamic property in that it has no natural frequencies in the range between 0 and 50 Hz, which should prove it more suitable for use in the bridge frequency measurement.

To identify further dynamic properties of the test cart under the moving conditions, the *ground dynamic tests* are conducted by pulling the cart slowly over a smooth concrete pavement at a speed of about 2 m/s. The time-history acceleration responses recorded for the three types of wheels/tires and the spectra obtained by FFT have been plotted in Figures 8.8 and 8.9, respectively.

As indicated by the time-history responses in Figures 8.8a and b, both the inflatable wheels and rubber wheels show similar acceleration amplitudes at about

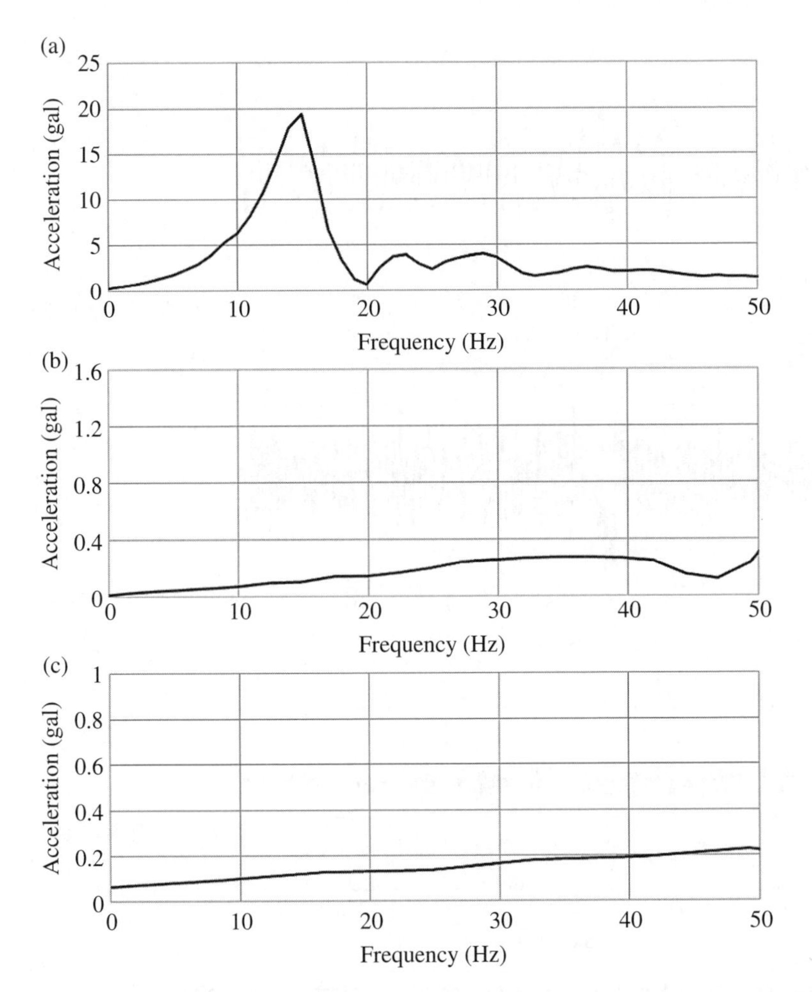


**Figure 8.7** Cart response in frequency domain under the free vibration test: (a) inflatable wheels; (b) rubber wheels; (c) PU wheels.

10 gal and larger, but the PU wheels shows a much smaller and even amplitude. Further, the frequency response in Figure 8.9 reveals that the test cart with inflatable wheels has a natural frequency peak at about 13 Hz, while the test cart with rubber wheels shows concentrated frequency peaks in the 30 to 35 Hz range. In contrast, the spectral content of the cart with PU wheels, as shown in Figure 8.9c, shows only a random, small, and even distribution of amplitudes throughout the range of frequencies of concern (i.e., up to 50 Hz), for which no natural frequencies are observed for the cart. The implication here is that when using the PU wheels, the cart response can *faithfully* reflect the smooth concrete pavement, of which the roughness profile is considered to be similar to the *white-noise random distribution*. For this reason, PU wheels are selected for further field tests for their relatively good performance in a later part of the study.

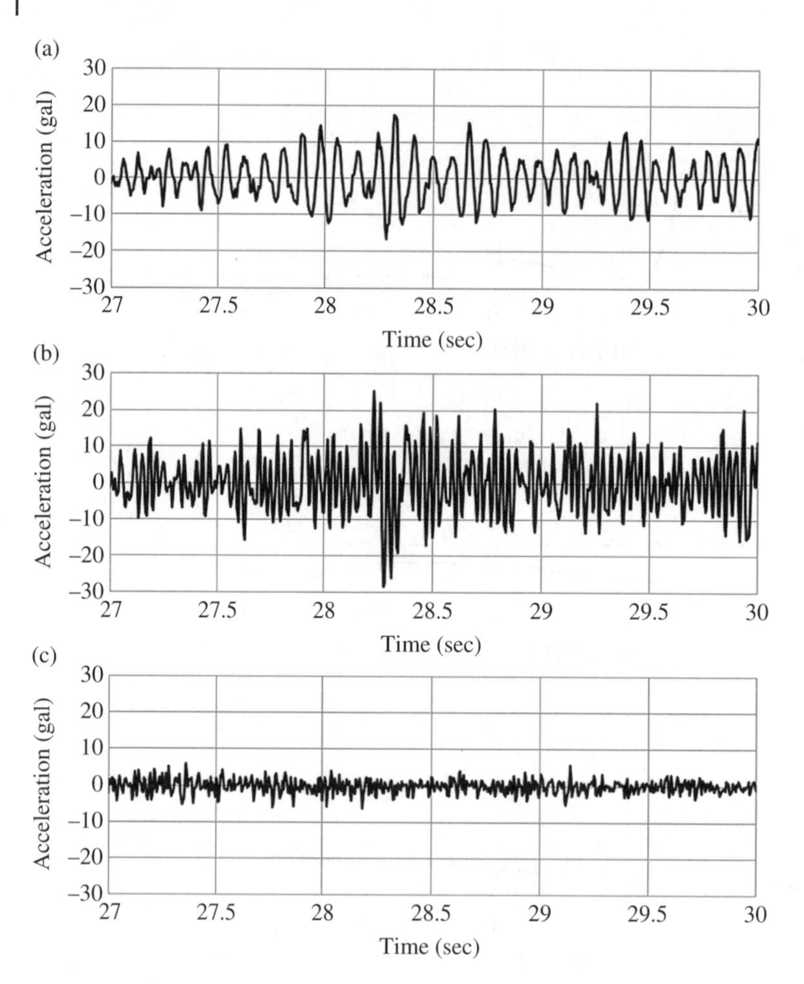


**Figure 8.8** Cart respond in time domain under the ground dynamic test: (a) inflatable wheels; (b) rubber wheels; (c) PU wheels.

## 8.4 Field Tests

Through the dynamic tests presented in the previous sections, PU wheels are selected for the field tests for its relatively good performance in reflecting the bridge frequencies. In this section, a series of field tests will be conducted for different bridges merely to avoid the influence of individual bridge characteristics. Most of the tests are done at midnight under virtually no traffic interference, unless otherwise noted. The test cart is pulled by a person at constant speed over (or near) the centerline of the bridge, with its dynamic response recorded during the passage. The advantage of using the human force is to reduce the effect of extra vibrations induced by the towing vehicle. The average speed of the test cart is calculated by dividing the distance of the travel on the bridge by the time used. The objective is to search for a better configuration of the test cart, such that it can be successfully applied to extracting the bridge frequencies from its dynamic response in field applications. The following are the key parameters to be

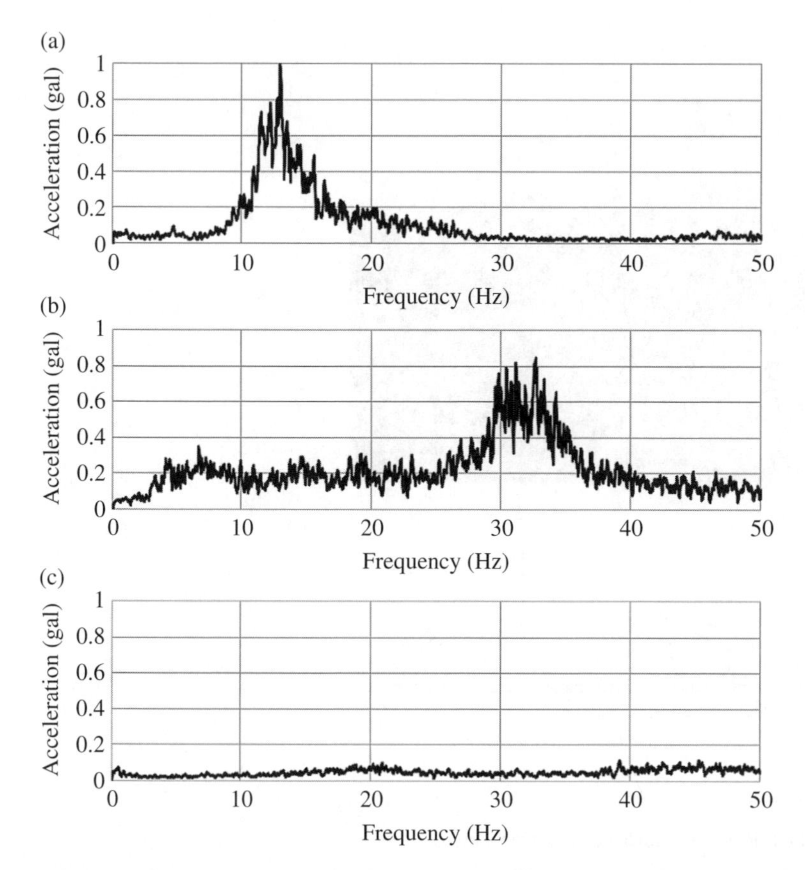


**Figure 8.9** Cart response in frequency domain under the ground dynamic test: (a) inflatable wheels; (b) rubber wheels; (c) PU wheels.

tested: (i) weight of the test cart; (ii) traveling speed of the test cart; and (iii) volume of existing traffic flows.

To this end, the stochastic subspace identification (SSI) method will be adopted to identify the natural frequency peaks of the bridge of concern. The SSI method is an output only signal identification technique, collocated with the household of FFT (Peeters and de Roeck 1999). By the SSI method, the original signal is divided into several parts to calculate the system poles. Then, based on the trend of increasing of the number of rows for the selected system poles, frequency peaks that are more regular can be identified. In other words, if a frequency peak is pointed out many more times by the number of rows, it is considered a reliable frequency of the system.

### 8.4.1 Effect of Cart Weight

In order to study the effect of cart weight, the cart has been designed such that extra steel plates can be inserted at a point near the center of gravity (i.e., right above the axle) of the test cart. The first bridge to be measured is the Ping-Pu Bridge located in the suburb of the Taipei City. This is a cable stayed bridge with a 95 m span completed in

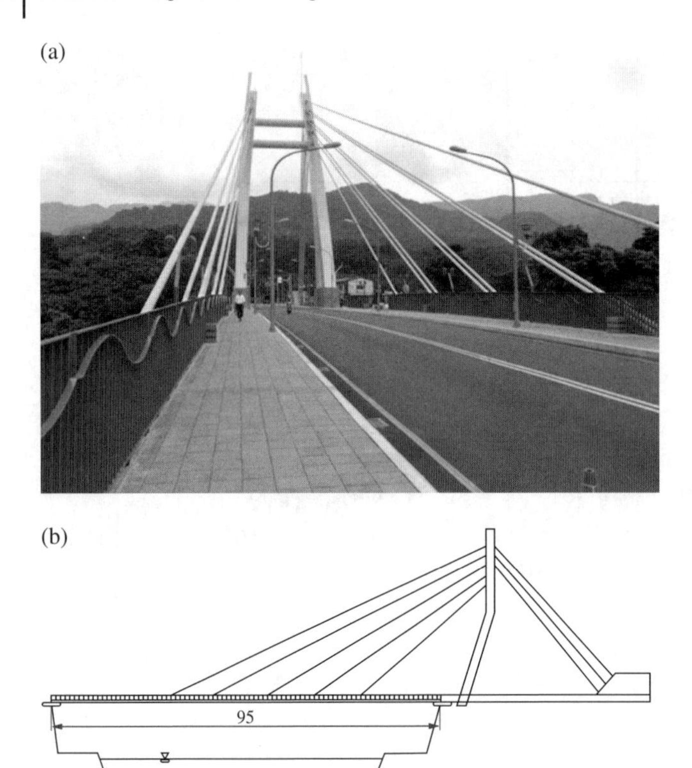


**Figure 8.10** Ping-Pu Bridge: (a) photo view; (b) side view.

February 2012, as shown in Figure 8.10. The test cart is pulled over the bridge with two different weights, 94 and 146 kg, at the speed of 2 km/h. The sampling rate for recording the data of vertical acceleration is 200 Hz.

The time-history acceleration responses recorded for the test cart during its passage over the bridge at the two speeds were plotted in Figure 8.11. Evidently, *the higher the cart weight, the smaller the cart response*. This means that a heavier cart is relatively insensitive to the road surface roughness and should be adopted in field applications. It is known that the road surface roughness may create a pollution or blurring effect on the test cart response, making it difficult to identify the bridge frequencies (Chang et al. 2010; Yang et al. 2012b).

The FFT/SSI results obtained for the measured response of the test cart with weights 94 and 146 kg have been presented in Figures 8.12a and b, respectively. Clearly, *a heavier cart can extract many more bridge frequencies than a lighter cart*, due to the fact that the pollution effect of road surface roughness has been largely suppressed by the cart weight. For instance, for the cart with 94 kg, one observes from Figure 8.12a that only the second and fifth frequencies of the bridge are identified. However, for the cart with 146 kg, one observes from Figure 8.12b that all the frequencies of the bridge from the second to the eighth can be identified. Specifically, the FFT/SSI result obtained for the

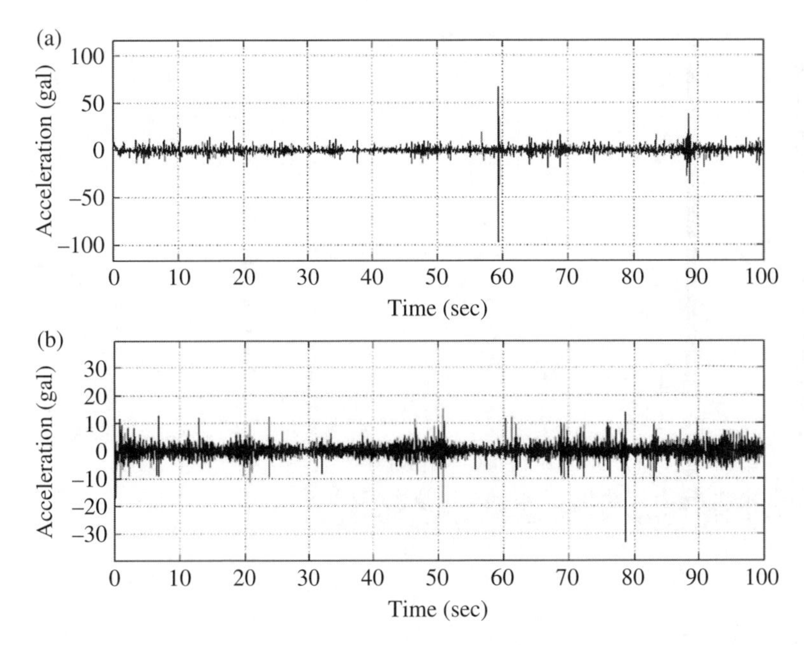


**Figure 8.11** Time-history response of the cart moving over the Ping-Pu Bridge at 2 km/h speed with weight of: (a) 94 kg; (b) 146 kg.

146 kg cart in Figure 8.12b was shown to be consistent with the one obtained directly from the ambient vibration test of the Ping-Pu Bridge under random traffic in Figure 8.13. It should be admitted, of course, that the first frequency of the bridge was not duly captured by the test cart with or without added weight, for which the reason remains to be investigated.

### 8.4.2 Effect of Various Traveling Speeds

The speed of the test cart is a crucial parameter to extracting the bridge frequencies. For the sake of efficiency, one would prefer a test cart that can be pulled at a speed as high as possible. There are two concerns herein, however. First, a higher moving speed for the test cart means that the *contact time* of the cart wheels/tires with the bridge surface is shortened, which may reduce the effectiveness of energy transmission from the bridge to the test cart. Second, a test cart moving at higher speeds tends to become less stable, in that it can be more easily excited by road surface roughness. To evaluate how the moving speed of the test cart affects the measured response, three testing speeds are considered: 2, 4, and 8 km/h. The bridge to be tested in this regard is the Jie-Shou Bridge, also located in the suburb of the Taipei City, which is a steel arch bridge of 84 m span, completed in July 2010, as shown in Figure 8.14.

The time–history acceleration responses recorded of the test cart moving over the Jie-Shou Bridge at the speeds of 2, 4, and 8 km/h have been plotted in Figures 8.15a–c, respectively. Figures 8.15a–c indicates that higher traveling speed will induce larger vibrations, which however is not good for bridge frequency identification, since the adverse effect of road surface roughness is going to play a larger role.

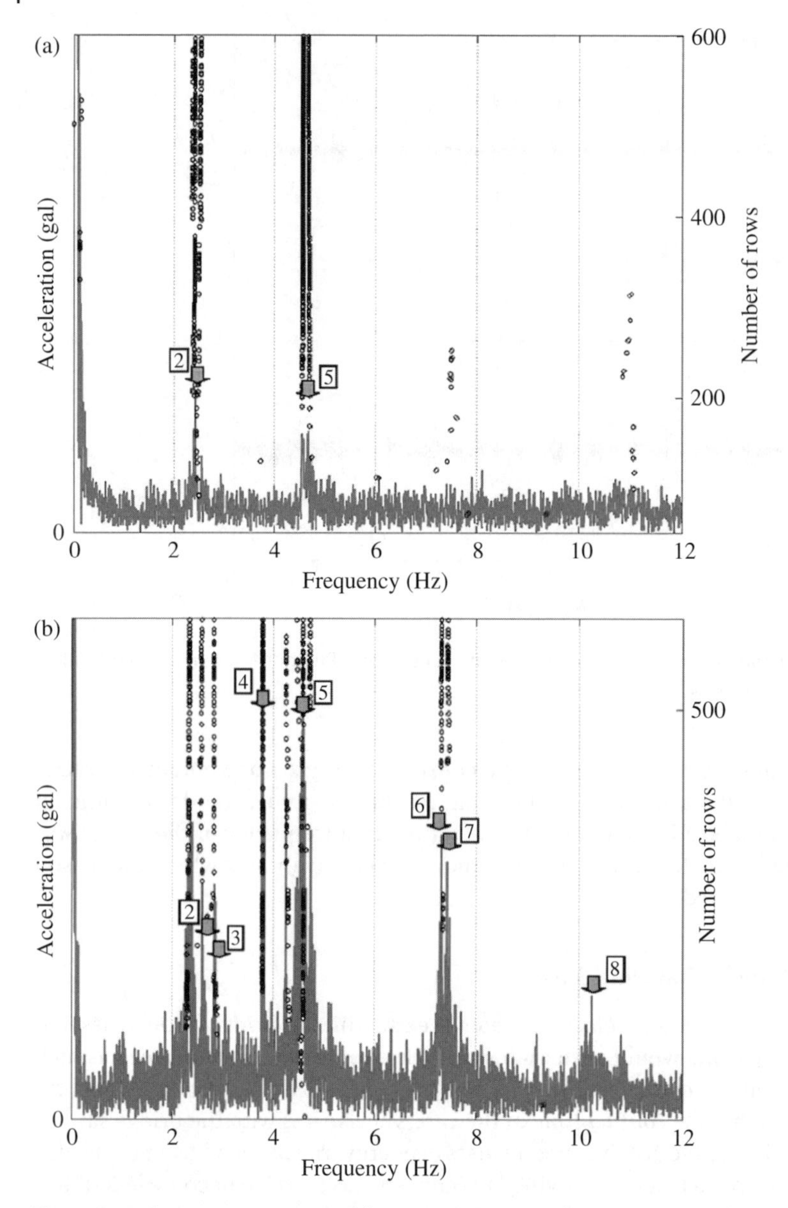


**Figure 8.12** Frequency response of the cart moving over the Ping-Pu Bridge at 2 km/h speed with weight of: (a) 94 kg; (b) 146 kg.

Correspondingly, the frequency responses obtained via the FFT/SSI processing for the three cart speeds have been plotted in Figures 8.16a–c. For comparison, the frequency response obtained from the ambient vibration test of the bridge was also plotted in Figure 8.17. From Figure 8.16a for the test cart moving at the lowest speed of 2 km/h, one observes that six frequency peaks of the bridge can be identified: 1.5, 2.2, 3.8, 4.7, 6.8, and 11 Hz, which coincide excellently with the ambient test result shown in Figure 8.17.

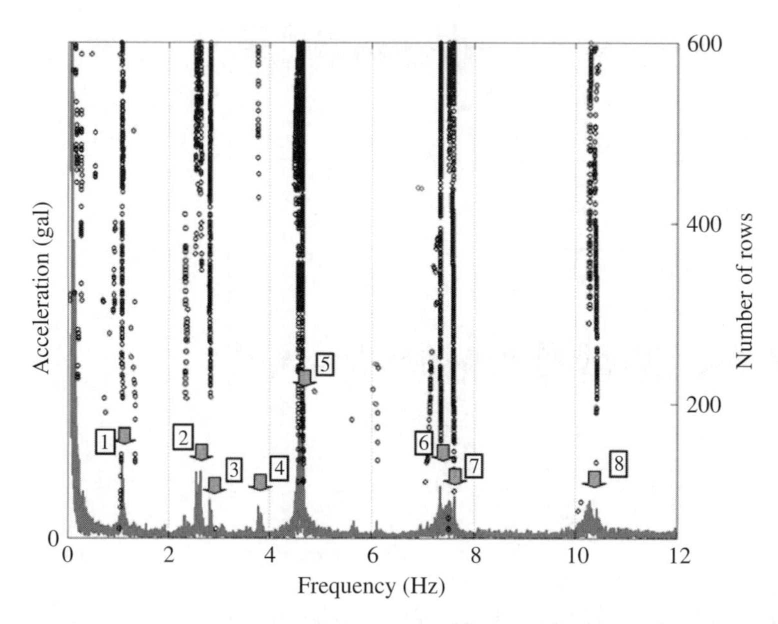


**Figure 8.13** Frequency response obtained from the ambient vibration test of the Ping-Pu Bridge.

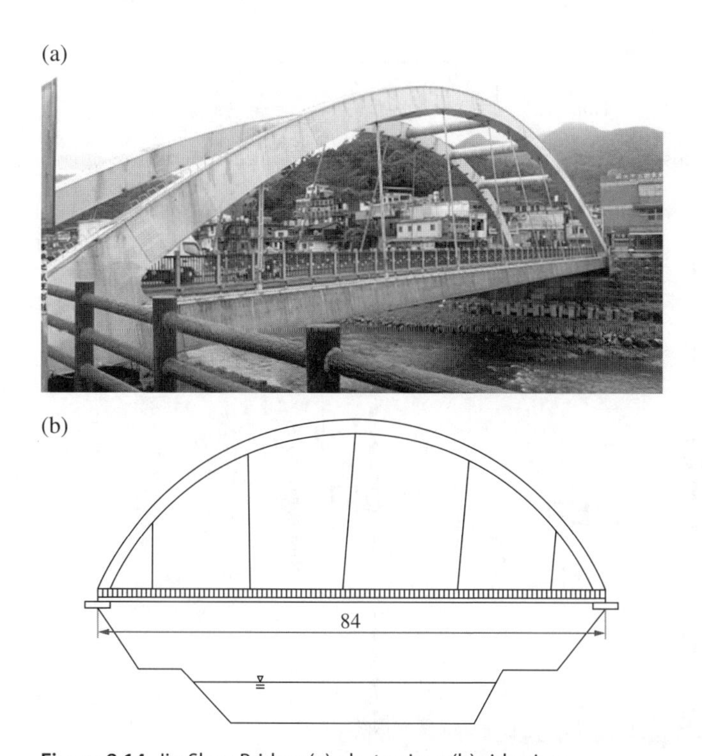


**Figure 8.14** Jie-Shou Bridge: (a) photo view; (b) side view.

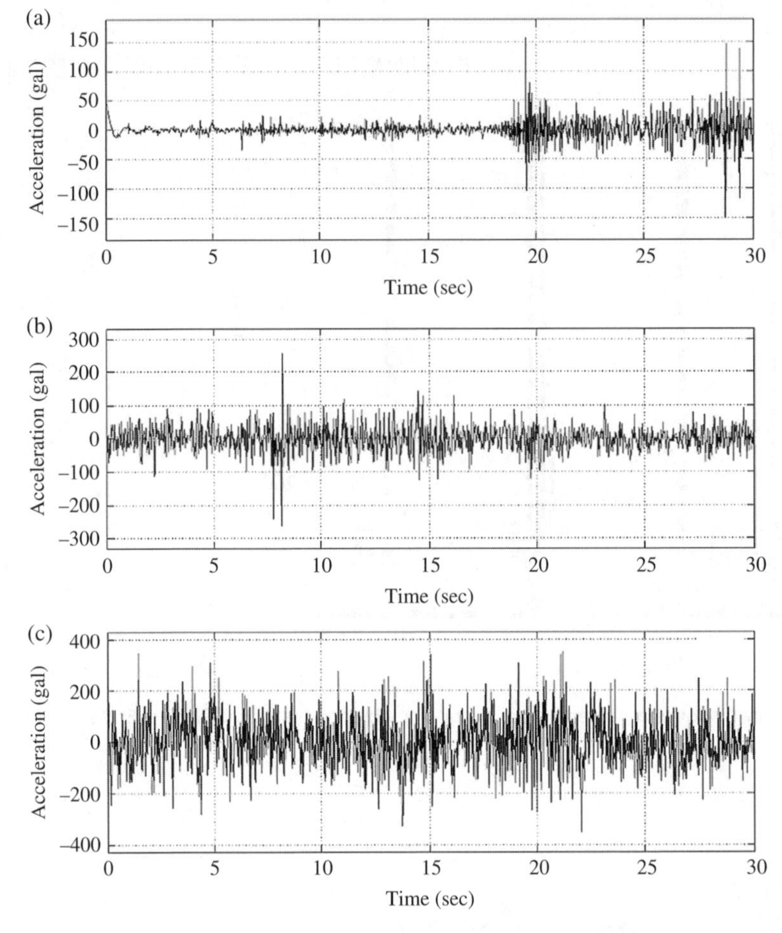


**Figure 8.15** Time-history response of the cart moving over the Jie-Shou Bridge at speed of: (a) 2 km/h; (b) 4 km/h; (c) 8 km/h.

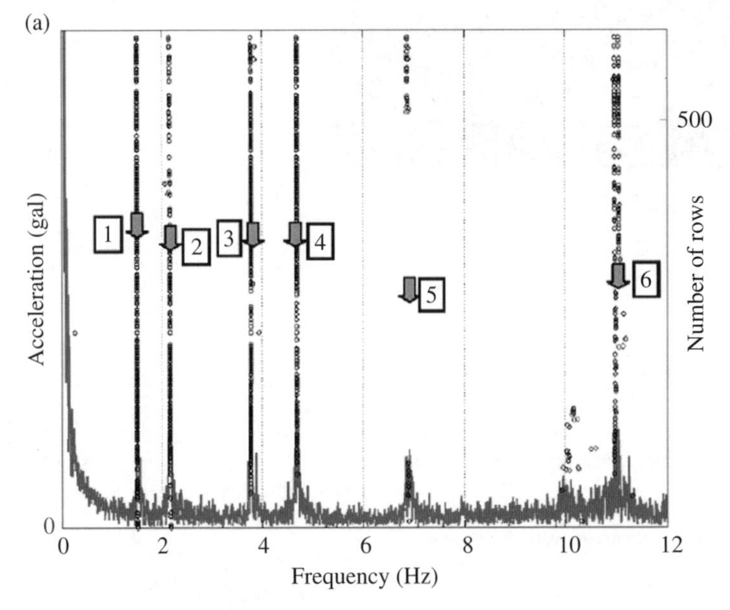


**Figure 8.16** FFT/SSI frequency response of the cart moving over the Jie-Shou Bridge at speed of: (a) 2 km/h; (b) 4 km/h; (c) 8 km/h.

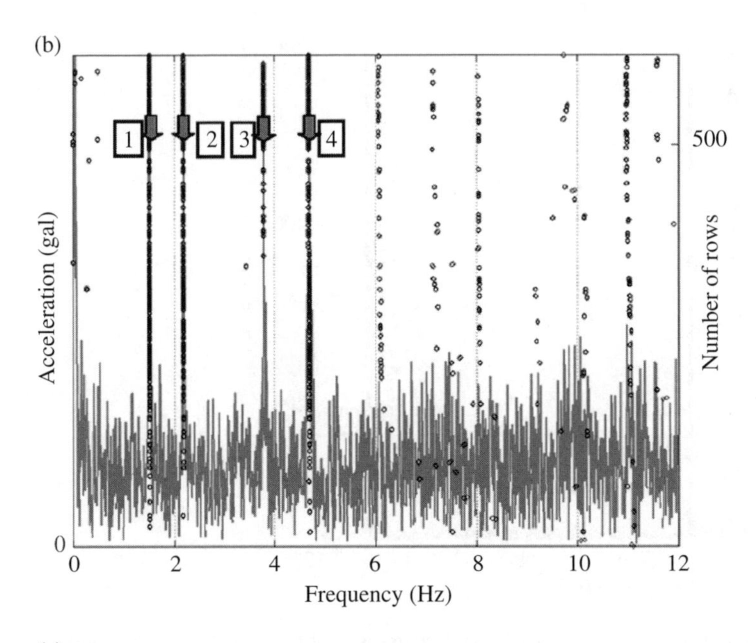


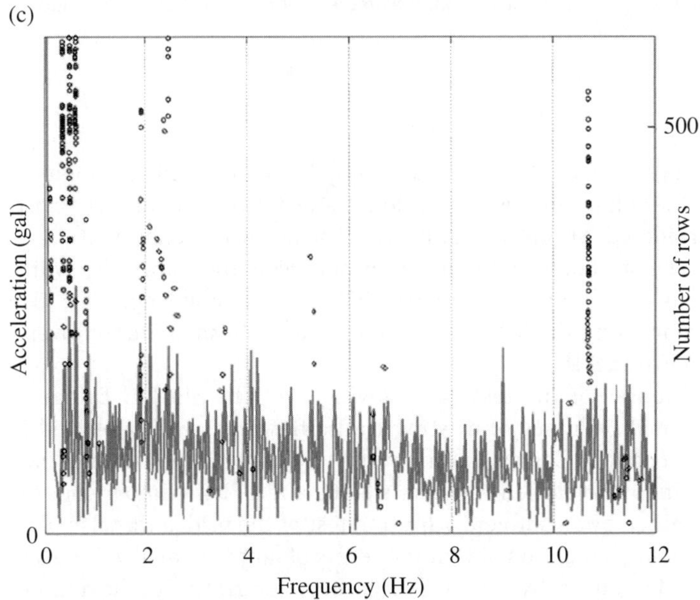


**Figure 8.16** (Continued)

However, the number of bridge frequencies that can be identified decreases as the cart speed increases to 4 km/h. For this case, only four bridge frequencies have been identified. Even worse, when the cart speed increases to 8 km/h, basically no bridge frequencies can be identified. As was mentioned previously, for a test cart moving at higher speeds, the contact time between the test cart and bridge is reduced, while the pollution effect of road surface roughness will be largely induced. Both factors are not good for the indirect extraction of bridge frequencies from the moving cart response.

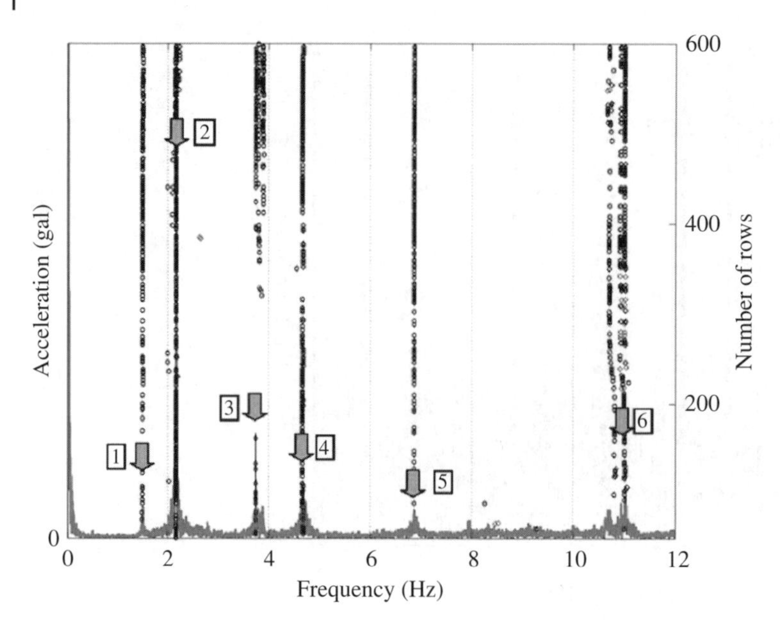


**Figure 8.17** FFT/SSI frequency response obtained from the ambient vibration test of the Jie-Shou Bridge.

### 8.4.3 Various Volumes of Existing Traffic Flows

An experiment has been designed to test the relationship between the identifiability of bridge frequencies and the volume of existing random traffic flows. Two different existing traffic flows are considered for the Shi-Lin Bridge. The test cart is allowed to move at the speed of 4 km/h. The Shi-Lin Bridge is an important link in the Taipei City, which is a four-span continuous steel bridge completed in 2005, with a total length of 113 m and a width of 17 m, connecting the MRT Shilin Station and Zhishan Station over a local creek, as shown in Figure 8.18.

The time-history responses of the test cart moving over the Shi-Lin Bridge at 4 km/h with large and small volumes of existing traffic flows have been plotted in Figures 8.19a and b, respectively. Correspondingly, the frequency responses obtained by the FFT/SSI processing were plotted in Figures 8.20a and b. For comparison, the frequency response obtained by an ambient vibration test of the bridge was also plotted in Figure 8.21. It should be mentioned that the terms of large or small volumes of traffic flows represent only a qualitative, not very rigorous, description of the number of vehicles moving over the bridge at the time of testing. To give some feeling, a large volume of traffic flow refers to the case where there is a continuous flow of vehicles moving at rather high speeds, say, over 50 km/h. The reverse is true for the small volume of traffic.

From the frequency response for the case of a *large volume* of existing traffic flow plotted in Figure 8.20a, one observes that virtually *all the first five frequencies* of the bridge can be identified. However, the case of a small volume of existing traffic flow, only the first, fourth, and fifth frequencies of the bridge can be roughly identified, i.e., at a

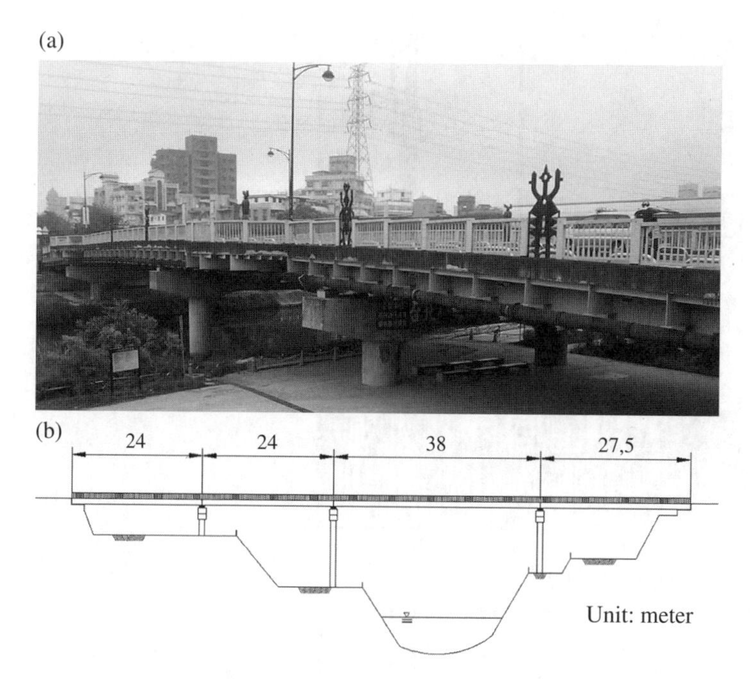


**Figure 8.18** The Shi-Lin Bridge: (a) photo view; (b) schematic diagram.

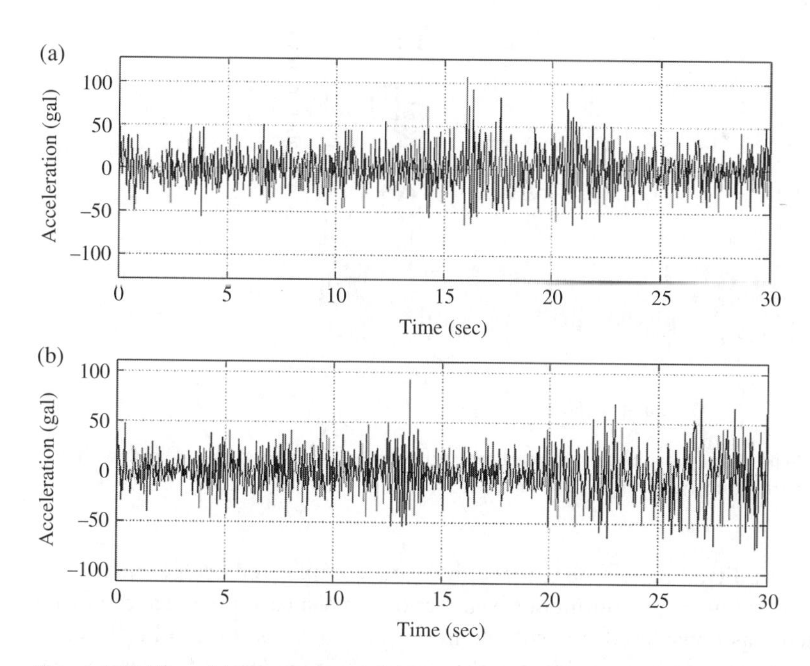


**Figure 8.19** Time–history response of the cart moving over the Shi-Lin Bridge at 4 km/h with volumes of existing traffic flows: (a) large; (b) small.

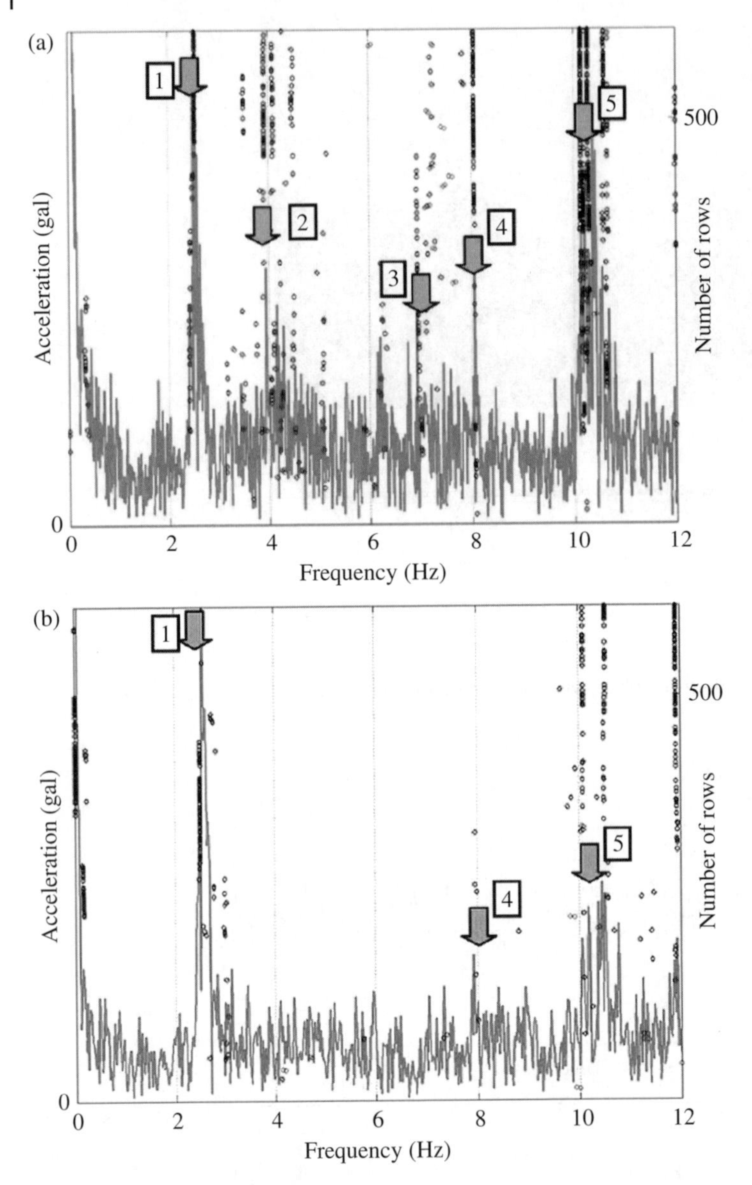


**Figure 8.20** FFT/SSI frequency response of the cart moving over the Shi-Lin Bridge at 4 km/h with volumes of existing traffic flows: (a) large; (b) small.

lesser degree of visibility. It should be realized that for a complicated bridge such as the Shi-Lin Bridge (i.e., a four-span continuous beam) and for a test cart moving at the speed of 4 km/h (which was considered unfavorable in the previous test of another bridge), such a result is very encouraging. Evidently, *the identifiability of bridge frequencies can be greatly enhanced by the existing traffic*, which is available almost in all bridges. Further, *the higher the volume of the existing traffic is, the better the resolution can be achieved.*

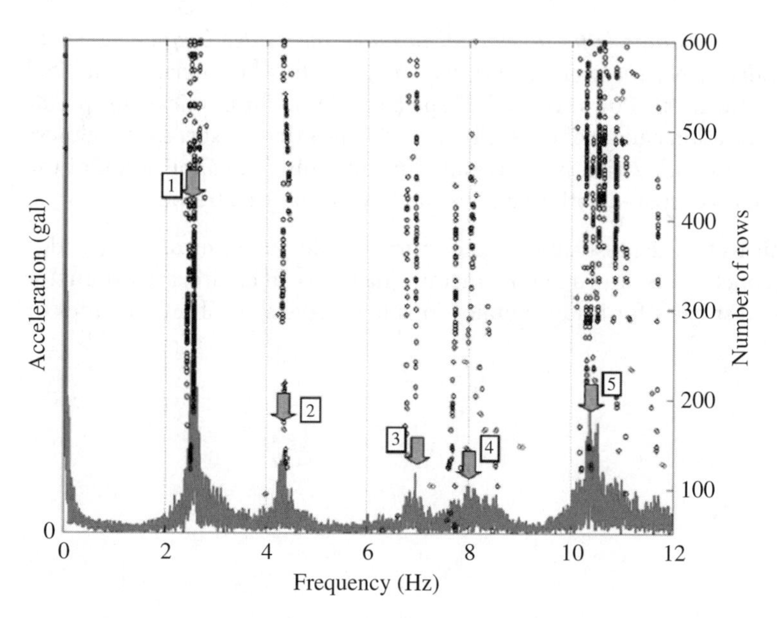


**Figure 8.21** FFT/SSI frequency response obtained from the ambient vibration test of the Shi-Lin Bridge.

This test has confirmed that the vibration energy transmitted by the existing traffic or accompanying vehicles onto the bridge is helpful for identifying the bridge frequencies from the dynamic response of the moving cart, and is consistent with previous findings (Lin and Yang 2005). One reason for this is that the vibration energy of the bridge associated with the road surface roughness becomes relatively small compared with that generated by the existing traffic (Yang et al. 2012a).

## 8.5 Concluding Remarks

Based on the test cart designed and the test program conducted for three bridges near Taipei City, the following conclusions can be drawn, concerning the applicability of the test cart in extracting bridge frequencies:

1) Of the three types of wheels/tires tested, the PU wheels are most suitable for use in the field test for extracting the bridge frequencies, since they show no particular self-frequencies in the range of frequencies of interest, mainly due to the better contact of the PU wheels with the road surface.
2) Higher moving speeds for the test cart will bring in extra interference from the road surface roughness, thereby making it difficult to identify the bridge frequencies.
3) *The heavier the test cart is, the smaller the cart response, and the higher the visibility can be achieved for the bridge frequencies.* A physical interpretation for this is that a heavier cart is relatively insensitive to road surface roughness, which therefore should be adopted in the field measurement.

4) *Larger volumes of existing traffic flows tend to make the bridge frequencies more visible as measured from the moving cart response,* which is beneficial to the field measurement as the traffic flows are available on all bridges. The physical interpretation is that the vibration energy of the bridge associated with road surface roughness becomes relatively *small,* which is always an annoying source to application of the indirect technique, compared with that generated by the existing traffic.

This study confirms that the test cart is a handy, feasible device for measuring bridge frequencies. The conclusions drawn herein, mainly qualitative in nature, are helpful for designing full-size test carts for bridge applications that serve as vital link in cities or highway systems.

# 9

# Theory for Retrieving Bridge Mode Shapes

This chapter presents a theoretical framework for constructing the bridge mode shapes from the dynamic response of a passing test vehicle. Compared with the conventional approaches that only measure a certain number of points (with sensors) on the bridge, the current approach offers much more spatial information, and therefore higher resolution, in retrieving the mode shapes, as the test vehicle can virtually touch each point along its path on the bridge. Basically, only one or few sensors need to be installed on the test vehicle. Factors that affect the accuracy of this approach are studied, including the vehicle speed, random traffic, and road surface roughness. Through numerical simulations, the approach is verified to be feasible under the condition of constant and low vehicle speeds. The materials presented in this chapter are based primarily on Yang et al. (2014) with some modifications according to Chapter 3.

## 9.1 Introduction

Measuring the mode shapes of a bridge is a crucial task in bridge engineering, since the measured mode shapes serve as a useful index for many applications related to bridges, such as numerical model calibration and updating (Brownjohn et al. 2001; Jaishi and Ren 2005; Altunisik et al. 2012), structural health monitoring and damage detection (Doebling et al. 1998; Farrar et al. 2001; Chang et al. 2003; Carden and Fanning 2004; Chrysostomou et al. 2008), and so on. Conventionally, to measure the mode shapes of a bridge, a certain number of sensors should be mounted on the bridge deck to record the dynamic responses of the bridge under external sources of vibration. Depending on the type and length of the bridge, the number of sensors required can be quite large. From the data recorded by the sensors, the mode shapes of the bridge can be identified by techniques such as system identification and/or other data processing techniques. Details for such modal identification approaches are available in many references, including the books by Ewins (2000) and Wenzel and Pichler (2005) and the papers by Farrar and James (1997) and Huang et al. (1999a). The number and locations of the sensors may vary from case to case, depending on the trade-off between the spatial resolution and experiment cost according to engineers' judgment. In general, a higher spatial resolution for mode shapes requires a larger number of sensors to be mounted, which is accompanied by higher consumption of cost, time, and labor.

*Vehicle Scanning Method for Bridges*, First Edition. Yeong-Bin Yang, Judy P. Yang, Bin Zhang and Yuntian Wu.

The vehicle scanning method proposed by Yang et al. (2004a) in the early stage works mainly for extraction of the bridge frequencies from the moving test vehicle's response. This approach requires only few sensors to be installed on the test vehicle, but no sensors on the bridge to be measured, which has the advantage for fast scanning the bridge frequencies. Since no deployment works need to be done on the bridge, such an approach was referred to as the indirect approach, in comparison with the conventional approach that works directly on the bridge, and therefore was referred to as direct approach. Recently, the vehicle scanning method have been extended to various applications on bridges, such as frequency extraction (Yang and Lin 2005; Yang and Chang 2009a), damage detection (Bu et al. 2006; McGetrick et al. 2009), cable tension loss detection (Yin and Tang 2011), and moving vehicle enhanced by a tapping device (Zhang et al. 2012).

Theoretically, the moving test vehicle can receive the vibration characteristics of virtually all the points along its path on a bridge, although the data for different points on the path are not synchronously taken. In contrast, using the conventional direct approaches, the vibration data are limited by the number of points installed with vibration sensors on the bridge, but they are synchronously recorded. Because of this, the vehicle scanning method is said to be richer in spatial information than the conventional approach, but it suffers from the drawback that the data are not taken synchronously; the latter is an issue to dealt with in this chapter.

The objective of this chapter is to present a solid, theoretical framework for constructing the mode shapes of a bridge with high spatial resolutions from the dynamic responses of the passing test vehicle. Central to the present study is the concept of instantaneous amplitudes obtained as a result of the Hilbert transform, which will be summarized in the beginning of the following section. Then, the theoretical formulation for the present approach will be given in detail, followed by the operating algorithms and constraints. Through four numerical cases, the feasibility of the present approach is illustrated and the factors that affect the accuracy of the retrieved mode shapes are evaluated.

## 9.2 Hilbert Transformation

Given a time series $s(t)$, the Hilbert transform (HT) of $s(t)$ is defined as (Bandat and Piersol 1986):

$$\hat{s}(t) = \mathbf{H}(s(t)) = \frac{1}{\pi} \boldsymbol{PV} \int_{-\infty}^{\infty} \frac{s(\tau)}{t-\tau} d\tau \tag{9.1}$$

where $\boldsymbol{PV}$ denotes the Cauchy principal value. The HT of $s(t)$ can be interpreted as the convolution of $s(t)$ with a unit impulse function of $1/\pi t$, thus it preserves most local information of $s(t)$. Then, $s(t)$ and $\hat{s}(t)$ can form an analytical function $z(t)$ as

$$z(t) = s(t) + i\hat{s}(t) \tag{9.2}$$

Mapping $z(t)$ from the complex Cartesian coordinates to the polar coordinates yields

$$z(t) = A(t)e^{i\theta(t)} \tag{9.3}$$

where

$$A(t) = \sqrt{s^2(t) + \hat{s}^2(t)} \tag{9.4}$$

and

$$\theta(t) = \arctan\left(\frac{\hat{s}(t)}{s(t)}\right) \tag{9.5}$$

The time-dependent function $A(t)$ is the *instantaneous amplitude function* of the original function $s(t)$, and $\theta(t)$ as the *instantaneous phase function*. In addition, the instantaneous amplitude function $A(t)$ can be regarded as the envelope function of $s(t)$. These definitions for the instantaneous amplitude and phase functions are physically meaningful only when the time series is *mono-component* or *narrow-band* (Huang et al. 1998, 1999b).

## 9.3 Theoretical Formulation

Figure 9.1 shows the mathematical model of a test vehicle moving on a simple bridge. In this model, the vehicle is simplified as a moving sprung mass $m_v$, supported by a spring of stiffness $k_v$; the bridge is a simply supported beam of length $L$, mass density $m^*$ per unit length, and bending rigidity $EI$. To focus on the physical meanings of the vehicle responses, five assumptions are adopted without losing the generality of the problem:

1) Road surface roughness is ignored in the derivation, but will be included in one of the numerical cases studied later to evaluate its influence.
2) The vehicle mass is negligibly small in comparison with the bridge mass.
3) Zero initial conditions are assumed for the bridge before the test vehicle starts to move on the bridge. This assumption is reasonable because the bridge vibrations caused by ambient excitations are generally small, in comparison with those by the moving vehicular loads.
4) Damping is neglected for both the vehicle and bridge. This assumption is also reasonable because the vibrations of both the vehicle and bridge under the moving vehicle's action are forced vibrations of the transient nature, for which the role of damping is insignificant.
5) The test vehicle travels at constant speed $v$ during its passage over the bridge.

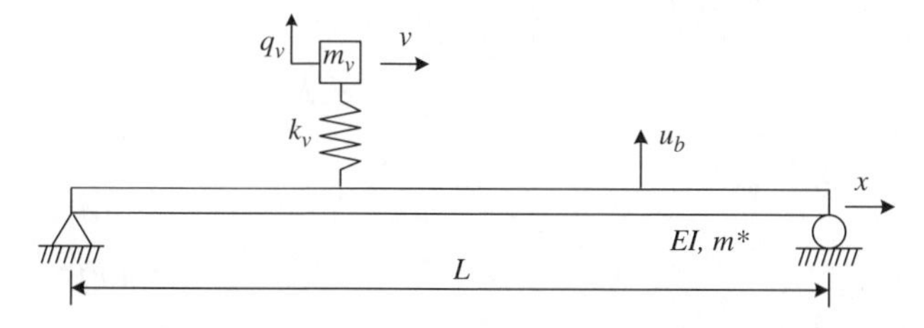


**Figure 9.1** Model of a test vehicle moving on a bridge.

The equations of motion can be written for the vehicle and bridge as follows:

$$m_v \ddot{q}_v(t) + k_v \left( q_v(t) - u(x,t)\Big|_{x=vt} \right) = 0 \tag{9.6}$$

$$m^* \ddot{u}(x,t) + EIu''''(x,t) = f_c(t)\delta(x - vt) \tag{9.7}$$

where $u(x,t)$ denotes the vertical displacement of the bridge, $q_v(t)$ the vertical displacement of the vehicle, measured from its static equilibrium position, and a dot and a prime represents the derivative with respect to time $t$ and longitudinal coordinate $x$, respectively. The contact force $f_c(t)$ can be expressed as

$$f_c(t) = -m_v g + k_v \left( q_v(t) - u(x,t)\Big|_{x=vt} \right) \tag{9.8}$$

where $g$ is the gravitational acceleration.

Using the modal superposition method, expressing the bridge displacement response $u(x,t)$ in terms of modal shapes $\sin(n\pi x/L)$ and generalized coordinates $q_{b,n}(t)$:

$$u(x,t) = \sum_{n=1}^{\infty} \sin \frac{n\pi x}{L} q_{b,n}(t) \tag{9.9}$$

we can obtain the solution of the displacement response of the test vehicle as follows (Yang and Lin 2005; Yang and Chang 2009a):

$$\begin{aligned} q_v(t) = \sum_{n=1}^{\infty} \Bigg[ & A_{1,n} + A_{2,n} \cos\left( \frac{2n\pi v}{L} \right) t + A_{3,n} \cos(\omega_v t) \\ & + A_{4,n} \cos\left( \omega_{b,n} - \frac{n\pi v}{L} \right) t + A_{5,n} \cos\left( \omega_{b,n} + \frac{n\pi v}{L} \right) t \Bigg] \end{aligned} \tag{9.10}$$

where the coefficient of each term is

$$A_{1,n} = \frac{\Delta_{st,n}}{2(1 - S_n^2)}, \tag{9.11}$$

$$A_{2,n} = \frac{\Delta_{st,n}}{2(1 - S_n^2)(4\mu_n^2 S_n^2 - 1)}, \tag{9.12}$$

$$A_{3,n} = \frac{2\Delta_{st,n} S_n^2 \mu_n^4 (2 + \mu_n^2 S_n^2 - \mu_n^2)}{(4\mu_n^2 S_n^2 - 1)\left[ 1 - 2\mu_n^2 (1 + S_n^2) + \mu_n^4 (1 - S_n^2)^2 \right]}, \tag{9.13}$$

$$A_{4,n} = -\frac{\Delta_{st,n} S_n}{2(1 - S_n^2)\left[ 1 - \mu_n^2 (1 - S_n)^2 \right]}, \tag{9.14}$$

$$A_{5,n} = \frac{\Delta_{st,n} S_n}{2(1 - S_n^2)\left[ 1 - \mu_n^2 (1 + S_n)^2 \right]}, \tag{9.15}$$

and the vehicle frequency $\omega_v$, bridge frequency $\omega_{b,n}$, vehicle-induced static deflection $\Delta_{st,n}$ of the bridge, the speed parameter $S_n$, and the ratio of bridge frequency $\omega_{b,n}$ to the vehicle frequency $\omega_v$ of the $n$th mode ($\mu_n$) are defined as

$$\omega_v = \sqrt{\frac{k_v}{m_v}} \tag{9.16}$$

$$\omega_{b,n} = \frac{n^2\pi^2}{L^2}\sqrt{\frac{EI}{m^*}} \tag{9.17}$$

$$\Delta_{st,n} = \frac{-2m_v g L^3}{n^4\pi^4 EI} \tag{9.18}$$

$$S_n = \frac{n\pi v}{L\omega_{b,n}} \tag{9.19}$$

$$\mu_n = \frac{\omega_{b,n}}{\omega_v} \tag{9.20}$$

The bridge response can also be solved in a similar manner (Yang and Lin 2005; Yang and Chang 2009a). However, it won't be presented herein since it is not of concern in this study.

Taking the derivative of the vehicle displacement response twice, one can obtain the vehicle acceleration response as

$$\begin{aligned}\ddot{q}_v(t) = \sum_{n=1}^{\infty}\Bigg[&\overline{\overline{A}}_{2,n}\cos\left(\frac{2n\pi v}{L}\right)t + \overline{\overline{A}}_{3,n}\cos(\omega_v t)\\ &+\overline{\overline{A}}_{4,n}\cos\left(\omega_{b,n} - \frac{n\pi v}{L}\right)t + \overline{\overline{A}}_{5,n}\cos\left(\omega_{b,n} + \frac{n\pi v}{L}\right)t\Bigg]\end{aligned} \tag{9.21}$$

with the coefficients as

$$\begin{aligned}&\overline{\overline{A}}_{2,n} = -\left(\frac{2n\pi v}{L}\right)^2 \times A_{2,n}, \overline{\overline{A}}_{3,n} = -\omega_v^2 \times A_{3,n},\\ &\overline{\overline{A}}_{4,n} = -\left(\omega_{b,n} - \frac{n\pi v}{L}\right)^2 \times A_{4,n}, \overline{\overline{A}}_{5,n} = -\left(\omega_{b,n} + \frac{n\pi v}{L}\right)^2 \times A_{5,n}\end{aligned} \tag{9.22}$$

Clearly, the vehicle response is dominated by five frequencies, i.e., the left-shifted driving frequency $(n-1)\pi v/L$, right-shifted driving frequency $(n+1)\pi v/L$, vehicle frequency $\omega_v$, left-shifted bridge frequency $\omega_{b,n} - n\pi v/L$, and right-shifted bridge frequency $\omega_{b,n} + n\pi v/L$.

To retrieve the mode shapes of the bridge, the component response corresponding to the bridge frequency of the $n$th mode should be singled out from the vehicle response by a feasible filtering technique. According to Eq. (9.21), the retrieved bridge component response $R_b$ associated with the $n$th mode is

$$R_b = A_l\cos\left(\omega_{b,n} - \frac{n\pi v}{L}\right)t + A_r\cos\left(\omega_{b,n} + \frac{n\pi v}{L}\right)t \tag{9.23}$$

where the coefficients $A_l$ and $A_r$ corresponding to the left- and right-shifted bridge frequencies are

$$A_l = \overline{\overline{A}}_{4,n} = \left(\omega_{b,n} - \frac{n\pi v}{L}\right)^2 \frac{S_n \Delta_{st,n} \omega_v^2}{2\left(1-S_n^2\right)\left(\omega_v - \omega_{b,n} + \frac{n\pi v}{L}\right)\left(\omega_v + \omega_{b,n} - \frac{n\pi v}{L}\right)} \tag{9.24}$$

$$A_r = \overline{\overline{A}}_{5,n} = -\left(\omega_{b,n} + \frac{n\pi v}{L}\right)^2 \frac{S_n \Delta_{st,n} \omega_v^2}{2\left(1-S_n^2\right)\left(\omega_v - \omega_{b,n} + \frac{n\pi v}{L}\right)\left(\omega_v + \omega_{b,n} - \frac{n\pi v}{L}\right)} \tag{9.25}$$

The bridge component response $R_b$ is a narrow-band time series and thus can be processed by the HT to yield its transformed pair:

$$\hat{R}_b(t) = \mathbf{H}\left[R_b(t)\right] = A_l \sin\left(\omega_{b,n} - \frac{n\pi v}{L}\right)t + A_r \sin\left(\omega_{b,n} + \frac{n\pi v}{L}\right)t \tag{9.26}$$

From Eq. (9.4), the instantaneous amplitude of $R_b$ can be obtained as

$$\begin{aligned} A(t) &= \sqrt{R_b^2(t) + \hat{R}_b^2(t)} \\ &= \sqrt{A_l^2 + A_r^2 + 2A_l A_r \cos\left(\frac{2n\pi vt}{L}\right)} \\ &= \sqrt{\left(A_l + A_r\right)^2 - 4A_l A_r \sin^2 \frac{n\pi vt}{L}} \end{aligned} \tag{9.27}$$

In general, the driving frequency $n\pi v/L$ is much smaller than the bridge frequency $\omega_{b,n}$. Accordingly, the coefficients $A_l$ and $A_r$ in Eqs. (9.24) and (9.25) reduce to the following:

$$A_l = \left(\omega_{b,n}\right)^2 \frac{S_n \Delta_{st,n} \omega_v^2}{2\left(1-S_n^2\right)\left(\omega_v - \omega_{b,n}\right)\left(\omega_v + \omega_{b,n}\right)} \tag{9.28}$$

$$A_r = -\left(\omega_{b,n}\right)^2 \frac{S_n \Delta_{st,n} \omega_v^2}{2\left(1-S_n^2\right)\left(\omega_v - \omega_{b,n}\right)\left(\omega_v + \omega_{b,n}\right)} \tag{9.29}$$

As can be seen, the two coefficients $A_l$ and $A_r$ are equal in magnitude, but opposite in sign, i.e., $A_l + A_r = 0$. Hence, Eq. (9.27) reduces to

$$A(t) = \sqrt{-4A_l A_r \sin^2 \frac{n\pi vt}{L}} = A_m \left|\sin \frac{n\pi vt}{L}\right| \tag{9.30}$$

where

$$A_m = \sqrt{-4A_l A_r} = \frac{\omega_{b,n}^2 S_n \Delta_{st,n} \omega_v^2}{\left(1-S_n^2\right)\left|\omega_v^2 - \omega_{b,n}^2\right|} \tag{9.31}$$

Replacing the variable $x$ with $vt$ in Eq. (9.30) yields

$$A\left(\frac{x}{v}\right) = A_m \left|\sin\frac{n\pi x}{L}\right| \tag{9.32}$$

The preceding equation shows that the instantaneous amplitude history $A(x/v)$ of the retrieved component response is the mode shape function $\sin(n\pi x/L)$ of the bridge (in absolute sense) multiplied by the coefficient $A_m$. A closer look at the coefficient $A_m$ in Eq. (9.31) reveals that it is a function of the speed parameter $S_n$, static deflection $\Delta_{st,n}$, vehicle frequency $\omega_v$, and bridge frequency $\omega_{b,n}$. All these variables are constants, as is the coefficient $A_m$. Particularly, the product of the mode shape function and any constant is known to remain a mode shape. The implication here is that once the component response corresponding to the bridge frequency of a certain mode can be decomposed from the recorded response of the test vehicle during its passage over the bridge, the instantaneous amplitude history of the decomposed component response is representative of the mode shape of concern of the bridge. Theoretically, the mode shapes of the bridge can be retrieved with a very high resolution since each point on the path of the bridge has been touched by the test vehicle.

## 9.4 Algorithms and Constraints

Substituting Eqs. (9.23) and (9.26) into Eq. (9.5), the instantaneous phase $\theta(t)$ can be derived as

$$\begin{aligned}\theta(t) &= \arctan\left(\frac{\hat{R}_b(t)}{R_b(t)}\right) = \arctan\left(-\cot\omega_{b,n}t\right) \\ &= \omega_{b,n}t - \frac{\pi}{2}\end{aligned} \tag{9.33}$$

By substituting the expressions for the instantaneous amplitude in Eq. (9.30) and instantaneous phase in Eq. (9.33) into Eq. (9.3), the analytical function can be expressed as follows:

$$z(t) = A(t)e^{i\theta(t)} = \left[\frac{\omega_{b,n}^2 S_n \Delta_{st,n}\omega_v^2}{\left(1-S_n^2\right)\left(\omega_v^2-\omega_{b,n}^2\right)}\left|\sin\frac{n\pi vt}{L}\right|\right] e^{i\left(\omega_{b,n}t-\frac{\pi}{2}\right)} \tag{9.34}$$

The preceding equation indicates that, in the dynamic response of the test vehicle during its passage over the bridge, the component response of the bridge frequency $\omega_{b,n}$ of the $n$th mode oscillates with a varying amplitude, but with a shape identical to that of the $n$th mode shape of the bridge in sinusoidal form. Therefore, the bridge component response will oscillate within the envelope formed by the mode shape of the bridge. To sketch the envelope of the bridge component response is equivalent to calculating the instantaneous amplitude.

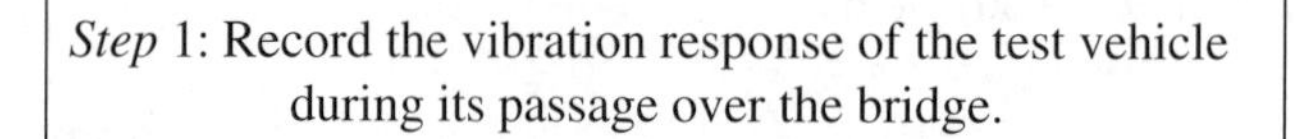
Step 1: Record the vibration response of the test vehicle during its passage over the bridge.
Step 2: Identify the vibration frequency of the bridge from the recorded vehicle responses.
Step 3: Distill the component response associated with a bridge frequency from the test vehicle response.
Step 4: Obtain the instantaneous amplitude history of the bridge component response for the particular mode.
Step 5: Recover the mode shape of the bridge from the instantaneous amplitude history.

**Figure 9.2** Flowchart for constructing the mode shapes.

The procedural steps proposed for retrieving the bridge mode shape from the dynamic response of a test vehicle are as follows (see also Figure 9.2):

*Step 1*: Record the vibration response $R(t)$ of the test vehicle during its passage over the bridge. In practice, the acceleration response of the test vehicle can be measured using the accelerometers (say, for seismic use) mounted on the vehicle. The recording interval should cover the whole duration of passage of the test vehicle, i.e., from the instant of entrance to the instant of departure from the bridge. The passing speed should be made as low as possible.

*Step 2*: Identify the vibration frequency $\omega_{b,n}$ of the bridge from the recorded vehicle responses $R(t)$. This step can be carried out using any feasible means of identification technique, such as the Fourier transform (Yang et al. 2004a; McGetrick et al. 2009) or Fourier transform collaborating with the empirical mode decomposition (EMD) pre-processing technique (Yang and Chang 2009a).

*Step 3*: Distill the component response associated with a bridge frequency from the test vehicle response. After the bridge frequency $\omega_{b,n}$ is made available, one can retrieve the bridge component response $R_{b,n}(t)$ associated with $\omega_{b,n}$ from the test vehicle response $R(t)$, by any feasible signal processing tools, such as the band-pass filters or singular spectrum analysis.

*Step 4*: Obtain the instantaneous amplitude history of the bridge component response for the particular mode. Performing the Hilbert transform, as defined in Eq. (9.1), to the decomposed bridge component response $R_{b,n}(t)$ yields its transformed pair $\hat{R}_{b,n}(t)$. Then, we can obtain the instantaneous amplitude history $A_n(t)$ of the bridge component response using Eq. (9.4).

*Step 5*: Recover the mode shape of the bridge from the instantaneous amplitude history. The curve of the instantaneous amplitude function $A_n(t)$ is representative of the mode shape of the bridge in absolute sense. The sign of the mode shape can be decided according to engineers' judgment or experience (Fang and Perera 2009). Note that a discontinuity may appear at common nodes where the signs at both sides of the nodes are forced to be opposite. Finally, the mode shape of the bridge obtained can be normalized or smoothed for any further engineering applications, in a way similar to those processed by other approaches.

With the procedural steps outlined above, the present approach for retrieving the mode shape of a bridge is subject to some restraints. One is imposed by the requirement of constant vehicle speed. If the vehicle speed is not constant, then the coefficient $A_m$ in Eq. (9.23) for the instantaneous amplitude history will not remain constant anymore. Consequently, the instantaneous amplitude calculated will deviate from the theoretical mode shape of the bridge in those nonconstant speed intervals. The level of deviation will be explored quantitatively as follows.

Suppose that the vehicle speed deviates from the original constant speed with a ratio of $\alpha$ at a certain instant, and so does the speed parameter $S_n$, the coefficient $A_m$ accordingly will vary as follows:

$$A_m(\alpha) = \frac{\alpha\, \omega_{b,n}^2\, S_n\, \Delta_{st,n} \omega_v^2}{\left(1-\alpha^2 S_n^2\right)\left(\omega_v^2 - \omega_{b,n}^2\right)} \tag{9.35}$$

The ratio $R_A$ of the varying coefficient $A_m(\alpha)$ to the original constant coefficient $A_m(1)$ can be computed as

$$R_A = \frac{A_m(\alpha)}{A_m(1)} = \frac{\alpha/\left(1-\alpha^2 S_n^2\right)}{1/\left(1-S_n^2\right)} = \frac{\alpha\left(1-S_n^2\right)}{\left(1-\alpha^2 S_n^2\right)} \tag{9.36}$$

Figure 9.3 shows the coefficient ratio $R_A$ with respect to the speed ratio $\alpha$ for the speed parameter of the following values: $S_n$ = 0.01, 0.05, 0.10, 0.15, and 0.20. It is observed that, for small $S_n$ values, say less than 0.15, $R_A$ is approximately equal to $\alpha$ within the speed with small variation, say, with $\alpha$ = 0.6–1.3. For this case, when the varying vehicle speed is $\alpha$ times the original constant speed, the coefficient $A_m$ is approximately equal to $\alpha$ times the theoretical value of the mode shape at that point. Such an observation is useful in retrieving the mode shape from the deviated mode shape, once the varying

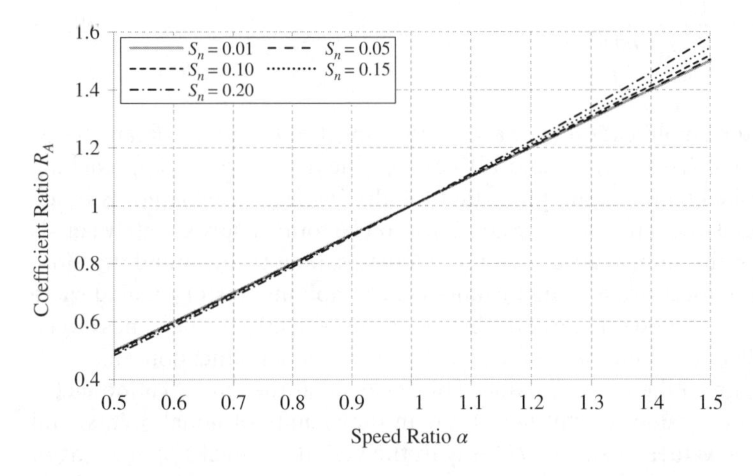


**Figure 9.3** Coefficient ratio $R_A$ vs. speed ratio $\alpha$.

speed history is recorded as well. Thus, the restraint on the constant vehicle speed can be relaxed and a slightly nonconstant vehicle speed is allowed.

Another restraint is imposed by the assumption that the driving frequency $n\pi v/L$ is negligibly smaller than the bridge frequency $\omega_{b,n}$. If the above assumption is violated, the instantaneous amplitude history given in Eq. (9.27) cannot be reduced to Eq. (9.30), accompanied by the fact that the instantaneous amplitude history is not a representative mode shape anymore. To evaluate the range of restraint quantitatively, the assumption is removed herein and Eq. (9.27) is re-derived into a form similar to Eq. (9.30) as follows:

$$\begin{aligned} A(t) &= \sqrt{(A_l + A_r)^2 - 4A_l A_r \sin^2 \frac{n\pi vt}{L}} \\ &= \sqrt{1 - \frac{(A_l + A_r)^2}{A_l A_r \sin^2 \frac{n\pi vt}{L}}} \sqrt{-4A_l A \sin^2 \frac{n\pi vt}{L}} \\ &= R_m A_m \left| \sin \frac{n\pi vt}{L} \right| \end{aligned} \tag{9.37}$$

where $A_m$ has been defined in Eq. (9.31) and $R_m$ is the amplification factor of the instantaneous amplitude history, given as

$$R_m = \sqrt{1 - \frac{(A_l + A_r)^2}{A_l A_r \sin^2 \frac{n\pi vt}{L}}} \tag{9.38}$$

If the assumption of $A_l + A_r = 0$ is valid, the amplification factor $R_m$ will reduce to unity and the instantaneous amplitude history is a representative mode shape of the bridge, which is the case discussed in Section 9.3. If the assumption is invalid, $R_m$ can be derived by substituting the expression of $A_l$ and $A_r$ in Eqs. (9.28) and (9.29), respectively, into Eq. (9.38), namely:

$$R_m = \sqrt{1 + \frac{16S_n^2}{\left(1 - S_n^2\right)^2 \sin^2 \frac{n\pi vt}{L}}} \tag{9.39}$$

Figure 9.4 shows the amplification factor $R_m$ with respect to the speed parameter $S_n$ for $\sin(n\pi v/L)$ taken as 0.2, 0.4, 0.6, 0.8, and 1.0. For the general case with $S_n$ approaching 0, as is the case implied by the assumption that the driving frequency $n\pi v/L$ is much smaller than the bridge frequency $\omega_{b,n}$, the amplification factor $R_m$ approximately equals unity. As $S_n$ increases, the amplification factor $R_m$ increases monotonously and therefore deviates from unity, indicating that the instantaneous amplitude history may deviate more from the theoretical mode shape of the bridge as the vehicle speed becomes larger. Moreover, the smaller the value of $\sin(n\pi v/L)$, the more the amplification factor $R_m$ deviates from unity, given the same $S_n$ value. This means that the amplification factor $R_m$ is sensitive for small values of $\sin(n\pi v/L)$, say in the vicinity of nodal points, and nonsensitive for large values of $\sin(n\pi v/L)$, say in the vicinity of peaks or troughs of the mode shape. To ensure accuracy of the solution, it is suggested that the test vehicle

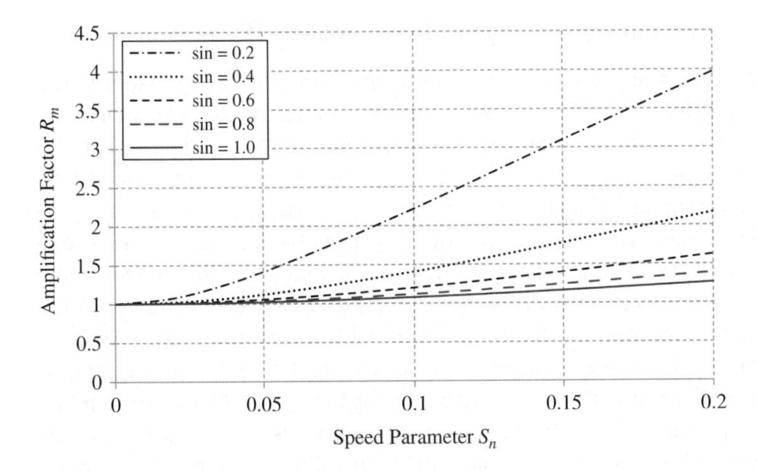


**Figure 9.4** Amplification factor $R_m$ vs. speed parameter $S_n$.

be allowed to move at low speeds. If the low-speed condition is not met, one can also retrieve the theoretical mode shape from the deviated mode shape making use of the calculated amplification factor.

## 9.5 Case Studies

In order to verify the feasibility of and to understand the restraints of the present approach for retrieving the mode shapes of a bridge from the dynamic responses of the passing test vehicle, four numerical cases are studied herein using the finite element simulations. The finite element simulating algorithm utilized in this study is the vehicle-bridge interaction element presented in Chang et al. (2010) and summarized in the Appendix, which is modified from the one by Yang and Yau (1997) and Yang et al. (2004b). For the cases considered, the following properties are adopted for the simply supported bridge: length $L$ = 30 m, Young's modulus $E$ = 27.5 GPa, moment of inertia $I = 0.175\,\text{m}^4$, and mass density $m^*$ = 1000 kg/m; and the following for the test vehicle: mass $m_v$ = 1000 kg and stiffness $k_v$ = 170 kN/m. The bridge is discretized into 20 identical beam elements, and the time step is selected as 0.001 second.

In the following, the acceleration responses of the test vehicle during its passage over the bridge will be processed with the procedural steps outlined previously to retrieve the mode shapes of the bridge. The accuracy of the retrieved mode shapes is evaluated by the *modal assurance criteria* (MAC) defined as

$$MAC = \frac{\phi_e^T \phi_t}{|\phi_e||\phi_t|} \tag{9.40}$$

where $\phi_e$ and $\phi_t$ denote the retrieved and theoretical mode shapes, respectively. Several factors, such as the test vehicle speed, random traffic, and road surface roughness, will be studied as well.

### 9.5.1 Test Vehicle Passing through a Bridge with Smooth Road Surface

The first case is to illustrate the present algorithm for retrieving the mode shapes of the bridge, while verifying its feasibility. In this case, a smooth road surface is considered for the bridge. In the first step, the acceleration response of the test vehicle during its passage of the bridge is generated, as shown in Figure 9.5. In the second step, the vehicle response is processed by the fast Fourier transform (FFT) to yield the frequency spectrum as shown in Figure 9.6. From this figure, the first three frequencies of the bridge can be identified as 3.87, 15.27, and 34.40 Hz. In the third step, by the conventional band-pass filters, the component responses that correspond to each of the identified bridge frequencies are decomposed from the test vehicle response, as shown in Figure 9.7. It is observed that the component responses oscillate with varying amplitudes in a way similar to the respective mode shapes of the bridge. In the fourth step, the Hilbert transform is performed to the bridge component responses to obtain the instantaneous amplitude histories. Finally, one can determine the signs of the mode shapes by engineering judgment, and obtain the mode shapes of the bridge as in Figure 9.8. In comparison with the theoretical mode shapes, the mode shapes identified herein show a high level of accuracy for the case studied.

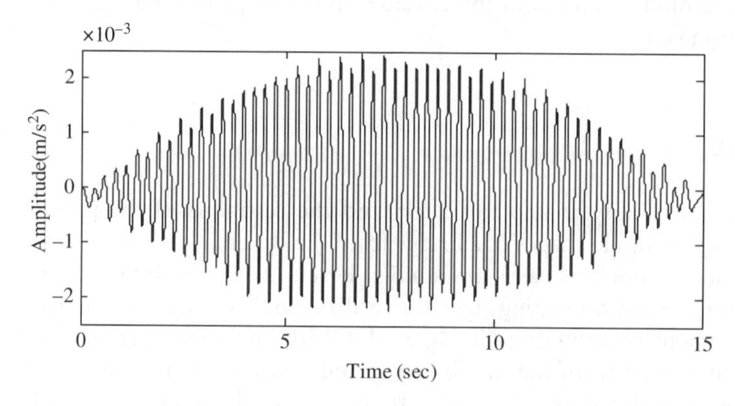


Figure 9.5 Acceleration response of the test vehicle.

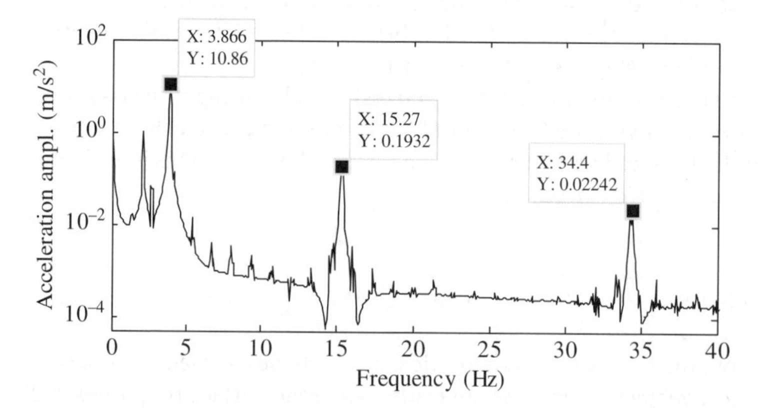


Figure 9.6 Acceleration spectrum of the test vehicle.

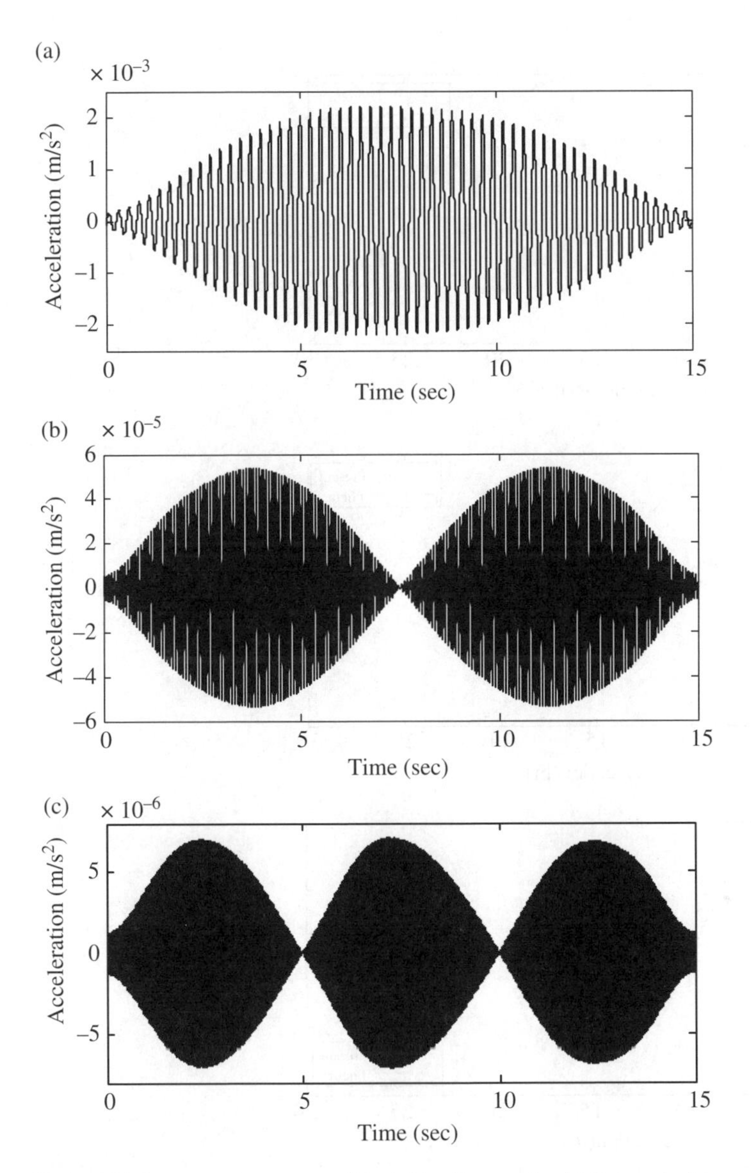


**Figure 9.7** Component responses of the bridge frequencies: (a) first mode; (b) second mode; (c) third mode.

### 9.5.2 Effect of Vehicle Speed

In this section, the effect of vehicle speed on the retrieved mode shapes of the bridge is studied for three different vehicle speeds: $v = 2$, 4, and 8 m/s. The bridge remains identical to the one previously studied. By following the same procedure, the mode shapes of the bridge can be extracted for each vehicle speed, as shown in Figure 9.9. In Table 9.1, the MAC values between the retrieved mode shapes and theoretical ones are listed. The effect of vehicle speed on the retrieved mode shapes can be clearly observed from Figure 9.9 and Table 9.1.

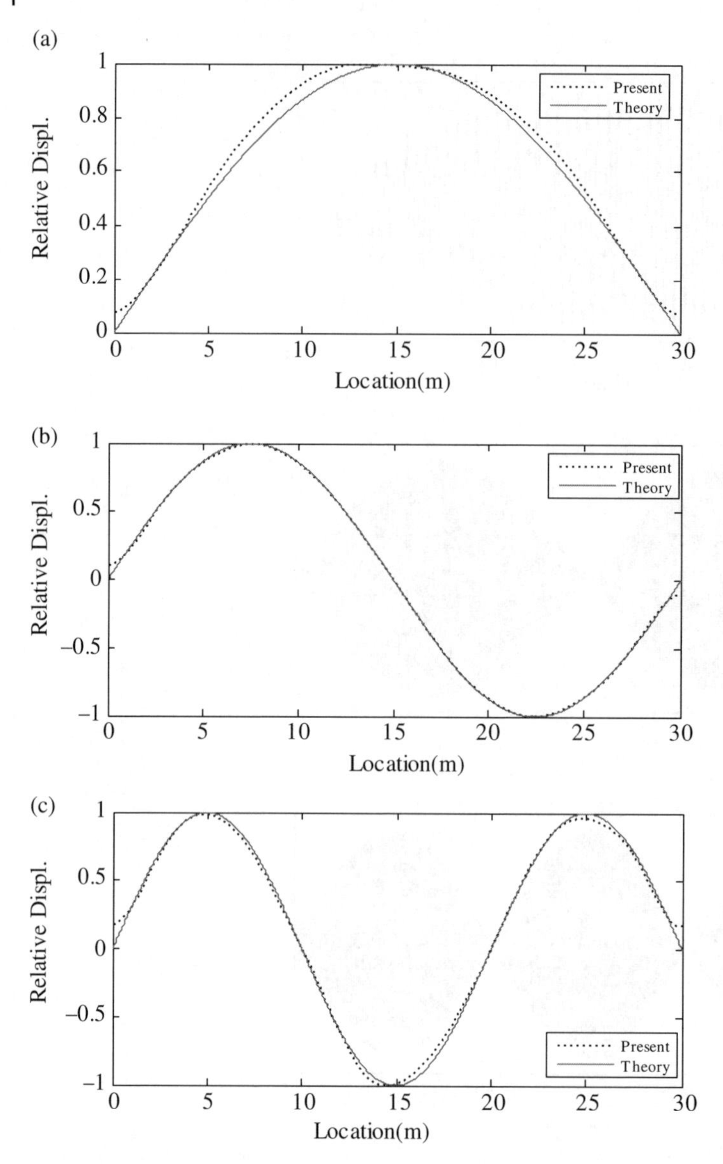


**Figure 9.8** Mode shapes of the bridge obtained by the present approach and theoretical formulae: (a) first mode; (b) second mode; (c) third mode.

If the vehicle moves at a low speed, say 2 m/s, the three retrieved mode shapes match very well with the theoretical ones, generally with MAC values over 0.998. As the vehicle speed is doubled to 4 m/s or twice doubled to 8 m/s, the first retrieved mode shapes show little deviation from the theoretical ones, still with MAC values over 0.998. However, the second retrieved mode shapes show slight deviations, with larger deviations concentrated at the end nodes, and the MAC value decreases as the vehicle speed increases. The third retrieved mode shapes show obvious deviations, also with decreasing MAC values as the vehicle speed increases. It is concluded herein that a lower vehicle

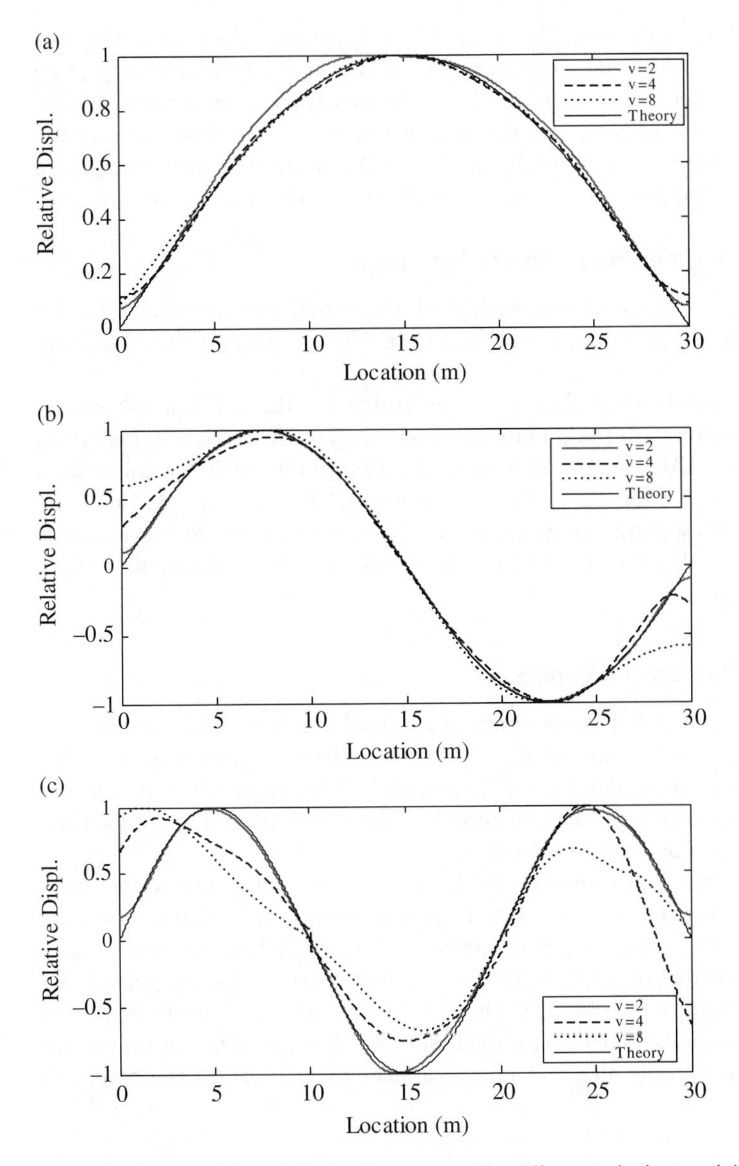


**Figure 9.9** Mode shapes of the bridge obtained for different vehicle speed: (a) first mode; (b) second mode; (c) third mode.

**Table 9.1** Modal assurance criteria (MAC) of the first three modes for different vehicle speeds.

| Speed (m/s) | First mode | Second mode | Third mode |
|---|---|---|---|
| 2 | 0.9984 | 0.9999 | 0.9981 |
| 4 | 0.9998 | 0.9955 | 0.8858 |
| 8 | 0.9997 | 0.9790 | 0.6824 |

speed guarantees higher accuracy of the retrieved mode shapes of a bridge, which is especially true for the higher modes. Such a conclusion is consistent with the suggestion made in the preceding section. In addition, the existence of larger discrepancy for the identified mode shapes at the supports is also consistent with the discussion of the preceding section: the theoretical amplification factor $R_m$ is sensitive in the vicinity of nodal points, and nonsensitive in the vicinity of peaks or troughs of the mode shape.

### 9.5.3 Test Vehicle Traveling along with Random Traffic

Except for the test vehicle, three accompanying vehicles moving at random speeds are adopted to simulate the effect of random traffic, whose dynamic properties are assigned as in Table 9.2.

The first three retrieved mode shapes of the bridge for this case are shown in Figure 9.10. In comparison with the theoretical mode shapes, the retrieved ones show high accuracy, with an MAC value of 0.9990 for the first mode, 0.999 for the second mode, and 0.9994 for the third mode. It is observed that the random traffic of the pattern presented in this section can hardly affect the accuracy of the retrieved mode shapes of the bridge, possibly because the total mass of the vehicles is too small to affect the bridge mode shapes.

### 9.5.4 Effect of Road Surface Roughness

In this case, the effect of road surface roughness is studied by letting the test vehicle pass through a bridge with rough road surface. The road surface roughness is generated according to the power spectrum density (PSD) curve of class A presented by International Organization for Standardization (1995). Figure 9.11 shows the generated road surface roughness profile that is adopted in this section.

First, we let the test vehicle pass over the bridge with surface roughness of the above profile. Then, we can process the dynamic responses of the test vehicle using the algorithm described. The mode shapes retrieved of the bridge have been plotted in Figure 9.12. Obvious distortions appear in the retrieved mode shapes, especially for the second and third modes. The MAC value between the retrieved mode shape and corresponding theoretical one is 0.9980 for the first mode, 0.9858 for the second mode, and 0.7957 for the third mode, all of which are smaller than those for the case with

**Table 9.2** Properties of random traffic.

| Vehicle Number | Mass (kg) | Stiffness (N/m) | Speed (m/s) | Initial spacing (m)[a] | Remark |
|---|---|---|---|---|---|
| 1 | 1000 | 170 000 | 2 | 0 | Test vehicle |
| 2 | 1000 | 170 000 | random | 1 | |
| 3 | 1000 | 170 000 | random | 3 | |
| 4 | 1000 | 170 000 | random | 5 | |

[a] Initial spacing denotes the spacing between the accompanying vehicle and test vehicle at the instant when the test vehicle enters the bridge.

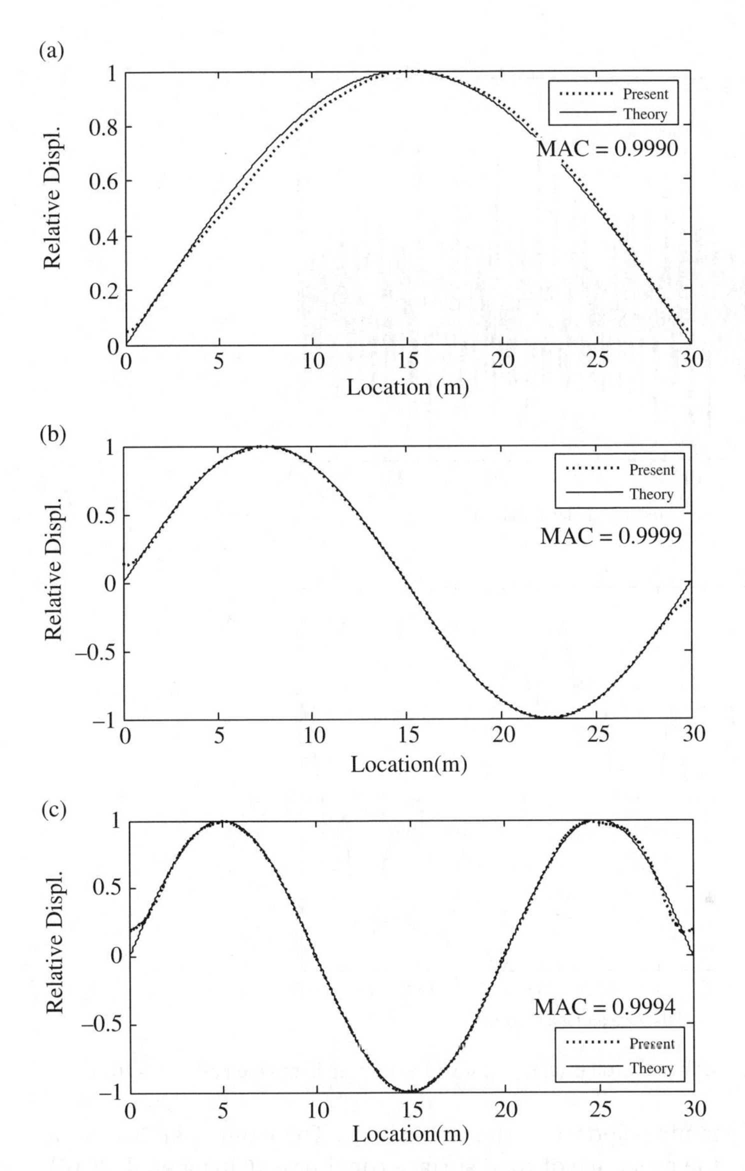


**Figure 9.10** Mode shapes of the bridge obtained from a test vehicle traveling with random traffic: (a) first mode; (b) second mode; (c) third mode.

smooth road surface. Therefore, it is concluded that the existence of road surface roughness has negative impact on retrieving the mode shapes of the bridge by the present approach, especially for the higher modes. In addition, discontinuity occurs at the nodes of inflection for the mode shapes, as can be observed from the values for the same nodes with different signs.

The reasons for the adverse effect of road surface roughness on the retrieval of mode shapes can be given as follows. First, the present approach for retrieving the mode shapes of the bridge is based on the assumption that the bridge frequencies have been

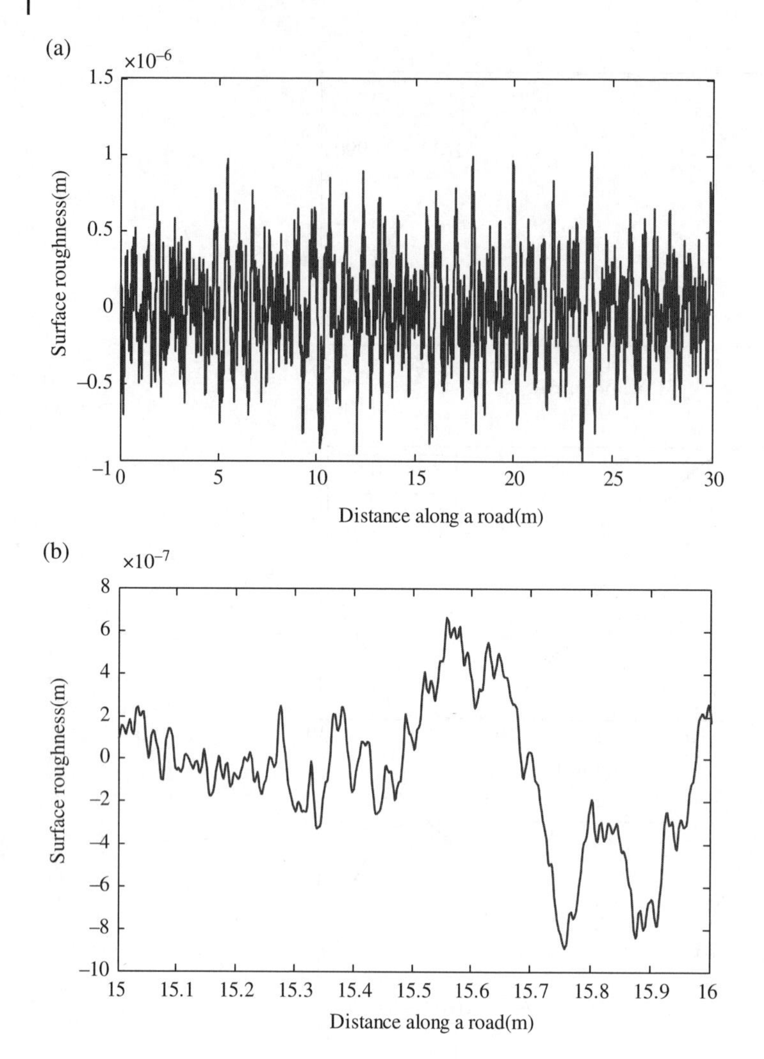


**Figure 9.11** Road surface roughness profile: (a) full span; (b) a close-up in the interval of 15–16 m.

identified from the dynamic response of the test vehicle. The latter is known to be negatively affected by the existence of road surface roughness (Chang et al. 2010). It follows that the former is also negatively affected. Second, during its passage over the bridge, the test vehicle is excited by both the vibrations of the bridge and road surface roughness. The bridge vibrates with fixed modal frequencies and shapes, while the road surface roughness injects a wide range of spatial frequency into the vehicle response. Clearly, the mode shapes of the vibrating bridge mingle with the surface roughness profile of the frequency identical or close to the bridge frequency of interest. The mingled mode shapes and surface roughness profile are hard to separate by the frequency-domain filtering techniques, thereby distorting the retrieved mode shapes. Third, such a distortion is more severe for higher modes since the amplitudes of higher modes are generally smaller than those of lower modes and thus easier to mingle with surface roughness profiles.

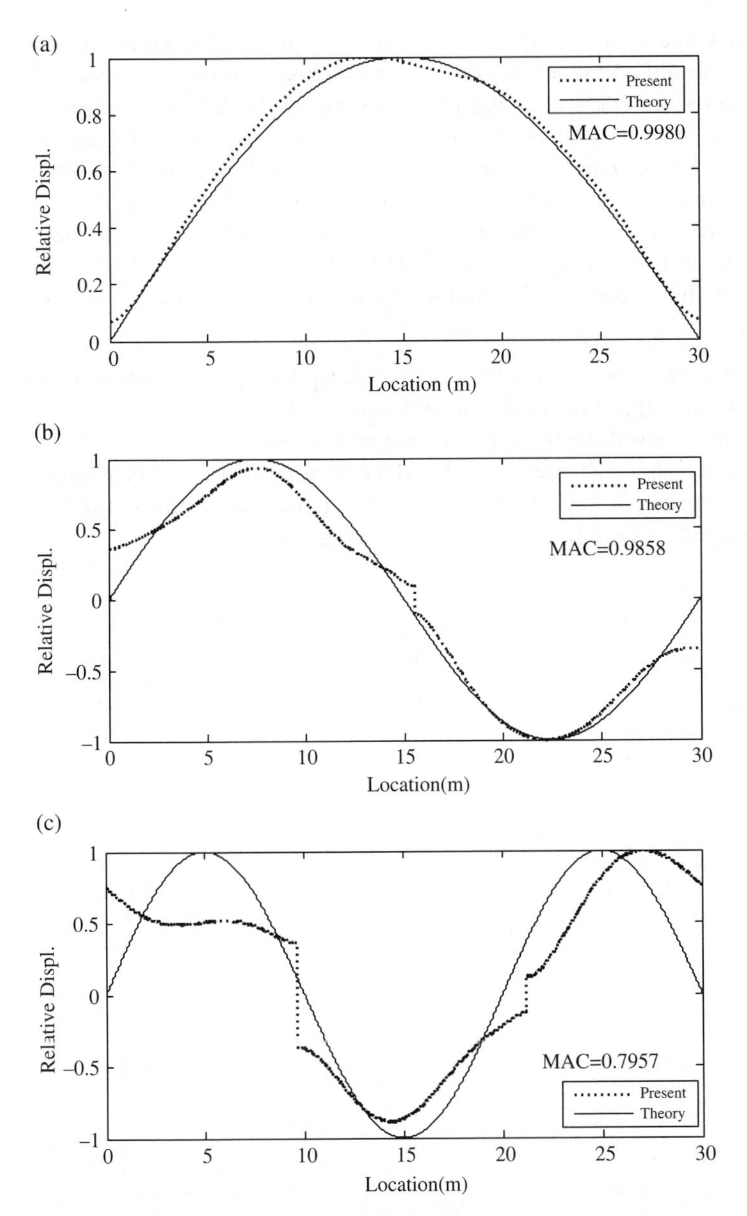


**Figure 9.12** Mode shapes of the bridge obtained from a test vehicle traveling on rough road surface: (a) first mode; (b) second mode; (c) third mode.

## 9.6 Concluding Remarks

This chapter described a theoretical framework for retrieving the mode shapes of a bridge from a passing test vehicle. Based on the theoretical formulation, we observed that the component response of the bridge frequency of each mode, when retrieved from the moving test vehicle's response, oscillates with a varying amplitude that is identical to the mode shape of the bridge of the mode of concern. Therefore, once a bridge frequency is identified and its corresponding component response is separated from the vehicle response, the instantaneous amplitude history of the retrieved component response can be regarded as a representative of the mode shape of the bridge. Theoretically, the mode shapes can be retrieved with a high resolution in space since each point along the path has been touched by the moving test vehicle.

Through the numerical case study, the present approach is verified to be feasible under the constraint that the vehicle speed is constant and low, say as low as 2 m/s. The chapter also evaluated the impact of the following factors on the accuracy of the retrieved mode shapes:

- *Vehicle speed.* Lower vehicle speeds generally guarantee higher accuracy of the mode shapes, which is especially true for the second and third modes.
- *Random traffic.* It can hardly affect the accuracy of the mode shapes.
- *Road surface roughness.* The existence of road surface roughness has negative effect on retrieval of the mode shapes, especially for the higher modes, partly due to the fact that road surface roughness in rich in high-frequency components.

# 10

# Contact-Point Response for Modal Identification of Bridges

The response of the contact point of the vehicle with the bridge, rather than the vehicle body itself, is a better parameter for modal identification of the bridge. In this chapter, approximate closed-form solutions were first derived for the vehicle and contact-point responses, and then verified by finite element solutions. The contact-point acceleration is born to be free of the vehicle frequency, an annoying effect that may overshadow the bridge frequencies in the presence of surface roughness. From the frequency response function (FRF) of the vehicle with respect to the contact point, it was shown that the contact-point response generally outperforms the vehicle response in retrieving the bridge frequencies in that it allows higher frequencies to be identified. In the numerical simulations, the contact-point response was compared with the vehicle response for various scenarios. It is concluded that in each case, say, for varying vehicle speeds or frequencies, for smooth or rough road surfaces, with or without existing traffic, the contact-point response outperforms the vehicle response in retrieving either the frequencies or mode shapes of the bridge. The materials presented in this chapter are based primarily on the paper by Yang et al. (2018a).

## 10.1 Introduction

The modal parameters of a bridge, including the frequency, mode shape, and damping, are the most useful information for engineering applications. For a newly completed bridge, one may like to measure the first few frequencies of the bridge for the purpose of model updating or calibration, so as to fill up the gap between the design model and the structure completed. An updated structural model is useful for vibration control or future design of similar bridges. For a bridge in use, a regular monitoring of the modal properties provides the crucial means for evaluating the degradation in stiffness, connections, supports, or material strength of the structure, and even for unveiling possible damages in the structure, say, due to overloading by trucks or natural hazards such as scouring or earthquakes.

Conventionally, methods such as the ambient vibration test, forced vibration test, and impact vibration test have been employed to identify the modal parameters of a bridge. These methods have been referred to as the *direct method* for bridge measurement, since they require the vibration sensors to be *directly* installed on the bridge so as to

*Vehicle Scanning Method for Bridges*, First Edition. Yeong-Bin Yang, Judy P. Yang, Bin Zhang and Yuntian Wu.

collect the bridge's response. Starting in the 1970s, a huge amount of research has been carried out along these lines. Some of the early works that adopt the ongoing traffic or controlled vehicular movement as the source of excitation have been presented by McLamore et al. (1971), Abdel-Ghaffar and Housner (1978), Ward (1984), Mazurek and DeWolf (1990), Casas (1995), Ventura et al. (1996), and Huang et al. (1999b), among others. A partial review of the relevant previous works along these lines was given by Carden and Fanning (2004). The direct approach is normally designed on a *one-system-per-bridge* basis. It requires quite a number of vibration sensors to be mounted on the bridge, if the mode shapes of the bridge are desired. Generally speaking, the on-site installation of sensors and equipment on the bridge is costly, time-consuming, and not maintenance free, while the monitoring system established for one bridge can hardly be transferred to another bridge. The other drawback is the *sea-like data* continuously generated by each bridge monitoring system, which cannot be efficiently digested.

To resolve the above problems, the idea for retrieving the bridge frequencies from a moving test vehicle was proposed by Yang et al. (2004a). At the time, only the first frequency of the bridge was attempted. The feasibility of the idea was validated immediately in a field test (Lin and Yang 2005). Later, Yang and co-workers proceeded to construct the mode shapes of the bridge using the concept of *instantaneous amplitude* based on the Hilbert transform of the component response of the bridge's frequency of concern (Yang et al. 2014). The technique of using the vehicle-based data was grossly referred to as the *vehicle scanning method* or *indirect method* for bridge measurement, which is featured by the fact that *no* sensors are required on the bridge, while only one or a few sensors are mounted on the test vehicle. Clearly, the vehicle scanning method has the advantage of mobility, economy, and efficiency, compared with the conventional direct approach.

In the past decade, Yang's research group has proceeded to tackle various aspects of the problem encountered in application of the vehicle scanning method for bridges. Specifically, the following issues have been investigated: the multi-mode effect of the bridge (Yang and Lin 2005), parametric study for relevant factors involved (Yang and Chang 2009b), application of empirical mode decomposition (EMD) technique to extract higher bridge frequencies (Yang and Chang 2009a), road surface roughness effect (Chang et al. 2010; Yang et al. 2012a), elimination of the road roughness effect by two connected vehicles (Yang et al. 2012b), design of a hand-drawn cart (Yang et al. 2013b), filtering techniques for removing the road roughness effect (Yang et al. 2013a), stochastic approach for noise removal (Yang and Chen 2016), and wave number-based technique for damage detection (Yau et al. 2017).

In the meantime, research has grown rapidly on the globe along the lines of using the vehicle-based data for modal identification and damage detection of bridges. For instance, the following issues have been investigated by researchers: bridge condition assessment (Bu et al. 2006), identification of bridge dynamic properties (McGetrick et al. 2009), multi-cracks detection of beam-like bridges (Nguyen and Tran 2010), identification of cable tension loss and deck damage in a cable-stayed bridge (Yin and Tang 2011), damage detection by mode shape squares using a test vehicle with tapping device (Zhang et al. 2012), use of a truck-trailer to monitor bridge damping (Keenahan et al. 2014), drive-by bridge inspection (Kim et al. 2014), identification of bridge mode shapes using short time frequency domain decomposition (Malekjafarian and OBrien 2014), drive-by damage detection using an apparent profile (Keenahan and OBrien 2014), among others.

In going through years of research on the indirect approach, Yang and co-workers have always used the *single-degree-of-freedom (SDOF) system* for the test vehicle, either in the pioneering theoretical study (Yang et al. 2004a), field tests (Lin and Yang 2005; Yang et al. 2013b), or in the following studies. But the same model was not exactly followed by other researchers. In the studies by Yang and co-workers, the SDOF vehicle model was not called a half-car model, but a *full-car model* with a *single axle and two wheels*. It is believed that *only when the test vehicle fully resembles the SDOF system used in the theoretical study*, i.e., the sprung mass system in Figure 10.1, *can the mechanical behavior revealed in the analytical counterpart be fully appreciated in the experimental and field tests*. The use of two-axle vehicles or more complicated vehicle models in the laboratory or in the field is likely to introduce the *linking action* or coupling effect between the front and rear axles of the vehicle, of which the effect has not been fully exploited yet.

This chapter can be regarded as an extension of Chapter 2 or the work by Yang et al. (2004a) in that the same SDOF model is used. However, it differs from Chapter 2 in that the response of the contact point of the moving vehicle with the bridge, instead of the vehicle response itself, will be used as the physical parameter for retrieving the modal dynamic properties of the bridge. Both the frequencies and mode shapes of the bridge are concerned. No consideration will be made for damage detection. To start, an analytical formulation for deriving the approximate closed-form solutions will be presented, based on which the physical difference between the vehicle response and contact-point response can be fully appreciated. Then, comprehensive numerical analyses will be conducted for various scenarios, namely, for varying vehicle speeds and frequencies, for smooth or rough road surface, with or without existing traffic, to show the superiority of using the contact-point response to identify the bridge modal properties, compared with the vehicle response.

## 10.2 Theoretical Formulation

Figure 10.1 shows the vehicle–bridge interaction (VBI) model adopted in this study. The vehicle is simplified as a lumped mass $m_v$ supported by a spring of stiffness $k_v$ and moving at constant speed $v$. An accelerometer is mounted on the vehicle to measure its vertical acceleration during its passage over the bridge. The bridge is modeled as a

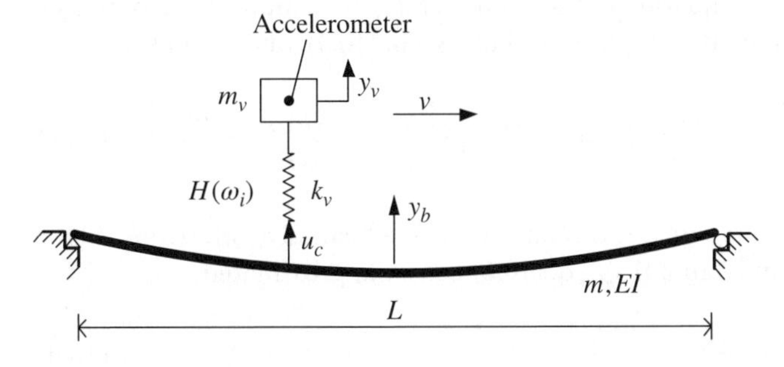


**Figure 10.1** Vehicle-bridge interaction (VBI) model.

simply supported beam of length $L$ with smooth surface. As already stated, the SDOF vehicle system adopted should not be interpreted as a half car model, but as a *vehicle model with single axle,* which has been successfully applied in the field investigations.

The equation of motion for the vehicle neglecting the damping effect is

$$m_v \ddot{y}_v + k_v \left( y_v - u_c \right) = 0, \tag{10.1}$$

where $y_v$ is the vertical displacement of the vehicle body; $u_c$ is the displacement of the *contact point* between the moving vehicle and the beam; and a dot indicates differentiation with respect to time $t$. Note that the vehicle displacement is measured from the static equilibrium position. The equation of motion for the beam neglecting the damping effect is

$$m\ddot{u} + EIu'''' = \left[ k_v \left( y_v - u_c \right) - m_v g \right] \delta \left( x - vt \right), \tag{10.2}$$

where $m$ is the unit mass, E the elastic modulus, $I$ the moment of inertia, $g$ the acceleration of gravity, $\delta$ the delta function, and a prime denotes differentiation with respect to coordinate $x$ of the beam.

The displacement $u$ of the beam can be represented as the sum of the modal responses (Biggs 1964):

$$u = \sum_n \left[ q_{bn} \left( t \right) \sin \frac{n\pi x}{L} \right], \tag{10.3}$$

where $q_{bn}$ denotes the modal displacement of the beam, and the contact-point displacement is included as $u_c = u|_{x = vt}$. For the derivations to follow, let us denote the vehicle frequency $\omega_v$ and the $n$th bridge frequency $\omega_{bn}$ as

$$\omega_v = \sqrt{\frac{k_v}{m_v}}, \qquad \omega_{bn} = \frac{n^2 \pi^2}{L^2} \sqrt{\frac{EI}{m}}. \tag{10.4a,b}$$

The VBI problem may be solved by various numerical methods. For the sake of physical interpretation, however, approximate closed-form solutions will be obtained for the contact-point and vehicle responses in the following.

### 10.2.1 Dynamic Response of the Vehicle–Bridge Contact Point

Substituting Eq. (10.3) into the equation of the beam in Eq. (10.2), multiplying both side by $\sin(n\pi x/L)$, integrating along the length $L$, and using Eq. (10.4), one obtains

$$\ddot{q}_{bn} \left( t \right) + \omega_{bn}^2 q_{bn} \left( t \right) - \frac{2 m_v \omega_v^2}{mL} \left( y_v - \sum_j q_{bj} \left( t \right) \sin \frac{j\pi vt}{L} \right) \sin \frac{n\pi vt}{L} = \frac{-2 m_v g}{mL} \sin \frac{n\pi vt}{L}. \tag{10.5}$$

Since the mass of the vehicle is *much less* than that of the beam, i.e., $m_v \ll mL$ (Fryba 1972; Yau et al. 1999; Yang and Yau 2015), Eq. (10.5) can be approximated as

$$\ddot{q}_{bn} \left( t \right) + \omega_{bn}^2 q_{bn} \left( t \right) = \frac{-2 m_v g}{mL} \sin \frac{n\pi vt}{L}. \tag{10.6}$$

Evidently, the problem of moving sprung mass has been reduced to one of moving load with static force $m_v g$, which can be easily solved. For zero initial conditions, the preceding equation can be solved to yield the modal displacement of the beam as

$$q_{bn}(t)=\frac{\Delta_{stn}}{1-S_n^2}\left(\sin\frac{n\pi vt}{L}-S_n\sin\omega_{bn}t\right), \tag{10.7}$$

where $\Delta_{stn}$ denotes approximately the $n$th modal static deflection of the beam under the gravity load $m_v g$ and the speed parameter $S_n$ is defined as the ratio of the half driving frequencies $n\pi v/L$ to the bridge frequencies $\omega_{bn}$, namely,

$$\Delta_{stn}=-\frac{2m_v gL^3}{EIn^4\pi^4},\quad S_n=\frac{n\pi v}{L\omega_{bn}}. \tag{10.8a,b}$$

By substituting Eq. (10.7) into Eq. (10.3), one obtains the contact-point displacement as

$$u_c=\sum_n\frac{\Delta_{stn}}{1-S_n^2}\left[\left(\sin\frac{n\pi vt}{L}-S_n\sin\omega_{bn}t\right)\sin\frac{n\pi vt}{L}\right], \tag{10.9}$$

which can be manipulated to yield

$$u_c=\sum_n\frac{\Delta_{stn}}{2\left(1-S_n^2\right)}\left\{1-\cos\frac{2n\pi vt}{L}-S_n\left[\cos\left(\omega_{bn}-\frac{n\pi v}{L}\right)t-\cos\left(\omega_{bn}+\frac{n\pi v}{L}\right)t\right]\right\}. \tag{10.10}$$

Differentiating Eq. (10.10) with respect to time $t$ twice yields the contact-point acceleration as

$$\ddot{u}_c=\sum_n\frac{\Delta_{stn}}{2\left(1-S_n^2\right)}\left\{\left(\frac{2n\pi v}{L}\right)^2\cos\frac{2n\pi vt}{L}\right.$$
$$\left.+S_n\left[\left(\omega_{bn}-\frac{n\pi v}{L}\right)^2\cos\left(\omega_{bn}-\frac{n\pi v}{L}\right)t-\left(\omega_{bn}+\frac{n\pi v}{L}\right)^2\cos\left(\omega_{bn}+\frac{n\pi v}{L}\right)t\right]\right\}. \tag{10.11}$$

As can be verified from Eq. (10.11), the response of the contact point is dominated by the driving frequency $2n\pi v/L$ and two shifted frequencies of the beam, i.e., $\omega_{bn}-n\pi v/L$, $\omega_{bn}+n\pi v/L$. Such a frequency composition is similar to the one obtained for the vehicle response presented in Chapter 3 or Yang and Lin (2005), except that *the vehicle frequency* $\omega_v$ *has been totally excluded*. Because of the exclusion of the vehicle frequency $\omega_v$, the contact-point response serves as a *better measurement* for extracting the modal frequencies and dynamic properties of the bridge using the moving vehicle technique.

### 10.2.2 Dynamic Response of the Moving Vehicle

Substituting the vehicle frequency in Eq. (10.4) into the equation of motion for the vehicle in Eq. (10.1), one obtains

$$\ddot{y}_v+\omega_v^2 y_v=\omega_v^2 u_c. \tag{10.12}$$

Then, substituting the response of the contact-point in Eq. (10.10) into Eq. (10.12), and using the zero initial conditions for vehicle, the vehicle responses $y_v$ can be solved as

$$y_v = \sum_n \frac{\Delta_{stn}}{2(1-S_n^2)} \left\{ 1 - \frac{\omega_v^2}{\omega_v^2 - \left(\frac{2n\pi v}{L}\right)^2} \cos\frac{2n\pi vt}{L} + A_n \cos(\omega_v t) \right.$$
$$\left. -S_n \left[ \frac{\omega_v^2}{\omega_v^2 - \left(\omega_{bn} - \frac{n\pi v}{L}\right)^2} \cos\left(\omega_{bn} - \frac{n\pi v}{L}\right)t - \frac{\omega_v^2}{\omega_v^2 - \left(\omega_{bn} + \frac{n\pi v}{L}\right)^2} \cos\left(\omega_{bn} + \frac{n\pi v}{L}\right)t \right] \right\}, \tag{10.13}$$

where

$$A_n = \frac{4S_n^2\mu_n^4(1-S_n^2)(2+\mu_n^2 S_n^2 - \mu_n^2)}{(4\mu_n^2 S_n^2 - 1)\left[1 - 2\mu_n^2(1+S_n^2) + \mu_n^4(1-S_n^2)^2\right]}, \quad \mu_n = \frac{\omega_{bn}}{\omega_v}. \tag{10.14a,b}$$

Differentiating Eq. (10.13) with respect to time $t$ twice yields the acceleration of the vehicle as follows:

$$\ddot{y}_v = \sum_n \frac{\Delta_{stn}}{2(1-S_n^2)} \left\{ \frac{\omega_v^2 \left(\frac{2n\pi v}{L}\right)^2}{\omega_v^2 - \left(\frac{2n\pi v}{L}\right)^2} \cos\frac{2n\pi vt}{L} + \overline{\overline{A}}_n \cos(\omega_v t) \right.$$
$$\left. +S_n \left[ \frac{\omega_v^2 \left(\omega_{bn} - \frac{n\pi v}{L}\right)^2}{\omega_v^2 - \left(\omega_{bn} - \frac{n\pi v}{L}\right)^2} \cos\left(\omega_{bn} - \frac{n\pi v}{L}\right)t - \frac{\omega_v^2 \left(\omega_{bn} + \frac{n\pi v}{L}\right)^2}{\omega_v^2 - \left(\omega_{bn} + \frac{n\pi v}{L}\right)^2} \cos\left(\omega_{bn} + \frac{n\pi v}{L}\right)t \right] \right\}, \tag{10.15}$$

where

$$\overline{\overline{A}}_n = -A_n \omega_v^2. \tag{10.16}$$

As it is clear from Eq. (10.15), the vehicle response also contains two shifted frequencies of bridge, $\omega_{bn} - n\pi v/L$ and $\omega_{bn} + n\pi v/L$. This is the theoretical basis for extracting the bridge frequencies from the vehicle response in most previous studies. It suffers from the drawback that *the vehicle frequency $\omega_v$ may have a spectral amplitude too large for the bridge frequencies to be identified.*

### 10.2.3 Procedure for Calculating the Contact-Point Response in a Field Test

In the field test, the vehicle acceleration can be recorded by the accelerometer installed on the vehicle, as shown in Figure 10.1. However, the acceleration of the contact point cannot be measured directly, but can be calculated using the procedure presented below.

Differentiating Eq. (10.1) for the moving vehicle with respect to time $t$ twice yields

$$k_v\left(\ddot{u}_c-\ddot{y}_v\right)=m_v\frac{d^2\ddot{y}_v}{dt^2}. \tag{10.17}$$

Using the definition for $\omega_v$ in Eq. (10.4), the contact-point acceleration can be computed as

$$\ddot{u}_c=\ddot{y}_v+\frac{d^2\ddot{y}_v}{\omega_v^2 dt^2}. \tag{10.18}$$

Considering that the vehicle accelerations recorded are discrete data, the term $d^2\ddot{y}_v/dt^2$ is calculated by the central difference method as

$$\frac{d^2\ddot{y}_v}{dt^2}=\frac{\left(\ddot{y}_v\big|_{i+1}-2\ddot{y}_v\big|_i+\ddot{y}_v\big|_{i-1}\right)}{\left(\Delta t\right)^2}, \tag{10.19}$$

where $\Delta t$ is the sampling interval and $i$ denotes the $i$th sampling point. Clearly, the contact-point acceleration $\ddot{u}_c$ can be calculated by Eq. (10.18), once the vehicle acceleration $\ddot{y}_v$ is made available, either recorded by the accelerometer in the field test or numerically computed in a simulation study.

### 10.2.4 Relationship Between the Contact-Point and Vehicle Responses

Some relations exist between the contact-point and vehicle responses, due to the transmission of vibrations from the beam to the contact point and then to the moving vehicle, as part of the phenomenon of VBI. The first observation is that all the frequency components of the contact-point response are transferred to the vehicle response. Accordingly, the responses are amplified for the displacement from Eq. (10.10) to Eq. (10.13) and for the acceleration from Eq. (10.11) to Eq. (10.15). By letting $\omega_i$ denote each of the frequencies in the contact-point response, i.e. $2n\pi v/L$, $\omega_{bn}-n\pi v/L$ and $\omega_{bn}+n\pi v/L$, the frequency response function (FRF) of the vehicle with respect to the contact point can be expressed as

$$H\left(\omega_i\right)=\frac{\omega_v^2}{\omega_v^2-\omega_i^2}. \tag{10.20}$$

The acceleration of the vehicle can be obtained by adding up the self-oscillation and the contact-point acceleration multiplied by the FRF, as revealed by comparing Eq. (10.11) with Eq. (10.15). Namely, the vehicle acceleration can be related to the contact-point acceleration as follows:

$$\ddot{y}_v=\ddot{u}_c * H\left(\omega_i\right)+F\left(\omega_v\right), \tag{10.21}$$

where $F(\omega_v)$ is the self-oscillation of the vehicle with frequency $\omega_v$, which can be obtained from Eq. (10.15) as

$$F(\omega_v) = \sum_n \frac{\Delta_{stn} \bar{\bar{A}}_n}{2(1 - S_n^2)} \cos(\omega_v t). \tag{10.22}$$

The FRF $H(\omega_i)$ of the vehicle with respect to the contact point is plotted as black solid lines against the frequency ratio $\omega_i/\omega_v$ in Figure 10.2. Clearly, for frequency ratio $\omega_i/\omega_v < \sqrt{2}$, the absolute value of FRF is larger than 1. Thus, the bridge-related properties contained in the vehicle response are stronger than those of the contact-point response. Therefore, *for the range of $\omega_i/\omega_v < \sqrt{2}$, which is quite narrow, it is easier to retrieve the bridge dynamic properties from the vehicle response than the contact-point response.* As a matter of fact, only the lowest frequency of long-span bridges may fall in this range, but the frequencies of most short- and medium-span bridges, as well as the second and higher frequencies of long-span bridges, will fall beyond this range. The pollution effect resulting from road roughness must also be considered in the extraction.

For the special case when $\omega_i/\omega_v = 1$, the FRF tends to infinity, *resonance* will occur on the vehicle, this condition of resonance has also been noted and verified in theoretical analysis and simulation for the first frequency of the bridge in Chapter 2 or Yang et al. (2004a).

For frequency ratio $\omega_i/\omega_v > \sqrt{2}$, the absolute value of the FRF is less than 1 and decreases rapidly, eventually reaching an asymptote of zero value, with the increase of $\omega_i/\omega_v$. It can be inferred that *for the range $\omega_i/\omega_v > \sqrt{2}$, which is the most general case to be encountered in practice, the bridge dynamic properties contained in the vehicle response is much less than that in the contact-point response.* As a result, the contact-point response serves as a better measure for extracting bridge frequencies of higher orders, as will be exploited in the later part of this study.

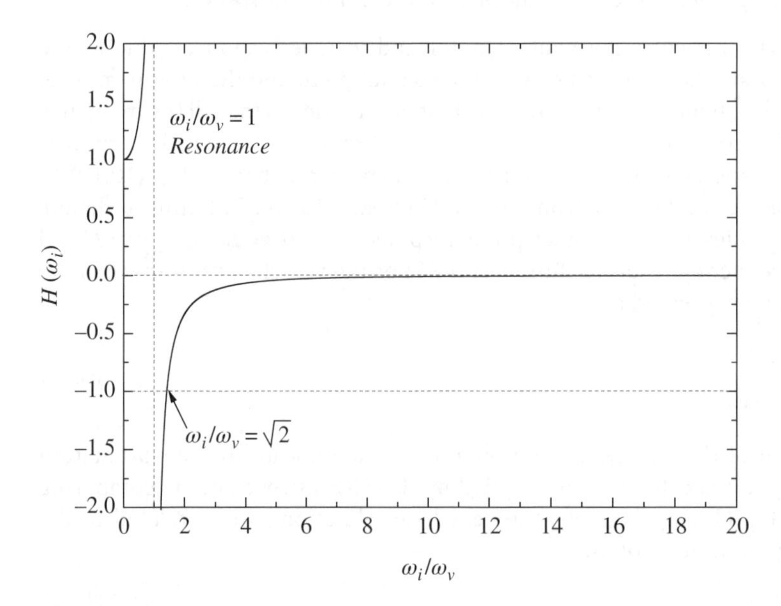


**Figure 10.2** Frequency response function (FRF) of the vehicle with respect to the contact point.

In order to show the amplitude of the bridge-related response more directly, let us compare the amplitudes of the bridge left-frequencies computed from the contact-point and vehicle responses. No consideration is made for the right-frequencies, as they are basically the same. As can be seen from Eq. (10.11), the amplitudes of the bridge left-frequencies $\omega_{bn} - n\pi v/L$ contained in the contact-point acceleration is

$$A(\ddot{u}_c) = \left| \frac{\Delta_{stn} S_n}{2(1-S_n^2)} \left( \omega_{bn} - \frac{n\pi v}{L} \right)^2 \right|. \tag{10.23}$$

Substituting Eq. (10.8) into Eq. (10.23) yields

$$A(\ddot{u}_c) = \frac{m_v g v L^2}{EI\omega_{bn} n^3 \pi^3} \left| \frac{\left( \omega_{bn} - \frac{n\pi v}{L} \right)^2}{(1-S_n^2)} \right|. \tag{10.24}$$

Noting that $n\pi v/L$ and $S_n^2$ are negligibly small and $\omega_{bn} = n^2 \omega_{b1}$, the preceding equation reduces to

$$A(\ddot{u}_c) = A_m \frac{1}{n}, \tag{10.25}$$

where $A_m$ is a constant equal to $m_v g v \omega_{b1} L^2 / EI\pi^3$.

Similarly, the amplitudes of the bridge left-frequencies can be obtained from the vehicle acceleration in Eq. (10.15) as

$$A(\ddot{y}_v) = A(\ddot{u}_c) \left| \frac{\omega_v^2}{\omega_v^2 - \left( \omega_{bn} - \frac{n\pi v}{L} \right)^2} \right|, \tag{10.26}$$

or

$$A(\ddot{y}_v) = A_m \left| \frac{1}{n(1-n^4 \mu_1^2)} \right|, \tag{10.27}$$

in terms of the ratio $\mu_1$ of the first bridge frequency to vehicle frequency, Eq. (10.14b).

Here, the amplitudes of the bridge frequencies retrieved from the contact point and vehicle responses, i.e. Eqs. (10.25) and (10.27), respectively, for three values of frequency ratios $\mu_1 = \omega_{b1} / \omega_v = 0.6, \sqrt{2}, 2$ will be compared. As can be seen from Figure 10.2, the corresponding amplitudes of the first bridge frequency extracted from the vehicle response for $\mu_1 = 0.6, \sqrt{2}, 2$ are respectively larger than, equal to, and less than those from the contact-point response. In Figure 10.3, the amplitudes of the $n$th bridge frequencies extracted from the contact-point and vehicle responses are compared for $\mu_1 = 0.6, \sqrt{2}, 2$. Aside from the fact observed for the first bridge frequency, as stated above, it is found that with the increase in the order of the bridge frequency, the amplitudes of the bridge responses extracted from the vehicle response decrease drastically and reach almost zero for the third mode. In contrast, *the amplitudes retrieved from the contact-point*

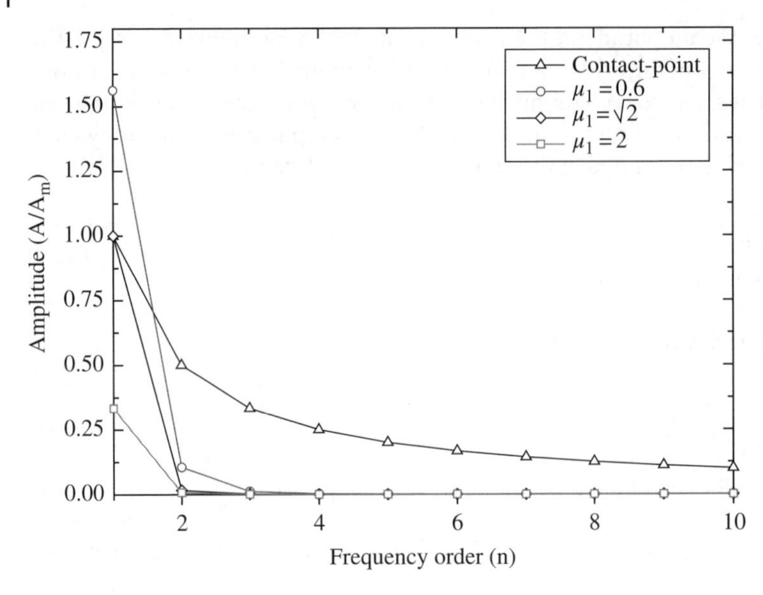


**Figure 10.3** Amplitudes of bridge frequencies extracted from contact-point and vehicle responses.

*response decrease rather slowly*. This explains the reason why only the first two frequencies of the bridge can be easily identified from the vehicle response in some previous studies. In fact, it was noted that when using the vehicle response, better resolution for bridge frequencies can be achieved if the condition $\omega_{b1}/\omega_v < 1$ is met (Yang et al. 2012a).

Another feature of the contact-point response is that *it is free of the vehicle frequency*, but only composed of the bridge related frequencies and the driving frequency, as revealed in Eq. (10.11). The problem encountered previously, e.g., in Yang et al. (2013a), in retrieval of bridge frequencies from the vehicle response due to the shadowing effect of the vehicle frequency itself on the bridge frequencies (rendering the latter too small to be identified) is just circumvented.

In short, compared with the vehicle response, the contact-point response is clear of the vehicle's self-oscillation and the transmission effect from the contact point to the vehicle, thereby enabling us to retrieve the bridge dynamic properties in an easy way. To further explore and confirm the above idea, numerical simulation will be conducted in the following using the finite element method.

## 10.3 Finite Element Simulation of VBI Problems

In the preceding section, the potential of using the contact-point response to identify the modal parameters of the bridge is theoretically outlined. To verify this, the finite element simulation will be presented in this section, by which most of the simplifications adopted in derivations of the preceding section are simply circumvented.

### 10.3.1 Brief on VBI Element

Figure 10.4 shows the VBI element for a vehicle modeled as a sprung mass passing through the bridge with rough surface, in which the vehicle is represented by a sprung

Figure 10.4 VBI element.

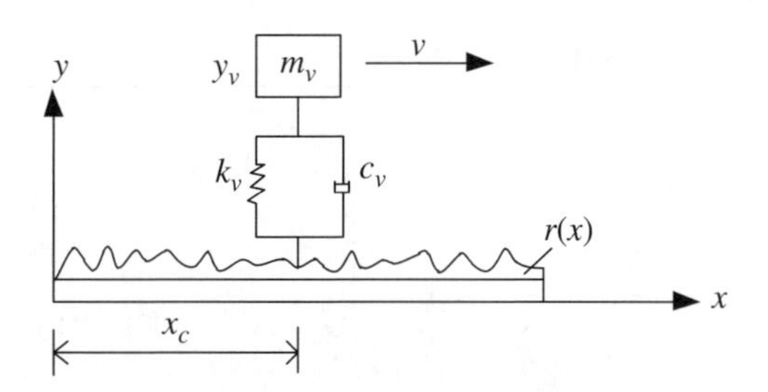


mass of magnitude $m_v$ supported by a spring of stiffness $k_v$ and a dashpot of damping coefficient $c_v$, and $x_c$ denotes the coordinate of the contact-point, $r(x)$ is the roughness profile of the bridge. The VBI element can be derived by combining the equations of motion for the sprung mass and the beam element in contact as (Appendix or Chang et al. 2010):

$$\begin{bmatrix} m_v & 0 \\ 0 & [m_b] \end{bmatrix} \begin{Bmatrix} \ddot{y}_v \\ \{\ddot{q}_b\} \end{Bmatrix} + \begin{bmatrix} c_v & -c_v\{N\}_c^T \\ -c_v\{N\}_c & [c_b]+c_v\{N\}_c\{N\}_c^T \end{bmatrix} \begin{Bmatrix} \dot{y}_v \\ \{\dot{q}_b\} \end{Bmatrix}$$
$$+\begin{bmatrix} k_v & -c_v v\{N'\}_c^T - k_v\{N\}_c^T \\ -k_v\{N\}_c & [k_b]+c_v v\{N\}_c\{N'\}_c^T + k_v\{N\}_c\{N\}_c^T \end{bmatrix} \begin{Bmatrix} y_v \\ \{q_b\} \end{Bmatrix} \tag{10.28}$$
$$= \begin{Bmatrix} c_v v r_c' + k_v r_c \\ -c_v v r_c'\{N\}_c - k_v r_c\{N\}_c - m_v g\{N\}_c \end{Bmatrix}.$$

where $[m_b]$, $[c_b]$, and $[k_b]$ denote the mass, damping, and stiffness matrices, respectively; $\{q_b\}$ is the displacement vector of the beam element; $\{N\}_c$ is the interpolation functions evaluated at the contact point $x_c$, and $r_c$ the roughness value at the contact point $x_c$.

By assembling the VBI element (with sprung mass) and the conventional beam elements for the parts of the bridge not directly acted on by the sprung mass, one can obtain the global finite element equation for the VBI system at time $t$. Such an equation can be augmented to include the step-by-step marching feature for time-history analysis, say, using the Newmark-$\beta$ method (with $\beta = 0.25$ and $\gamma = 0.5$ for unconditional stability) via updating of the acting position of the sprung mass moving at speed $v$ (Yang and Yau 1997). The response of the contact point calculated in the time–history analysis will be used to identify the modal parameters of the bridge, and compared with those based on the vehicle response.

### 10.3.2 Verification of the Theoretical Solution

As for now, the responses of the moving vehicle and contact-point have been made available in closed form, and the VBI element and procedure for solving the time–history response of the VBI problem have been briefed. The accuracy of the analytical solution and the feasibility of using the contact-point response for identifying the modal properties of the bridge will be numerically investigated in this section.

The following data are adopted for the simply supported bridge: length $L = 25$ m, mass per unit length $m = 2000$ kg/m, elastic modulus $E = 27.5$ GPa, and moment of inertia

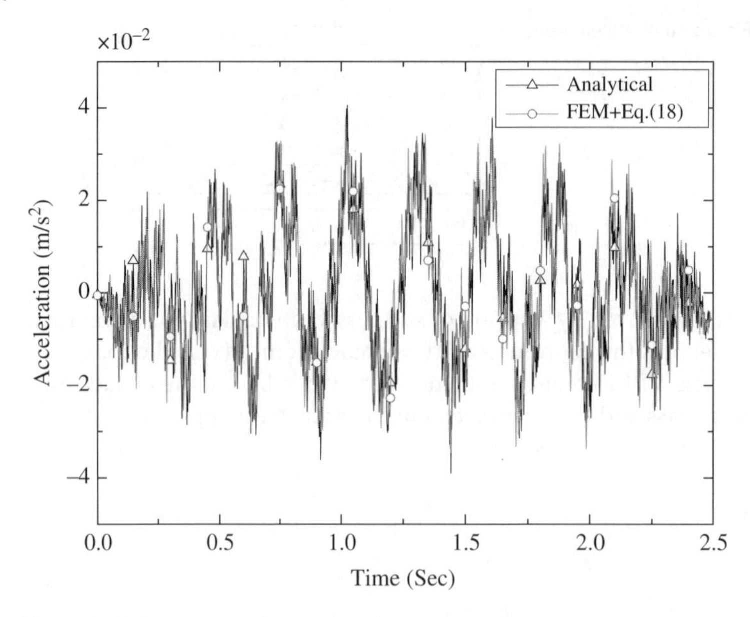


**Figure 10.5** Comparison of contact-point responses.

$I = 0.15\,\text{m}^4$, smooth surface $r(x) = 0\,\text{m}$, and the following for the vehicle, $m_v = 1000\,\text{kg}$, $c_v = 0\ \text{kN m/s}$, and $k_v = 200\,\text{kN/m}$. Accordingly, the vehicle frequency calculated is $\omega_v = 2.25\,\text{Hz}$ and the first bridge frequency is $\omega_{b1} = 3.61\,\text{Hz}$, implying that the condition $\omega_i/\omega_v > \sqrt{2}$ is met. In the finite element analysis, 30 beam elements are used for the bridge.

For the case with a vehicle speed of $v = 10\,\text{m/s}$, the acceleration of the contact-point *derived from the vehicle response* in Eq. (10.18) and the closed-form solution in Eq. (10.11) have been plotted in Figure 10.5. Clearly, the contact-point response calculated from the vehicle response, i.e. from the "measured response" of the instrumented vehicle in Eq. (10.18) is generally *reliable* for further applications.

In addition, the acceleration of the vehicle obtained by the finite element method and the analytical solution in Eq. (10.15) have been compared in Figure 10.6. As revealed by Figures 10.5 and 10.6, the contact-point and vehicle accelerations obtained by the two methods are in good agreement each with the analytical solution. Further, a comparison of Figure 10.5 with Figure 10.6 indicates that the contact-point acceleration is of *higher* amplitude than the vehicle acceleration, but roughly of the same order. *The contact-point acceleration oscillates more drastically* than the vehicle acceleration and contains more high frequency components, which are mainly dominated by bridge frequencies of higher orders, as will be investigated later on.

## 10.4 Retrieval of Bridge Frequencies

As was shown in the previous sections, the contact-point response is a better measure than the vehicle response for retrieving the bridge frequencies, especially those of higher orders. In this section, such a point will be numerically evaluated. For our

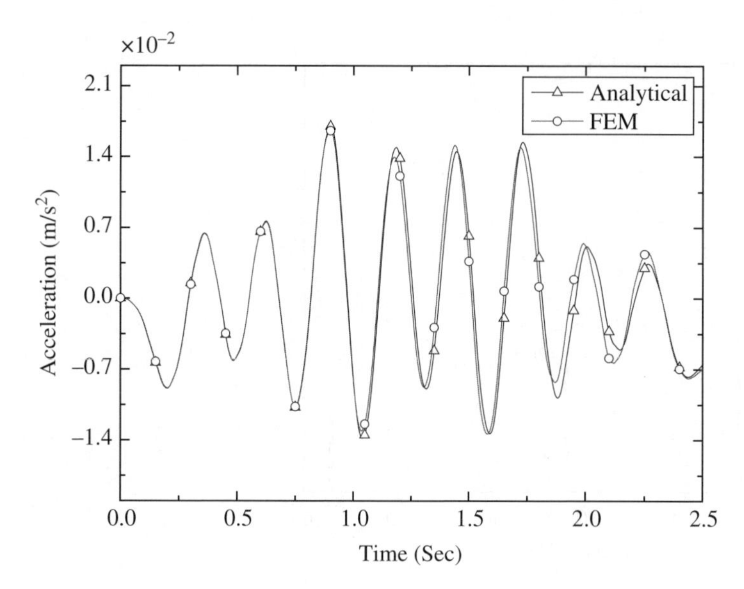


**Figure 10.6** Comparison of vehicle responses.

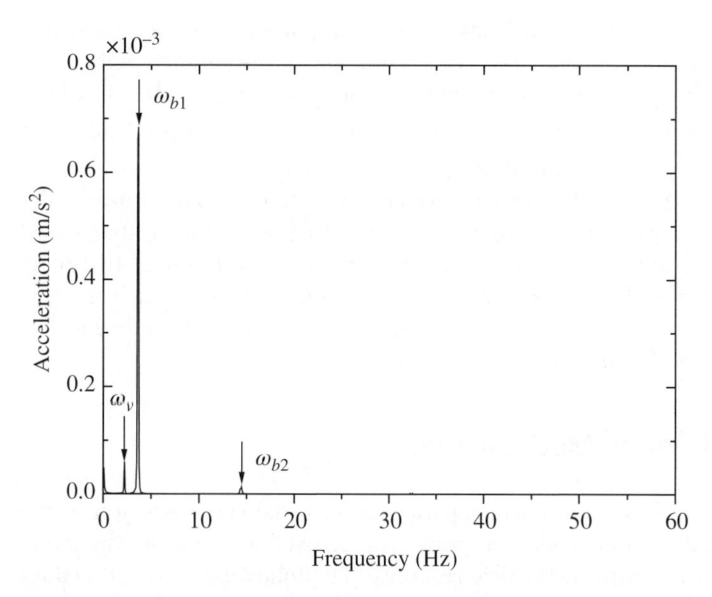


**Figure 10.7** Acceleration spectrum of the vehicle (sprung mass).

purposes, the bridge and vehicle models adopted are the same as in the previous example. The vehicle speed is set at $v = 2\,\text{m/s}$ for better resolution in the spectral results.

The FFT spectra of the vehicle and contact-point accelerations have been plotted in Figures 10.7 and 10.8, respectively. As can be seen, aside from the vehicle frequency, only two bridge frequencies can be identified from the vehicle spectrum, while the second frequency is merely visible. In contrast, the contact-point spectrum contains only bridge frequencies, and four bridge frequencies can be clearly identified. Besides,

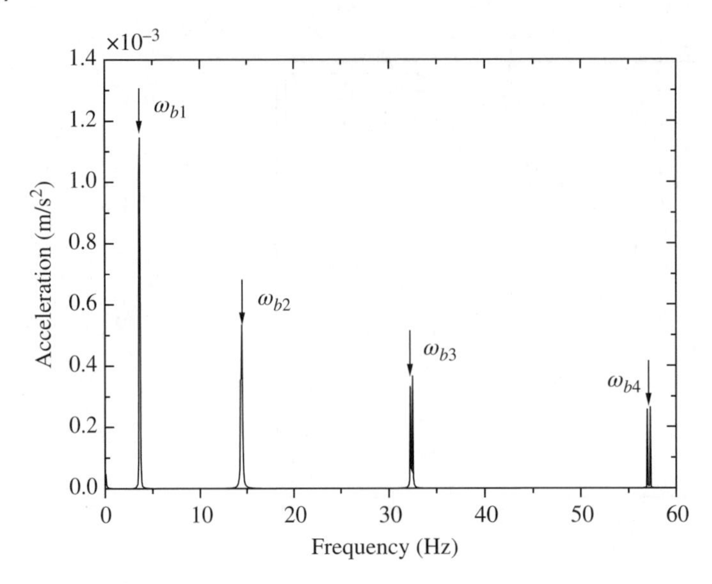


**Figure 10.8** Acceleration spectrum of the contact point.

in the vehicle spectrum (Figure 10.7), the bridge spectral amplitude decreases drastically against the increase in frequency, while the same is not true in the contact-point spectrum (Figure 10.8). Such a phenomenon can be explained using Figure 10.2, in which the FRF value of the vehicle decreases drastically for $\omega_i > \sqrt{2}\omega_v$, noting that with $\omega_v = 2.25$ Hz and $\omega_{b1} = 3.61$ Hz, the condition $\omega_i > \sqrt{2}\omega_v$ is met.

By comparing Figure 10.7 with Figure 10.8, we also note that the amplitude of the first bridge frequency in the vehicle response is less than that in the contact-point response. This phenomenon can be explained by Figure 10.3 for $\mu_1$ equal to 1.6 (as implied by $\omega_{b1} = 3.61$ Hz and $\omega_v = 2.25$ Hz). From the above discussion, it is clear that *the contact-point acceleration is a better measure for identifying the bridge frequencies, especially those of higher orders.*

## 10.5 Retrieval of Bridge Mode Shapes

The mode shapes of a bridge are important parameters for damage detection of the bridge. In this chapter, the contact-point response will be used to construct the mode shape of a bridge compared with the vehicle response. The following is the procedure presented in Chapter 9 or Yang et al. (2014) for constructing the mode shapes of a bridge using the acceleration data recorded or computed of the vehicle during its passage over the bridge:

1) Identify the frequencies of the bridge from the vehicle response using the FFT.
2) Distill the component response associated with the frequency of the bridge of interest from the vehicle response by feasible signal processing schemes, such as band-pass filters.
3) Obtain the instantaneous amplitude history of the bridge component response for the particular mode of concern using the Hilbert transform.

4) Recover the mode shape of the bridge from the instantaneous amplitude history by engineering judgment such that opposite signs exist at the two sides of a discontinuity at a common node are adjusted, along with the mode shape normalized.

Our focus in this chapter is to use the contact-point response, rather than the vehicle response, for constructing the mode shapes.

### 10.5.1 Effect of Moving Speed

In Chapter 9 and Yang et al. (2014), it was shown that the mode shapes can be constructed accurately from the vehicle response with slow moving speeds. The bridge and vehicle models used remain identical to those in Section 10.3.2. To study the effect of moving speed on the construction of mode shapes using the contact-point response, three moving speeds are considered, 2, 6, and 10 m/s (or 7.2, 21.6, 36 km/h). For illustration, let us consider the third mode shape of the bridge constructed by using the vehicle and contact-point responses in Figures 10.9 and 10.10, respectively. As revealed by Figure 10.9, for the very low speed of 2 m/s, the mode shape retrieved from the vehicle response agrees well with the theoretical one. But for the other higher speeds, serious distortion occurs with the mode shapes. In contrast, the third mode shape retrieved from the contact-point response in Figure 10.10 appears to be good for all the speeds considered.

To verify the range of applicability of the contact-point response for retrieving the bridge mode shapes, the vehicle speed has been increased up to 20 m/s (or 72 km/h), and the result was plotted in Figure 10.11. Clearly, the mode shape recovered agrees well with the theoretical one, even for the high speed of 20 m/s. The above analysis indicates that *the contact-point response is a better measure than the vehicle response for retrieving the bridge mode shapes for the range of vehicle speeds considered*, while the vehicle response may be used only for very low speeds.

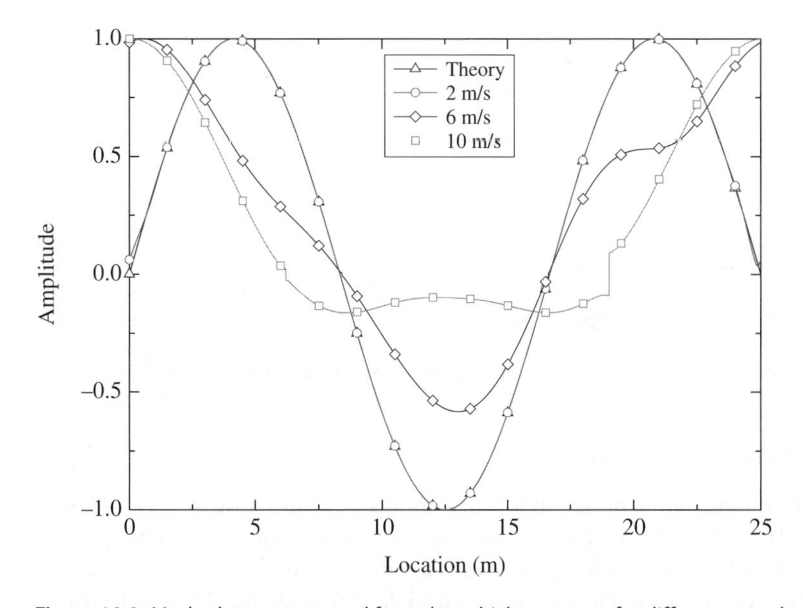


**Figure 10.9** Mode shapes recovered from the vehicle response for different speeds.

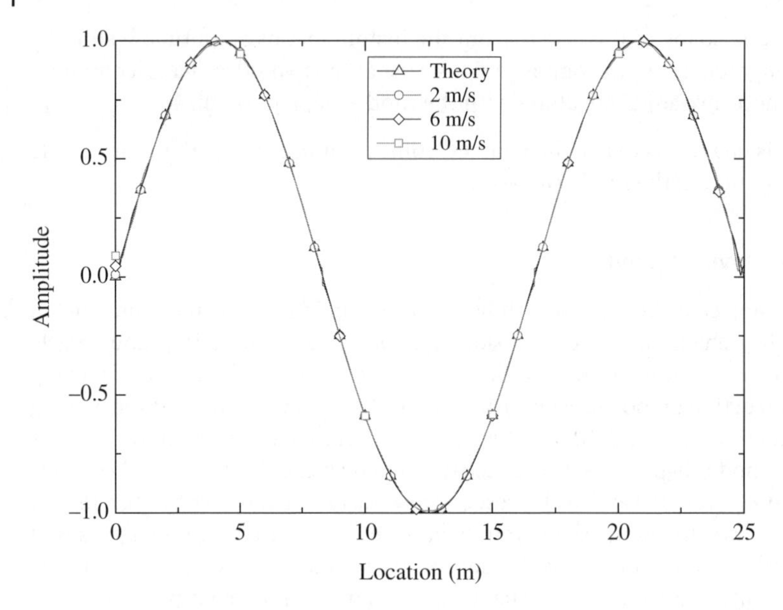


**Figure 10.10** Mode shapes recovered from the contact-point response for different speeds.

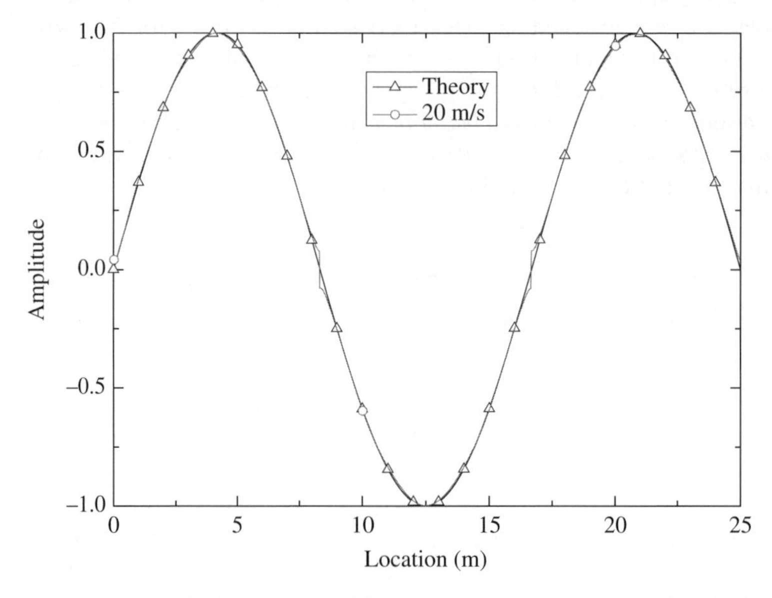


**Figure 10.11** Mode shape recovered from the contact-point response for vehicle speed of 20 m/s.

### 10.5.2 Effect of Vehicle Frequency

In practice, the frequencies of a bridge to be measured, and whether the bridge frequencies are close to the frequency of the test vehicle, are not known beforehand. To this end, how the vehicle frequency affects the construction of bridge mode shapes will be investigated. Three vehicle frequencies are considered: 2.25, 2.76, and 3.18 Hz, to which the corresponding stiffness of the vehicle are 200, 300, and 400 kN/m. The other parameters of the vehicle and the bridge are the same as those used in Section 10.4.

The first mode shape retrieved from the vehicle and contact-point responses are plotted in Figures 10.12 and 10.13, respectively. As revealed by Figure 10.12, when the vehicle frequency gets closer to the bridge frequency (3.61 Hz), the mode shape retrieved from the vehicle response gets worse. Since the contact-point response does

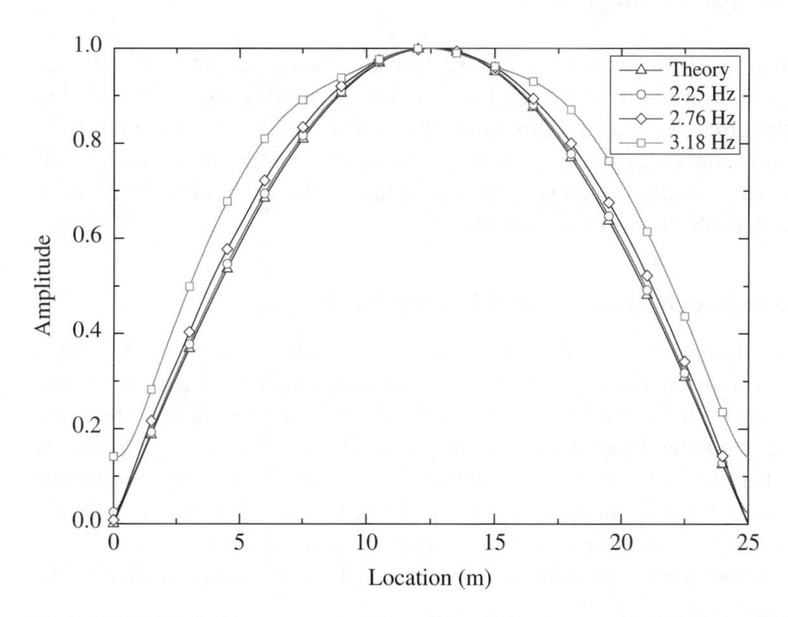


**Figure 10.12** Mode shapes recovered from the vehicle response for different vehicle frequencies.

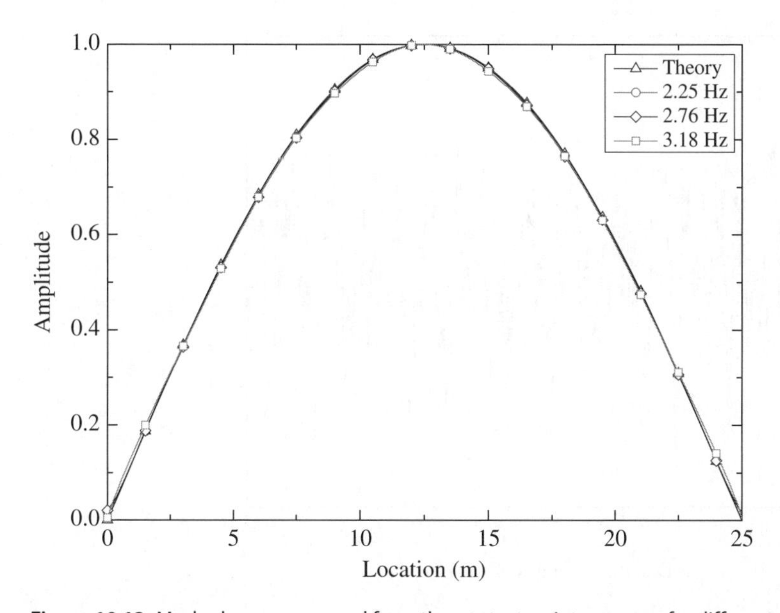


**Figure 10.13** Mode shapes recovered from the contact-point response for different vehicle frequencies.

not contain the vehicle frequency, it is natural that *the mode shape retrieved from the contact-point response is not affected by the vehicle frequency*, as indicated in Figure 10.13. This is another advantage of using the contact-point response for identification of bridge mode shapes.

## 10.6 Effect of Road Roughness

For a bridge with smooth surface, it was shown that the contact-point response outperforms the vehicle response in retrieving frequencies and mode shapes of the bridge. In this section, the effect of road roughness on the performance of the contact-point response will be investigated. For the present purposes, the surface roughness profile of the bridge generated using the power spectral density (PSD) function of ISO 8608 (1995) with Class A is plotted in Figure 10.14.

### 10.6.1 Bridge with Rough Surface Free of Existing Traffic

To simulate the vibration behavior of the vehicle passing over a bridge with rough surface, the VBI element given in Eq. (10.28) is used. The acceleration spectra of the vehicle and contact-point are plotted in Figure 10.15, in which the black dashed line denotes the vehicle response and the red line the contact-point response. As can be seen, only the vehicle frequency and the first bridge frequency can be identified from the vehicle response, similar to the result identified from the field experiment in Chapter 4 and Lin and Yang (2005). The first bridge frequency is distinguished in the contact-point spectrum, which has an amplitude much higher than that identified from the vehicle response, similar to the

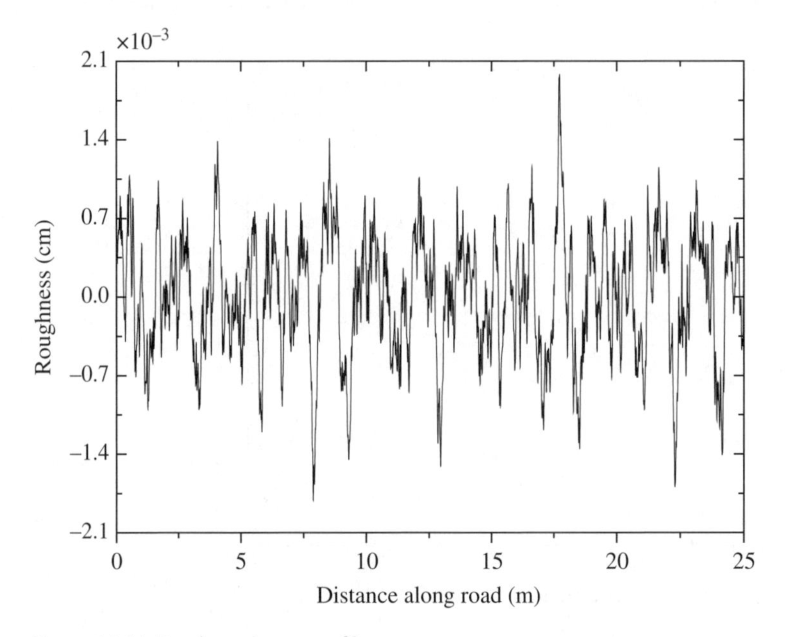


**Figure 10.14** Road roughness profile.

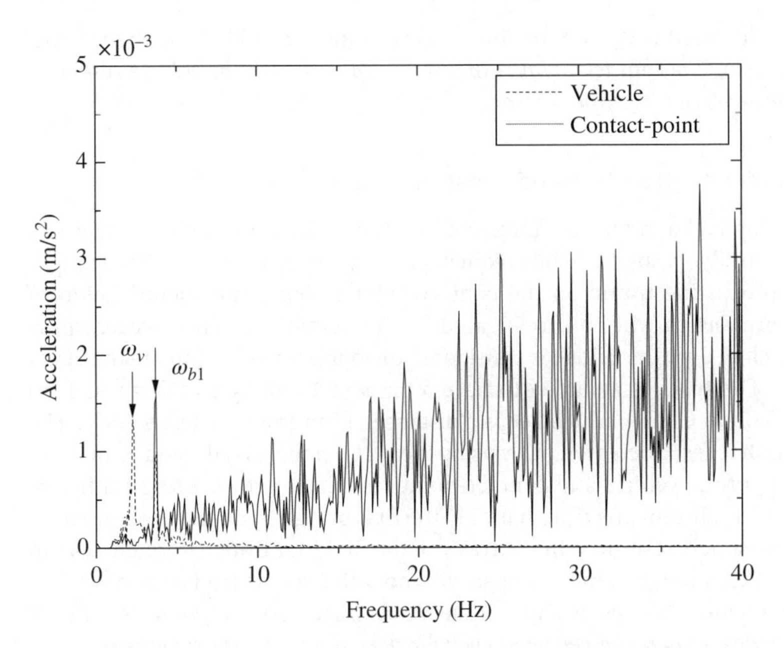


**Figure 10.15** Acceleration spectra of vehicle and contact-point for the bridge with roughness.

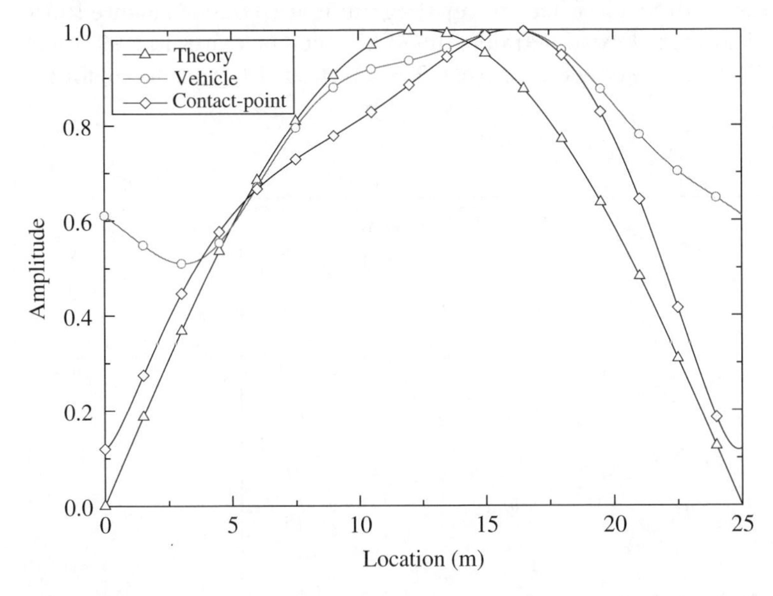


**Figure 10.16** Mode shapes recovered from the contact-point and vehicle responses for the bridge with roughness.

observation made in Section 10.4. In addition, the wide range of high-amplitude, high-frequency response for the contact point is induced by road roughness.

The first mode shapes of the bridge constructed from the contact-point and vehicle responses are shown in Figure 10.16, for which the MAC (modal assurance criterion)

(See Chapter 9) calculated is 0.9766 for the contact point and 0.9121 for the vehicle. *The advantage of using the contact-point response, compared with the vehicle response, in retrieving the bridge's mode shape is clear.*

### 10.6.2 Bridge with Rough Surface under Existing Traffic

It was shown in Chapter 4 and Lin and Yang (2005) that existing traffic on the bridge can help amplifying the bridge response, while reducing the disturbing effect of surface roughness. The effect of existing traffic on the contact-point response for identification of bridge dynamic properties will be investigated in this section. For the present purposes, a truck traveling over the bridge is considered to approximately simulate the effect of existing traffic. The truck is assumed to have a mass of 15 000 kg and stiffness 1350 kN/m (Model NG80B in China), which keeps a distance of 5 m from the test vehicle. The same procedure as that described previously using the VBI elements is adopted in analysis.

The FFT spectra of the vehicle and contact-point responses with existing traffic are plotted in Figure 10.17. It remains true that only the first bridge frequency can be easily identified from the vehicle response. In contrast, all the first four bridge frequencies can be identified from the contact-point response with no difficulty, which is free of vehicle frequency. Theoretically, this result shows that *the contact-point response is a better measure for the bridge frequencies retrieval even for the case with road roughness.*

The first mode shapes of the bridge constructed from the contact-point and vehicle responses for the case with rough surface and existing traffic are plotted in Figure 10.18. Due to the beneficial effect of existing traffic, the MAC value increases to 0.9971 and 0.9947 (Figure 10.18) from the previous case of 0.9766 and 0.9121 (Figure 10.16) for the

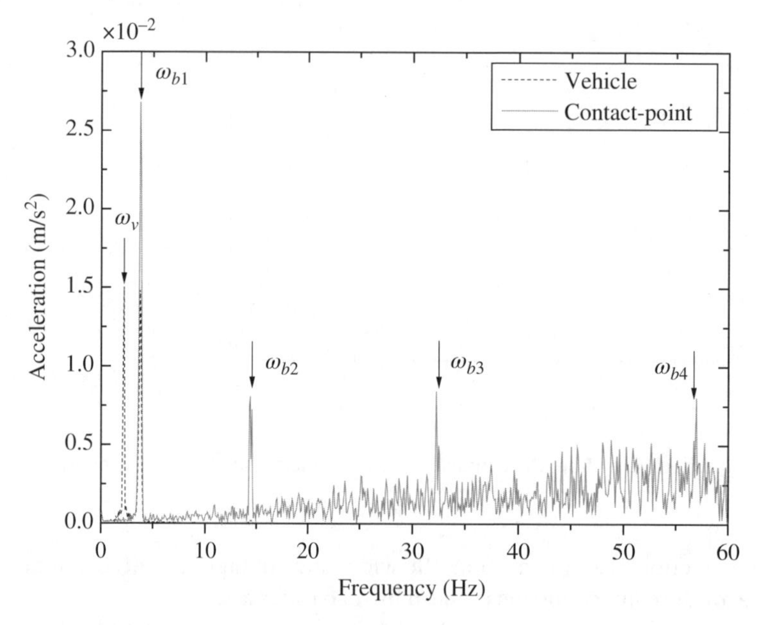


**Figure 10.17** Acceleration spectra of vehicle and contact-point for the bridge with roughness under existing traffic.

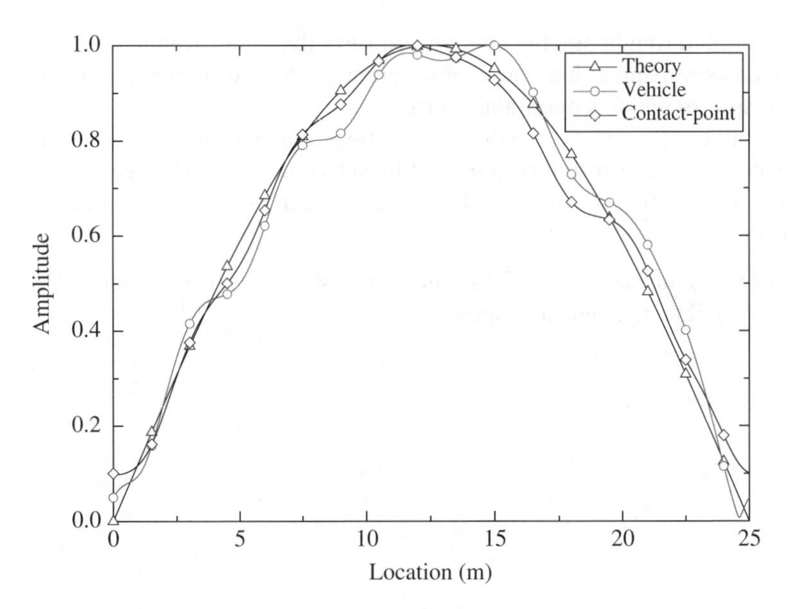


**Figure 10.18** Mode shapes recovered from the contact-point and vehicle responses for the bridge with roughness under existing traffic.

contact-point and vehicle responses, respectively. Therefore, it has been demonstrated that *a better result can be identified for the bridge modal properties using the contact-point than the vehicle response. Overall, the existing traffic is beneficial to application of such a technique.*

## 10.7 Concluding Remarks

In this theoretical study, the response of the *contact point* between the vehicle and bridge, rather than the vehicle itself, is firstly proposed to identify the modal properties of the bridge. As indicated by the approximate closed-form solutions and finite element simulations, the contact-point acceleration is featured by the fact that it is *free of* the vehicle frequency, a disturbing effect in retrieval of bridge frequencies for rough road surface. To facilitate the field application, a procedure was presented for computing the contact-point acceleration from the measured acceleration of the moving vehicle. From the FRF of the vehicle with respect to the contact point, it was found that only for the case with $\omega_i/\omega_v < \sqrt{2}$ and no road roughness, the vehicle response works well for retrieving the bridge frequencies. For the more general case with $\omega_i/\omega_v > \sqrt{2}$, the contact-point response outperforms the vehicle response in retrieving the bridge frequencies, in that more frequencies can be identified.

Based on the structural data adopted in numerical simulations, the following conclusions were drawn:

1) For the case with smooth road surface, more bridge frequencies can be retrieved from the contact-point than the vehicle response. In recovering the bridge mode shapes, the contact-point response is by nature not affected by the vehicle speed and frequency, unlike the vehicle response.

2) In the presence of road roughness, the ability to retrieve the bridge frequencies and mode shape decreases using either the contact-point or vehicle response, but the contact-point response remains marginally better.
3) When coupled with existing traffic, the disturbing effect of road roughness is greatly alleviated. The contact-point response can be utilized to retrieve bridge frequencies of the first few modes, while only one bridge frequency can be identified from the vehicle response.

In addition, better quality is achieved for the mode shapes retrieved using the contact-point response than the vehicle response.

# 11

# Damage Detection of Bridges Using the Contact-Point Response

To further the technique of vehicle scanning method, the contact-point response of a moving test vehicle is adopted for the damage detection of bridges. First, the contact-point response of the vehicle moving over the bridge is derived both analytically and in central difference form (for field use). Then, the instantaneous amplitude squared (IAS) of the driving component of the contact-point response is calculated by the Hilbert transform (HT), making use of its narrow-band feature. The IAS peaks serve as the key parameter for damage detection. In the numerical simulation, a damage (crack) is modeled by a hinge-spring unit. The feasibility of the proposed method to detect the location and severity of a damage or multi damages of the bridge is verified. Also, the effects of surface roughness, vehicle speed, measurement noise and random traffic are studied. In the presence of ongoing traffic, the damages of the bridge are identified from the repeated or invariant IAS peaks generated for different traffic flows by the same test vehicle over the bridge. The materials presented in this chapter are based primarily on the paper by Zhang et al. (2018).

## 11.1 Introduction

As parts of the land transportation networks, the bridge has been playing an irreplaceable role in ensuring the free and safe passage of passengers and cargoes. Regardless of its vital role in lifelines, a bridge may suffer from varying degrees of damages due to degradation of stiffness in structural members, connections, supports, or material strength, caused by overloaded vehicles, weathering, or natural disasters, such as earthquakes and typhoons, etc. Obviously, there is a strong demand to develop efficient and mobile techniques for the damage detection of bridges so as to enhance the quality of routine management and maintenance.

To monitor the operational and/or damage conditions of the bridges, vibration-based methods have been adopted for half a century or longer, see the review works by Salawu (1997), Doebling et al. (1998), Carden and Fanning (2004), and Fan and Qiao (2011). In general, these methods require the installation of quite a number of sensors on the bridge to detect the dynamic properties of bridge, such as frequencies, mode shapes, and damping coefficients. They were referred to as the *direct approach* in that the vibration responses *directly* measured from the bridge were used to retrieve the physical

*Vehicle Scanning Method for Bridges*, First Edition. Yeong-Bin Yang, Judy P. Yang, Bin Zhang and Yuntian Wu.

properties of concern. An enormous volume of researches has been carried out along these lines using the ambient vibration, traffic vibration, forced vibration, impact vibration, etc. (Mclamore et al. 1971; Ward 1984; Mazurek and DeWolf 1990; Casas 1995; Soyoz and Feng 2009; Kim et al. 2016; Ni and Zhang 2016). One drawback with the direct approach is that it usually requires numerous sensors to be installed on the bridge along with a data acquisition system, for which the deployment and maintenance cost is generally high. Another drawback is that the vast amount of data generated, the so-called *sea-like data*, may not be effectively used. It should be added that the monitoring system tailored for one bridge can hardly be transferred to another bridge to do the same work of monitoring, a problem known as the *lack of mobility*.

The vehicle scanning method for bridge measurement was proposed by Yang et al. (2004a) mainly to circumvent the drawbacks of the direct approach. This method was known in the early days of development as the *indirect approach* to distinguish it from the direct approach conventionally used. With this technique, the vibration data collected by one or few sensors installed on the moving test vehicle were used to retrieve the physical parameters of the supporting bridge, as schematically shown in Figure 11.1. No sensors are needed on the bridge. Compared with the direct approach, the vehicle scanning method shows great potential in economy, mobility, and efficiency, although further research in software and hardware is required to enhance its robustness in field applications.

The idea of the vehicle-based measurement proposed by Yang et al. (2004a) was validated in the field tests (Lin and Yang 2005). Since then, numerous researches have been inspired and conducted along these lines, including the theoretical studies (Yang and Lee 2017), damage assessment (Bu et al. 2006; Kim and Kawatani 2008; Yin and Tang 2011), techniques for improving recognition results (Nasrellah and Manohar 2010), experiments (McGetrick et al. 2015), mode shape construction (Yang et al. 2014; Malekjafarian and OBrien 2014), damping identification (Gonzalez et al. 2012; Keenahan et al. 2014), damage detection of the bridges (Zhang et al. 2012; OBrien 2017a), and reviews of relevant works (Malekjafarian et al. 2015; Yang and Yang 2018).

From the above review, it is known that the vehicle (body) response has been used frequently for identifying the modal parameters and damages of the bridge. This is mainly due to the fact that the vehicle response can be easily recorded by sensors mounted on the test vehicle. In this chapter, however, the response of the vehicle's contact point with the bridge will be used instead, which is shown to be a better measure for detecting the damages of the bridge, as it is born to be free of the vehicle frequency. For illustration of the idea involved, only simply supported beams will be considered herein.

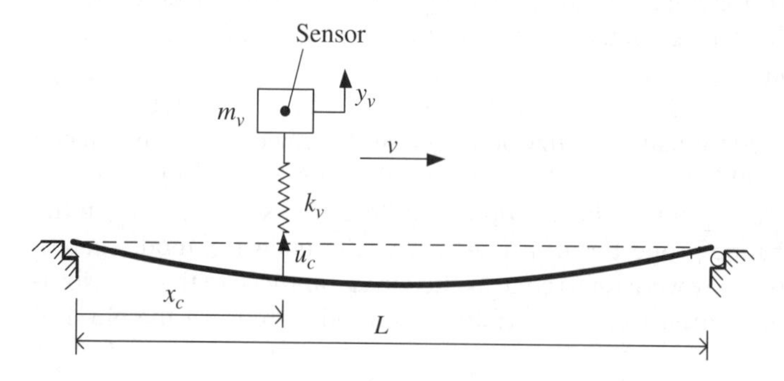


**Figure 11.1** Schematic of the vehicle-bridge interaction (VBI) model.

The contents of the chapter are outlined as follows. First, the moving vehicle's contact-point response (acceleration) is derived both analytically and in central-difference form for field use. Central to the damage detection is the calculation of the instantaneous amplitude squared (IAS) for the low-frequency driving component of the contact-point response by the Hilbert transform (HT), making use of its narrow-band feature. In the finite element simulation, a damage (crack) is modeled by a hinge-spring unit. Using the IAS peaks, it is demonstrated that the location and severity of a damage or multi damages of the bridge can be clearly identified. Also, a parametric study is conducted for the effects of the key factors, such as surface roughness, vehicle speed, measurement noise and random traffic, on damage detection. In the presence of ongoing traffic, the damages of the bridge are identified from the repeated IAS peaks generated for different traffic flows by the same test vehicle over the bridge.

## 11.2 Dynamic Response of the Vehicle-Bridge System

As shown in Figure 11.1, an undamped *single-axle* test vehicle is modeled as a lumped mass $m_v$ supported by a spring of stiffness $k_v$, passing through a simple beam of length $L$. Such a *single-degree-of-freedom* (SDOF) model was the one used in the theoretical and field studies (Yang et al. 2004a; Lin and Yang 2005), which should not be regarded as a half or quarter vehicle, but as the trailer of a tractor-trailer system. All the dynamic properties of the test vehicle, including the frequency, are assumed to have been made available by an ambient or impact test in practice. The equation of motion for the vehicle is

$$m_v \ddot{y}_v + k_v \left( y_v - u_c \right) = 0, \tag{11.1}$$

where $y_v$ = the vertical displacement of the vehicle measured from the static equilibrium position, $u_c$ = the displacement of the vehicle's contact point on the beam, or of the beam under the vehicle's wheels, and $(\dot{\square}) = \mathrm{d}(\ )/\mathrm{d}t$. In practice, the contact-point response can also be interpreted as the bridge response under the action of the vehicle's wheels. The equation of motion for the beam is

$$\bar{m}\ddot{u} + EIu'''' = f_c(t)\delta(x - vt), \tag{11.2}$$

where $\bar{m}$ = mass per unit length, $E$ = elastic modulus, $I$ = moment of inertia, $u$ = displacement of the beam, $(\ )' = \mathrm{d}(\ )/\mathrm{d}x$, and $\delta$ = delta function.

For simple beams, the displacement $u$ can be expressed as the sum of each ($j$th) modal contribution via the modal shape $\phi_j(x)$ and modal coordinate $q_{bj}(t)$ as follows (Biggs 1964):

$$u(x,t) = \sum_{j=1}^{n} \left[ q_{bj}(t) \sin \frac{j\pi x}{L} \right]. \tag{11.3}$$

For the case of constant speed $v$, the response $u_c$ of the contact point at $x_c$ (=$vt$) is $u_c = u(x, t)|_{x = vt}$. The contact force $f_c(t)$ between the beam and the vehicle is the sum of the elastic (suspension) force and gravity of the vehicle,

$$f_c(t) = k_v \left( y_v - u_c \right) - m_v g, \tag{11.4}$$

where $g$ = acceleration of gravity.

### 11.2.1 Contact-Point Response: Analytical Solution

To solve for the vehicle response, the response of the beam should be obtained first. Substituting Eqs. (11.3) and (11.4) into Eq. (11.2), multiplying both sides of the equation by $\sin(j\pi x/L)$ and integrating with respect to $x$ over the length $L$ of the beam, one obtains

$$\ddot{q}_{bj}(t)+\omega_{bj}^2 q_{bj}(t)-\frac{2m_v\omega_v^2}{\bar{m}L}\left(y_v-\sum_{k=1}^{n} q_{bk}(t)\sin\frac{k\pi vt}{L}\right)\sin\frac{j\pi vt}{L}$$

$$=\frac{-2m_v g}{\bar{m}L}\sin\frac{j\pi vt}{L} \tag{11.5}$$

where the vehicle frequency $\omega_v=(k_v/m_v)^{1/2}$, and the $j$th frequency of the beam $\omega_{bj}=(j^2\pi^2/L^2)(EI/\bar{m})^{1/2}$. By neglecting the third term on the LHS of Eq. (11.5), since $m_v \ll \bar{m}L$, and by adopting zero initial conditions, the modal coordinates $q_{bj}(t)$ can be solved. Further, by substituting the modal coordinates $q_{bj}(t)$ into Eq. (11.3), along with $u_c = u(x,t)|_{x=vt}$, one obtains the contact-point displacement $u_c$ as

$$u_c=\sum_{j=1}^{n}\frac{\Delta_{stj}}{2\left(1-S_j^2\right)}\left[1-\cos\frac{2j\pi vt}{L}-S_j\cos\left(\omega_{bj}-\frac{j\pi v}{L}\right)t\right.$$

$$\left.+S_j\cos\left(\omega_{bj}+\frac{j\pi v}{L}\right)t\right]. \tag{11.6}$$

where $\Delta_{stj}$ = $j$th modal static deflection caused by the vehicle and $S_j$ = nondimensional speed parameter:

$$\Delta_{stj}=-\frac{2m_v gL^3}{EIj^4\pi^4},\qquad S_j=\frac{j\pi v}{L\omega_{bj}}. \tag{11.7a,b}$$

Differentiating Eq. (11.6) twice yields the contact-point acceleration as

$$\ddot{u}_c=\sum_{j=1}^{n}\frac{\Delta_{stj}}{2\left(1-S_j^2\right)}\left[\left(\frac{2j\pi v}{L}\right)^2\cos\frac{2j\pi vt}{L}+S_j\left(\omega_{bj}-\frac{j\pi v}{L}\right)^2\cos\left(\omega_{bj}-\frac{j\pi v}{L}\right)t\right.$$

$$\left.-S_j\left(\omega_{bj}+\frac{j\pi v}{L}\right)^2\cos\left(\omega_{bj}+\frac{j\pi v}{L}\right)t\right]. \tag{11.8}$$

Evidently, the contact-point acceleration $\ddot{u}_c$ consists of two parts. One is the *driving component* with frequencies $2j\pi v/L$ due to the vehicle's motion, and the other is the *bridge component* with frequencies $\omega_{bj}\pm j\pi v/L$, where the sign $\pm$ is interpreted as the Doppler's effect. Unlike most previous studies focused on the bridge component, attention will be paid to the driving component in this chapter.

### 11.2.2 Contact-Point Response: For Use in Field Test

In practice, the vehicle acceleration can be easily measured by accelerometers mounted on the vehicle, while the contact-point acceleration is relatively difficult to measure due to its moving nature. In this chapter, the contact-point acceleration will be computed indirectly from the vehicle acceleration. Namely, by differentiating Eq. (11.1) twice with respect to time and by replacing the term $\mathrm{d}^2\ddot{y}_v/\mathrm{d}t^2$ by its central difference, which is known to be reliable for expressing the discrete measured data $\ddot{y}_v$ in the test environment, one can compute the contact-point acceleration as

$$\ddot{u}_c = \ddot{y}_v + \frac{\left(\ddot{y}_v\big|_{s+1} - 2\ddot{y}_v\big|_s + \ddot{y}_v\big|_{s-1}\right)}{\omega_v^2\left(\Delta t\right)^2}. \tag{11.9}$$

where $s$ = $s$th sampling point and $\Delta t$ = sampling interval. Clearly, given the vehicle's frequency and recorded acceleration history during its passage over the bridge, the contact-point acceleration can be computed accordingly.

## 11.3 Algorithm for Damage Detection

### 11.3.1 Hilbert Transformation

The HT has been used to unveil the instantaneous attributes of a time series. The contact-point response generated by a moving test vehicle will be processed by the HT to detect the damage(s) of a bridge. The HT of a signal $s(t)$ is defined as its convolution with a unit impulse function of $1/\pi t$ (Hahn 1966):

$$h(t) = H\left[s(t)\right] = \frac{1}{\pi}\int_{-\infty}^{\infty}\frac{s(\tau)}{t-\tau}d\tau. \tag{11.10}$$

which enables the local properties to be highlighted. It can also be interpreted as a new signal with a phase delay of $\pi/2$ from the original one, but no change in signal amplitude. The two series $s(t)$ and $h(t)$ form an analytical function $z(t)$ as

$$z(t) = s(t) + ih(t) = A(t)\mathrm{e}^{i\theta(t)}, \tag{11.11}$$

where $A(t)$ = instantaneous amplitude and $\theta(t)$ = instantaneous phase, defined as

$$A(t) = \pm\sqrt{\left[s(t)\right]^2 + \left[h(t)\right]^2}, \quad \theta(t) = \arctan\left[\frac{h(t)}{s(t)}\right]. \tag{11.12a,b}$$

The amplitude $A(t)$ can be regarded as the envelope of the original signal, changing with time. The two parameters $A(t)$ and $\theta(t)$ are physically meaningful only in certain circumstances, namely, when the time series $s(t)$ is of a mono component or narrowband (Huang et al. 1998, 1999b). In the next section, the HT will be manipulated to yield messages that are useful for damage detection of bridges.

### 11.3.2 Strategy for Damage Detection

The driving component of the contact-point response can be obtained from Eq. (11.8):

$$R_d(t) = \sum_{j=1}^{n} A_j \cos\frac{2j\pi vt}{L}, \tag{11.13}$$

where the coefficient $A_j$ is

$$A_j = \frac{2\Delta_{stj}\left(j\pi v\right)^2}{L^2\left(1-S_j^2\right)}. \tag{11.14}$$

For a bridge to be measured, the coefficient $A_j$ represents a set of constants that do not vary with time. The frequency $v/L$ is the only frequency parameter in the driving component of the contact-point response, because all the others are its integral multiples. The driving component is one of *low frequency* and *narrow-band* in the spectrum, since the values of $v/L$ and its multiples are relatively small. Such a signal of multi-order frequencies contains dense information of the bridge, and is sensitive to changes in bridge stiffness, as will be exploited later on.

Being narrow-banded, the driving component of the contact-point acceleration can be well treated by the HT (Chondros et al. 1998). Substituting Eq. (11.13) into Eq. (11.10) yields the HT of the driving component as

$$H_d\left(t\right) = H\left[R_d\left(t\right)\right] = \sum_{j=1}^{n} A_j \sin\frac{2j\pi vt}{L}. \tag{11.15}$$

Then, the IAS of the driving component can be obtained by substituting Eqs. (11.13) and (11.15) into Eq. (11.12) as

$$A^2\left[R_d\left(t\right)\right] = \sum_{j=1}^{n} A_j^2 + 2\sum_{j=1}^{n}\sum_{k=2}^{n} A_j A_k\left[\cos\frac{(k-j)2\pi vt}{L}\right], \quad j<k. \tag{11.16}$$

For a simply supported beam, the mode shape is $\phi_j(x) = \sin(j\pi x/L)$. With this and using $x = vt$, the IAS of the driving component in Eq. (11.16) can be rewritten with respect to the mode shapes squared as

$$A^2\left[R_d\left(x\right)\right] = \sum_{j=1}^{n} A_j^2 + 2\sum_{j=1}^{n}\sum_{k=2}^{n} A_j A_k\left[1-2\phi_{k-j}^2\left(x\right)\right], \quad j<k. \tag{11.17}$$

As revealed in the preceding equation, the IAS of the driving component is dominated by the mode shapes $\phi_j(x)$, which will be abnormal at the damage position when a damage occurs (Chondros et al. 1998; Khaji et al. 2009). Such a property will be used to detect the damage(s) of a bridge. In comparison with the vehicle-body acceleration, the contact-point acceleration is more sensitive to structural damage or discontinuity, since the latter is a function of the second derivative of the former, as implied by Eq. (11.9).

The procedure for bridge damage detection by a moving test vehicle using the IAS can be summarized as follows:

1) Obtain the contact-point acceleration using Eq. (11.9), in which the vehicle acceleration is obtained either by field test or numerical simulation.
2) Obtain the fast Fourier transform (FFT) for the contact-point acceleration in frequency domain.
3) Identify the narrow band of the low-frequency driving component enclosing the first few driving frequencies $2j\pi v/L$, for given speed $v$ and span length $L$.

4) Extract the driving component response by the band-pass filter or other signal processing technique.
5) Obtain the IAS of the driving component as the square of the Hilbert transform.
6) Identify the damages of the bridge by the peaks of the IAS.

The procedure is simple and straightforward for detecting bridge damage. To further verify the above ideas and procedure, comprehensive studies will be conducted in the following using the finite element method.

## 11.4 Finite Element Simulation of the Problem

To verify the feasibility of the present technique, a finite element simulation will be conducted for the damage detection using the response generated of a vehicle during its passage over a damaged bridge. With the finite element simulation, most of assumptions adopted in deriving the closed-form solutions for the vehicle and contact-point accelerations are virtually circumvented.

### 11.4.1 Damage Element for Beams

In this study, the damage of a beam will be represented by an internal hinge with zero-length springs inserted between two connected beam segments, instead of reducing finite-length beam stiffness (Pandey et al. 1991; Lee et al. 2002). As shown in Figure 11.2, a *damage element* in its general form consists of a rotational spring of stiffness $K_r$ and a vertical spring of stiffness $K_v$ at the location of damage.

The stiffness matrix of the damage element can be written as

$$\mathbf{K_d} = \begin{bmatrix} \mathbf{K_v} & 0 & -\mathbf{K_v} & 0 \\ 0 & \mathbf{K_r} & 0 & -\mathbf{K_r} \\ -\mathbf{K_v} & 0 & \mathbf{K_v} & 0 \\ 0 & -\mathbf{K_r} & 0 & \mathbf{K_r} \end{bmatrix}. \tag{11.18}$$

When the stiffnesses $K_r$ and $K_v$ are very large or infinite, the segments on the two sides of the damage combine to form an intact beam. On the other hand, when the stiffnesses $K_r$ and $K_v$ are zero, the two segments become separate members. For our purposes, we shall consider only *rotational discontinuity* over the internal hinge, with the displacement discontinuity excluded. Namely, an infinitely large value is taken for the vertical spring stiffness $K_v$ and a finite value for the rotational spring stiffness $K_r$.

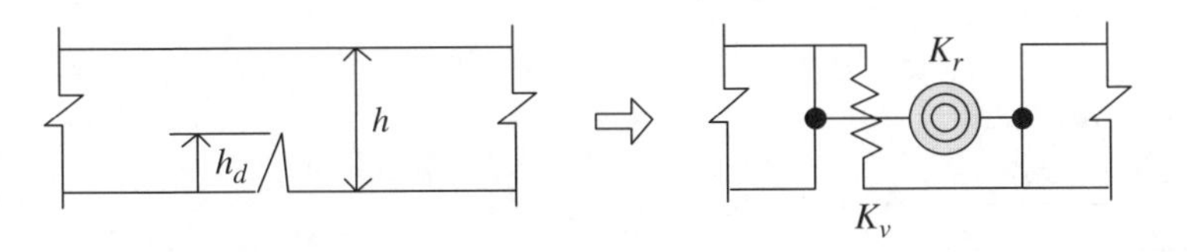


**Figure 11.2** Damage element for beams.

For a beam of a rectangular cross section, the equivalent rotational stiffness for a crack can be estimated as follows (Tada et al. 2000):

$$K_r = \left[ \frac{2h}{EI} \left( \frac{\alpha}{1-\alpha} \right)^2 \left( 5.93 - 19.69\alpha + 37.14\alpha^2 - 35.84\alpha^3 + 13.12\alpha^4 \right) \right]^{-1}, \tag{11.19}$$

As shown in Figure 11.2, the *damage parameter* $\alpha$ is defined as the ratio of the crack depth $h_d$ to beam height $h$, i.e., $\alpha = h_d/h$ indicating the severity of a damage.

### 11.4.2 Brief on Vehicle–Bridge Interaction (VBI) Element Used

The *VBI element* adopted is shown in Figure 11.3, in which the vehicle moving at speed $v$ is modeled as a sprung mass $m_v$ supported by a spring of stiffness $k_v$ and dashpot of damping coefficient $c_v$. By combining the equations of motion for the vehicle and the beam element in contact, the VBI element is derived as:

$$\begin{bmatrix} \mathbf{m_v} & \mathbf{0} \\ \mathbf{0} & [\mathbf{m_b}] \end{bmatrix} \begin{Bmatrix} \ddot{\mathbf{y}}_\mathbf{v} \\ \{\ddot{\mathbf{q}}_\mathbf{b}\} \end{Bmatrix} + \begin{bmatrix} \mathbf{c_v} & -\mathbf{c_v}\{\mathbf{N}\}_\mathbf{c}^\mathbf{T} \\ -\mathbf{c_v}\{\mathbf{N}\}_\mathbf{c} & [\mathbf{c_b}] + \mathbf{c_v}\{\mathbf{N}\}_\mathbf{c}\{\mathbf{N}\}_\mathbf{c}^\mathbf{T} \end{bmatrix} \begin{Bmatrix} \dot{\mathbf{y}}_\mathbf{v} \\ \{\dot{\mathbf{q}}_\mathbf{b}\} \end{Bmatrix}$$

$$+ \begin{bmatrix} \mathbf{k_v} & -\mathbf{c_v v}\{\mathbf{N'}\}_\mathbf{c}^\mathbf{T} - \mathbf{k_v}\{\mathbf{N}\}_\mathbf{c}^\mathbf{T} \\ -\mathbf{k_v}\{\mathbf{N}\}_\mathbf{c} & [\mathbf{k_b}] + \mathbf{c_v v}\{\mathbf{N}\}_\mathbf{c}\{\mathbf{N'}\}_\mathbf{c}^\mathbf{T} + \mathbf{k_v}\{\mathbf{N}\}_\mathbf{c}\{\mathbf{N}\}_\mathbf{c}^\mathbf{T} \end{bmatrix} \begin{Bmatrix} \mathbf{y_v} \\ \{\mathbf{q_b}\} \end{Bmatrix}$$

$$= \begin{Bmatrix} \mathbf{c_v v r'_c} + \mathbf{k_v r_c} \\ -\mathbf{c_v v r'_c}\{\mathbf{N}\}_\mathbf{c} - \mathbf{k_v r_c}\{\mathbf{N}\}_\mathbf{c} - \mathbf{m_v g}\{\mathbf{N}\}_\mathbf{c} \end{Bmatrix}, \tag{11.20}$$

where $[\mathbf{m_b}]$, $[\mathbf{c_b}]$, $[\mathbf{k_b}]$ denote the mass, damping, stiffness matrices, respectively, $\{\mathbf{q_b}\}$ the displacement vector of the VBI element, and $\{\mathbf{N}\}_\mathbf{c}$ is the vector of cubic Hermitian interpolation functions evaluated at the contact point.

By assembling the VBI element, damage element, and conventional beam elements used to represent the parts of the beam not directly acted upon by the vehicle, the global stiffness equation of motion of the VBI system can be established. Then, the response of the coupled system can be solved step-by-step by the Newmark $\beta$ method (with $\beta = 0.25$ and $\gamma = 0.5$ for unconditional stability) considering the vehicle's movement. The dynamic acceleration of the contact point calculated will be used to detect the damage of the bridge.

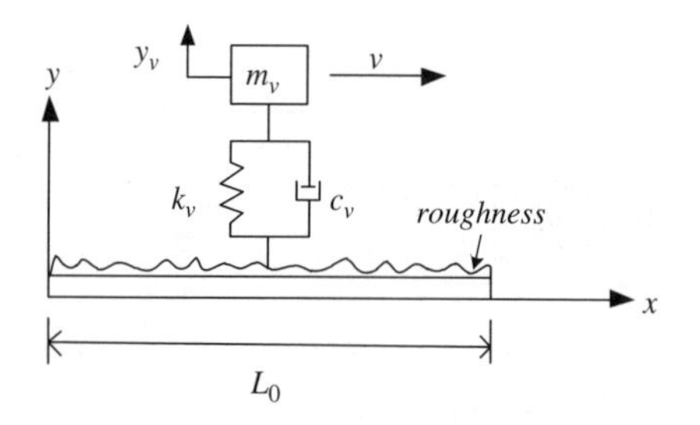


**Figure 11.3** Vehicle–bridge interaction element.

## 11.5 Detection of Damages on a Beam

As a further proof, the finite element simulation will be conducted by letting a vehicle move over a cracked beam. The properties adopted for the beam are close to those of the I-girders used in highways (Lin and Yang 2005), which has length $L$ = 35 m, width $b$ = 1 m, height $h$ = 2 m, area A = 2 m$^2$, moment of inertia I = 0.67 m$^4$, mass density $\rho$ = 2500 kg/m$^3$, and elastic modulus $E$ = 32.5 GPa; and the following for the vehicle, mass $m_v$ = 1500 kg and stiffness $k_v$ = 550 kN/m. A damping ratio of 0.1 is taken for the vehicle and 0.05 for the bridge (assumed to be of the Rayleigh type). The beam is discretized into 30 elements. Unless specified otherwise, the crack is assumed to be located at the midspan with a damage coefficient $\alpha$ = 0.35 and the vehicle speed of 1 m/s is adopted.

### 11.5.1 Detection of Damage Location on the Beam

In damage detection, the first task is to identify the existence and location of the damage. For the problem considered, the vehicle acceleration, contact-point acceleration, and the IAS result of the driving component of the contact-point acceleration have been plotted in Figures 11.4a–c, respectively. Although from the vehicle response or the contact-point

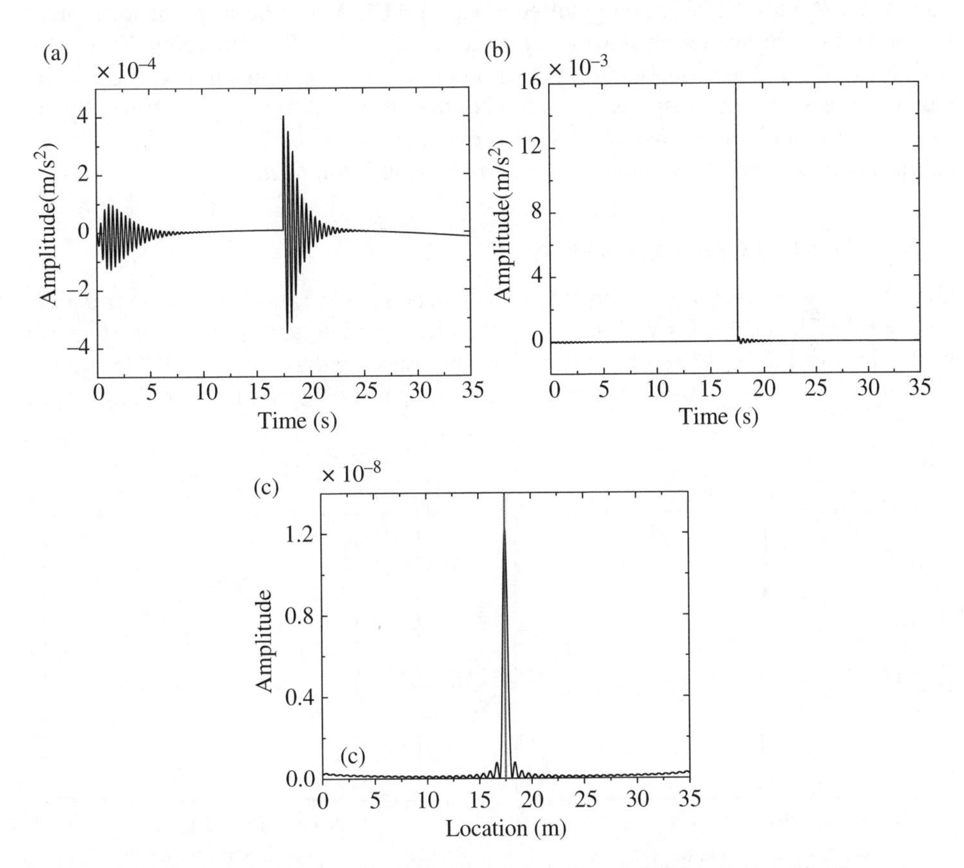


**Figure 11.4** Responses for beam with a midspan damage of α = 0.35: (a) vehicle acceleration; (b) contact-point acceleration; and (c) IAS based on contact-point response.

response in Figures 11.4a and b, the damage location can be visibly observed without further processing. This won't be true for the more general case with noises or for the test vehicle moving at higher speeds. All we like to say is that the IAS result to be investigated below has much higher sensitivity compared with the above two parameters for the more general case for damage identification.

From Figure 11.4c for the IAS result based on the contact-point response, it is clear that the outstanding peak indicates the occurrence of a damage with its location coincident with the damage location (red dotted line). It should be noted that for a healthy bridge, no peaks will occur in Figure 11.4. In general, the IAS is extremely small in amplitude, since it is obtained as the *square* of the instantaneous amplitude, which is very small by itself. This example demonstrates the capability of the vehicle-generated contact-point acceleration in detecting the damage of a bridge. The present technique is featured by the fact that *no baseline or reference is required for damage detection,* unlike those previously used for moving vehicle/load problems (Zhang et al. 2012; He and Zhu 2016).

Further, the present technique is tested against the variation of damage location, by allowing the damage to exist either at $1/3L$ or $1/6L$ of the beam. The IAS results of the driving component of the contact-point acceleration calculated for the damage locations at $1/3L$ and $1/6L$ have been plotted in Figure 11.5. As can be seen, the location of the damage can be accurately detected, no matter where it is. By comparing Figure 11.4 with Figure 11.5, one observes that for different damage locations, but with the same damage level $\alpha$, the IAS response generated of the beam is different, as revealed by the varying peak amplitudes. Nevertheless, it is confirmed that *the location of damage can be well identified regardless of the variation in IAS peak amplitudes.*

### 11.5.2 Detection of Damage Severity

To test the present technique's sensitivity to damage severity, another two crack depths are considered, i.e., $\alpha = 0.1$ and 0.2, while the other parameters of the VBI system are kept unchanged. The IAS results obtained for the crack depths $\alpha = 0.1$ and 0.2 have been plotted in Figures 11.6a and b, respectively. By comparing Figure 11.4 with Figure 11.6,

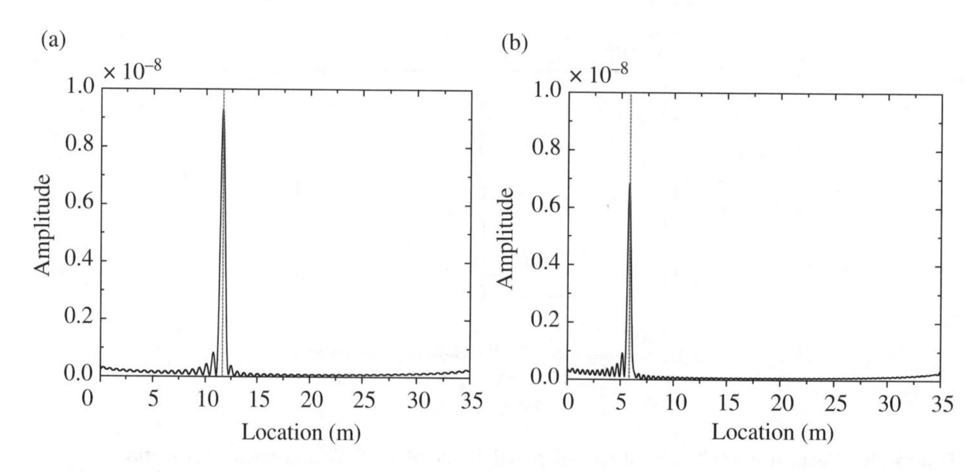


**Figure 11.5** IAS results of the beam with damage at: (a) 1/3 span; and (b) 1/6 span.

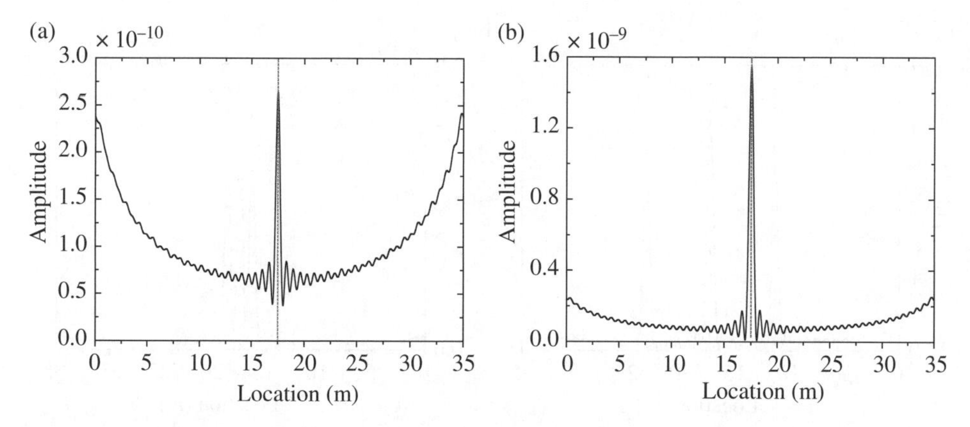


**Figure 11.6** IAS results of the beam with damage parameter of: (a) $\alpha = 0.1$; and (b) $\alpha = 0.2$.

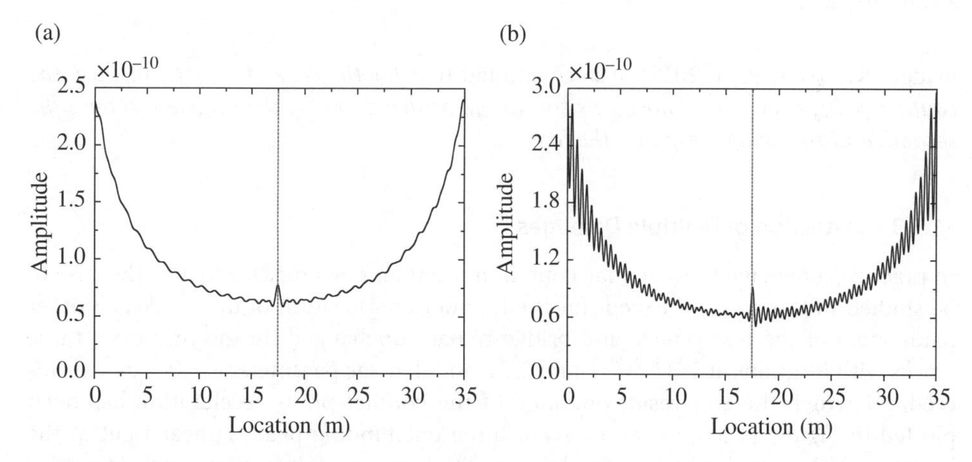


**Figure 11.7** IAS results of the beam with damage $\alpha = 0.03$ based on: (a) contact-point acceleration; and (b) vehicle acceleration.

one observes that the peak value of the IAS increases rapidly as the damage extent increases, i.e., *the more serious the damage is, the greater the peak value of the IAS.*

In order to test the sensitivity of the present method to minor damages, the damage parameter $\alpha$ has been reduced to 0.03, which is regarded as a very slight damage. For comparison, the IAS results of the beam based on the contact-point acceleration and "vehicle acceleration" have been plotted in Figures 11.7a and b, respectively. From Figure 11.7a, it is evident that even for this slight damage, the contact-point IAS response remains distinguishable in detecting the damage. As for the vehicle's IAS in Figure 11.7b, the strong oscillations caused by the boundary effect near the two ends may dilute the resolution of the result in real test. It is therefore concluded that *the contact-point acceleration is a better measure than the vehicle acceleration for detecting bridge damages as slight as of $\alpha = 0.03$ for the case of smooth surface.* Moreover, by comparing specifically with the following works for bridges' damage identification using: a fixed displacement sensor on the bridge (Zhu and Law 2006), fixed displacement sensor on the moving vehicle (Nguyen and Tran 2010) and moving displacement sensor on the

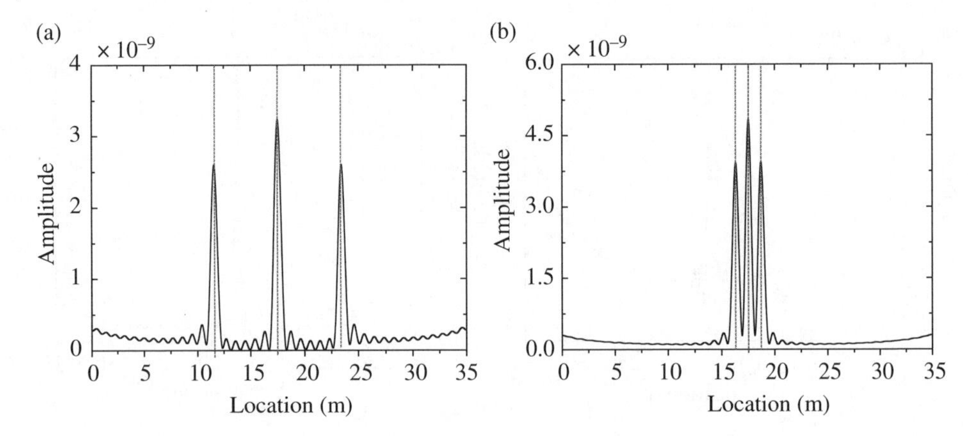


**Figure 11.8** IAS results of the beam with damages at: (a) 1/3, 1/2, and 2/3 span; and (b) 14/30, 15/30, and 16/30 span.

bridge (Khorram et al. 2012), it is concluded that *for the case of smooth surface, the contact-point response is most sensitive to the bridge damage, the vehicle response the second, and the bridge response the last.*

### 11.5.3 Detection of Multiple Damages

In practice, a damaged bridge may contain more than one cracks, which is the case to be studied in this section. Except for the number and location of the cracks, all other parameters of the test vehicle and bridge remain unchanged. In the first case, three cracks with locations at $1/3L$, $1/2L$ and $2/3L$ and damage parameter $\alpha = 0.3$ are considered, for which the IAS result obtained of the contact-point acceleration has been plotted in Figure 11.8a. As can be seen, three outstanding peaks appear right at the locations of damages. It is confirmed that *multi damages of the bridge can be identified from the IAS contact-point acceleration with no difficulty for beams with smooth surface.*

Previously, it was noted that damages of multi cracks cannot be properly identified by the wavelet-based crack detection method when the crack spacing is small (Zhu and Law 2006). To test the performance of the present method, three cracks at $14/30L$, $15/30L$ and $16/30L$ with damage parameter $\alpha = 0.35$ are assumed. The IAS result obtained for the contact-point acceleration has been plotted in Figure 11.8b. As can be seen, three distinct peaks corresponding to the three locations of cracks can be identified. It is hence concluded that *the present method works well for detecting multi-crack damages, even when the cracks are closely spaced.*

## 11.6 Parametric Study

In the field test, the measured response of the VBI system may be effected by the vehicle moving speed, measurement noise, roughness of the bridge deck, and random traffic during the measurement process. All these factors will be evaluated below to explore the feasibility of the method presented.

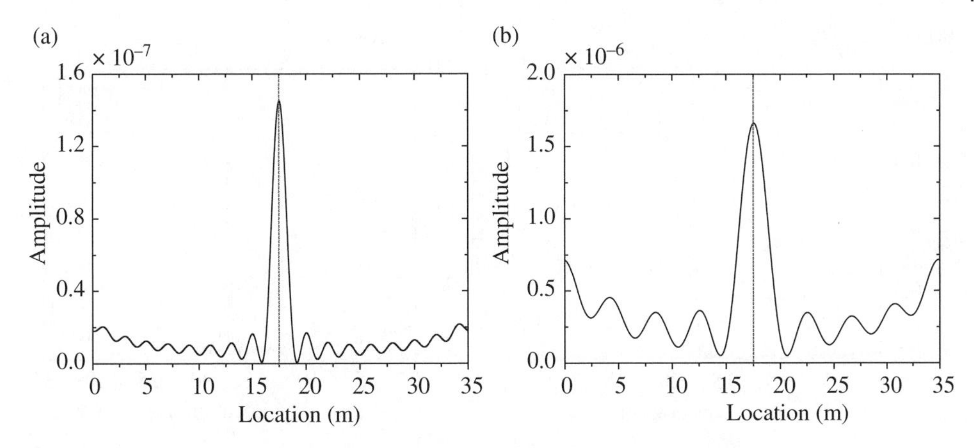


**Figure 11.9** IAS results of the beam with vehicle speed of: (a) 3 m/s; and (b) 7 m/s.

### 11.6.1 Effect of Test Vehicle Speed

In this section, the effect of moving speed of the test vehicle on the damage detection will be investigated by allowing the test vehicle to pass the cracked beam at three speeds 3 and 7 m/s. The IAS results of the beam with vehicle speeds of 3 and 7 m/s have been plotted in Figures 11.9a and b, respectively. By comparing Figure 11.4 (for 1 m/s) with Figures 11.9a and b, one observes that the peak value of the IAS increases dramatically with the increase in moving speed. Another phenomenon is the occurrence of some small peaks at the undamaged regions of the beam for speed up to 7 m/s, though it does not affect the result of damage detection. Such a phenomenon can be attributed to the fact that for a bridge with fixed length, the faster the vehicle's speed is, the less the amount of data collected, thereby resulting in more spectral leakage and lower resolution in frequency domain. All the small peaks in Figures 11.9a and b are caused by the spectral leakage. It should be noted that the occurrence of a damage is judged by the major peak's relative magnitude to the remainders, but not its absolute magnitude. Consequently, *a test vehicle moving at lower speeds permits better damage detection using the present approach.*

### 11.6.2 Effect of Measurement Noise

In practice, the data taken by the accelerometers installed on the test vehicle will be inevitably polluted by measurement noise. To study the effect of measurement noise, white noise of various levels will be hypothetically superimposed on the vehicle acceleration, resulting in the polluted acceleration $\ddot{y}_p$ as

$$\ddot{y}_p = \ddot{y}_v + E_p N_s \sigma_{\ddot{y}_v}, \tag{11.21}$$

where $E_p$ is the noise level, $N_s$ the standard normal distribution, and $\sigma$ the standard deviation. The reason why the noise is added to the vehicle acceleration, rather than to the contact-point acceleration, is that the former is the time–history response actually taken in the field test. It is realized that the noise existing on the vehicle acceleration will be carried over to the contact-point acceleration by Eq. (11.9).

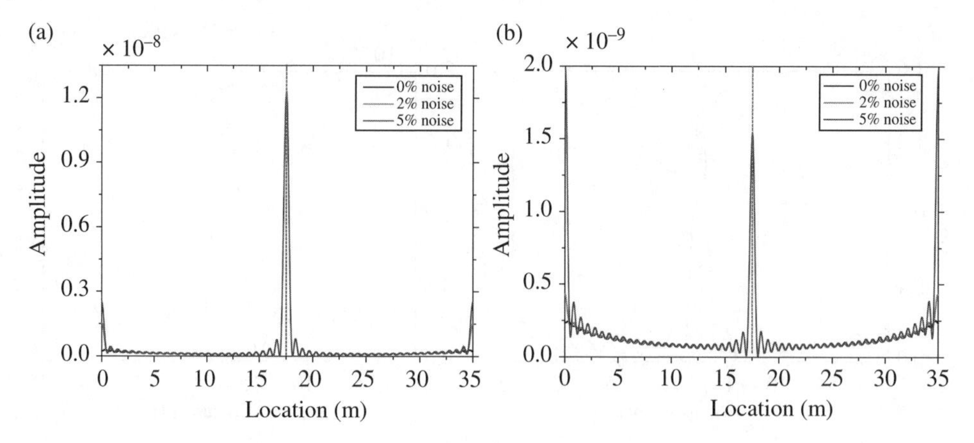


**Figure 11.10** IAS results of the beam with various levels of measurement noise and with damage parameter of: (a) $\alpha = 0.35$; and (b) $\alpha = 0.2$.

Figure 11.10a and b show the IAS results of the beam affected by 0%, 2%, and 5% of measurement noise for two damage levels $\alpha = 0.35$ and $\alpha = 0.2$, respectively. Clearly, *the damage on a beam can be located by the present approach even in the presence of certain noise.* The relatively good performance of the present approach for detecting damages in the noise environment can be attributed to two reasons. First, the second derivatives involved in Eq. (11.9) for computing the contact-point acceleration preclude the effect of low-frequency components of the noise. Second, the noise interfering the IAS result is by nature of narrow band, as implied by the algorithm for damage detection in Section 11.3.2.

### 11.6.3 Bridge with Rough Surface Free of Random Traffic

The capability of the present method to work on bridges with rough surface will be investigated herein. For the present purposes, the surface roughness profile adopted (see Figure 11.11) is constructed using the power spectral density (PSD) function of ISO 8608 (1995) with Class A and frequency range 0.011–2.83/m.

A single-axle test vehicle (i.e., trailer) towed by a tractor was used by Lin and Yang (2005) in the field test, as schematically shown in Figure 11.12. The trailer is representative of the SDOF system used in the closed-form analytical formulation, by which the technique for extracting the bridge frequencies from the moving test vehicle has been verified to be feasible (Lin and Yang 2005; Yang et al. 2013b). In this section, the tractor is simulated by a moving load of 15 tons, and the trailer as a moving sprung mass of 1.5 tons, spaced at $d = 5\,\text{m}$.

Figure 11.13 shows the IAS result for the contact-point acceleration of the test vehicle towed by the tractor over the bridge with rough surface. As can be seen, two additional peaks appear in the IAS result, aside from the peak of the test vehicle indicative of the damage location. This can be explained as follows. When the tractor reaches the damage location (17.5 m), the dynamic response of the beam caused by slope discontinuity caused by the damage will be transmitted to the those of the test vehicle and the contact point at the instant, thereby producing the peak at 12.5 m. Similarly, when the tractor

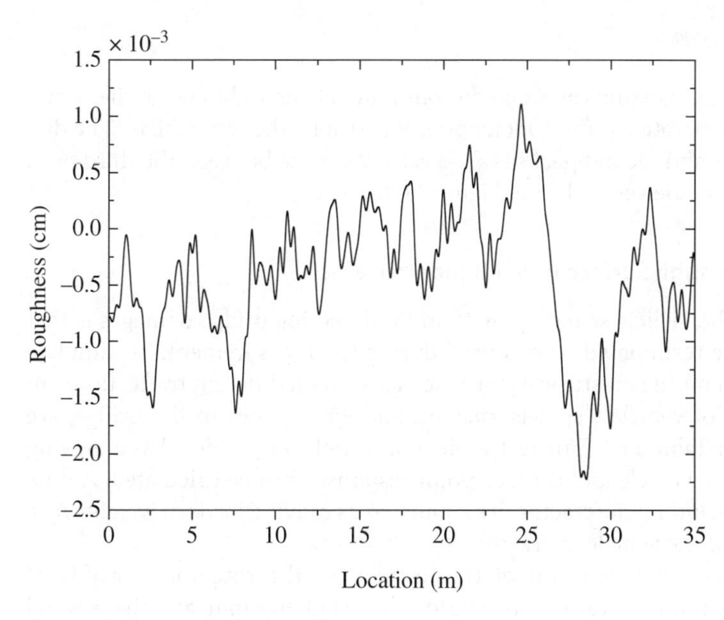

**Figure 11.11** Road surface profile.

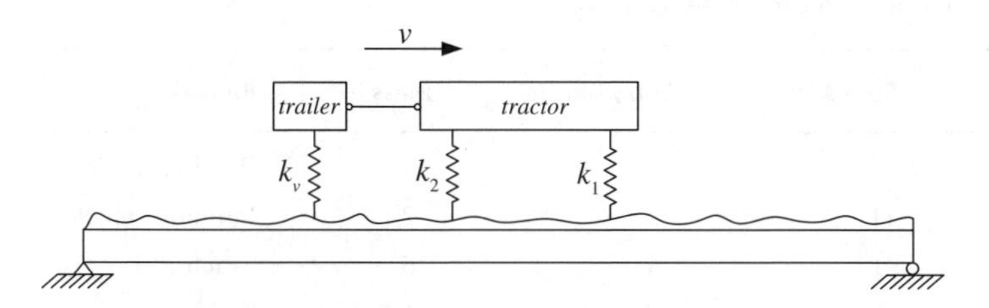

**Figure 11.12** Tractor-trailer model.

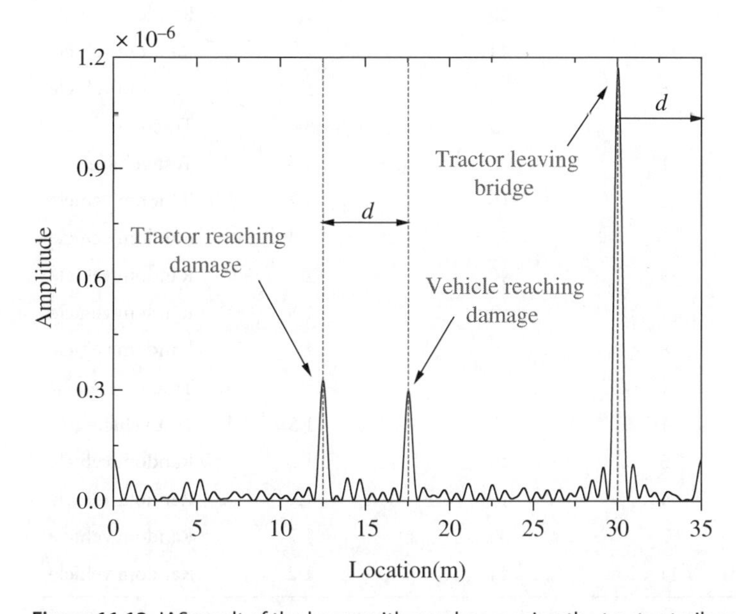

**Figure 11.13** IAS result of the beam with roughness using the tractor-trailer system.

departs from the bridge, crossing the slope discontinuity at the right end of the beam, another peak will be generated at the contact point of 30 m by the test vehicle. The distance between the first and second peaks is $d$, so is the distance between the third peak and the beam end, as can be verified from Figure 11.13.

### 11.6.4 Bridge with Rough Surface under Random Traffic

A distinct feature of the vehicle scanning method for detecting bridge damages is that the traffic need not be terminated or detoured during the measurement. To simulate the random traffic in a more realistic way, three scenarios with different traffic flows, in terms of the number of vehicles, speeds, masses, and entry times to the bridge, are considered as listed in Table 11.1. The test vehicle (i.e., trailer) is modeled as a moving sprung mass so that the vehicle and contact-point responses can be calculated. All the passing-by vehicles, including the tractor, are simulated as moving loads of weights $m_i g$, where $m_i$ is the mass of the vehicle or tractor.

Figure 11.14a shows the IAS result of the beam with the roughness profile of Figure 11.11 under the random traffic of Scenario I. A first glance indicates that several

**Table 11.1** Properties of random traffic for three scenarios.

| Scenario | Vehicle | Speed (m/s) | Entry time (s) | Mass ($t$) | Remark |
|---|---|---|---|---|---|
| | 1 | 7 | 0 | 1.8 | Random vehicle |
| | 2 | 11 | 4 | 2 | Random vehicle |
| | 3 | 1 | 5 | 15 | Tractor |
| I | 4 | 1 | 10 | 1.5 | Test vehicle |
| | 5 | 8 | 17 | 1 | Random vehicle |
| | 6 | 7 | 30 | 1.7 | Random vehicle |
| | 7 | 10 | 44 | 2 | Random vehicle |
| | 1 | 8 | 0 | 1.5 | Random vehicle |
| | 2 | 1 | 3 | 15 | Tractor |
| II | 3 | 1 | 8 | 1.5 | Test vehicle |
| | 4 | 10 | 16 | 1.2 | Random vehicle |
| | 5 | 7 | 29 | 1.4 | Random vehicle |
| | 6 | 9 | 40 | 2 | Random vehicle |
| | 1 | 11 | 0 | 1.5 | Random vehicle |
| | 2 | 8 | 4 | 1.7 | Random vehicle |
| | 3 | 1 | 7 | 15 | Tractor |
| III | 4 | 1 | 12 | 1.5 | Test vehicle |
| | 5 | 6 | 16 | 1 | Random vehicle |
| | 6 | 9 | 19 | 1.4 | Random vehicle |
| | 7 | 7 | 35 | 1.7 | Random vehicle |
| | 8 | 11 | 44 | 1.2 | Random vehicle |

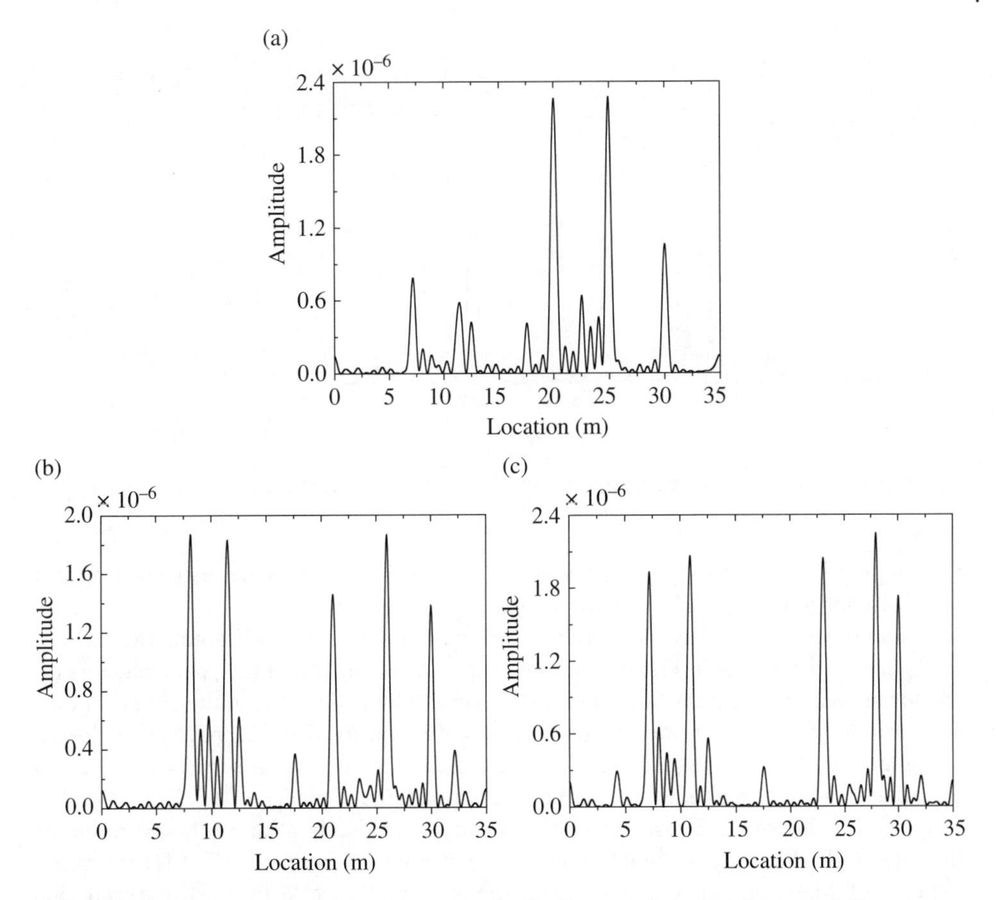


**Figure 11.14** IAS result of the beam under various traffic flows: (a) Scenario I; (b) Scenario II;, and (c) Scenario III.

extra peaks mingle with those induced by the test vehicle and tractor, thereby rendering it difficult to identify which peak is the one corresponding to the damage location. Given the damage location at 17.5 m and the tractor-trailer interval of $d = 5$ m, one concludes that the peaks at the position of 12.5, 17.5, and 30 m are those induced by the tractor and test vehicle, and the remaining ones by the random vehicles when they enter or depart from the bridge, or cross the damage location. Though the tractor-trailer interval $d$ is known beforehand, the distance between each of the random vehicles and the damage location remains unknown in practice. Evidently, the random traffic will bring in extra peaks in the IAS result, rendering it difficult to identify the damage location in a straightforward way.

In order to resolve the extra peaks problem brought by random traffic, a simple strategy is adopted herein. The idea is that in spite of the randomness of the traffic, the damage location of the bridge is fixed, so is the interval $d$ between the tractor and trailer. Hence, by allowing the tractor-trailer to travel over the bridge under *different traffic flows*, some peaks corresponding to the damage location and the effects of slope discontinuity caused by the damage will *invariably* appear in the IAS result for each traffic flow.

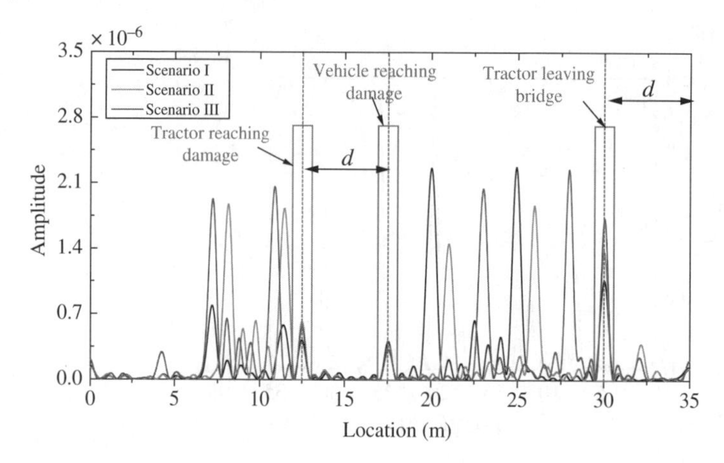


**Figure 11.15** IAS results of the beam under traffic flows of all three scenarios.

Such repeated or invariant peaks provide the clue for detecting the damage location and slope discontinuities associated with damages on the bridge.

To demonstrate the above idea, two additional scenarios with different traffic flows are considered as in Table 11.1, where the properties of the trailer (i.e., test vehicle) and tractor, as well as their interval *d*, remain the same. The IAS results of the contact-point acceleration of the beam under the traffic flows of Scenarios II and III have been plotted in Figures 11.14b and c, respectively. At a first glance, the IAS results in Figure 11.14 do not appear to be the same, when judged by the positions of peaks.

By overlapping the IAS results for the contact-point acceleration of the beam under the three traffic flows, one obtains the combined result in Figure 11.15. It is interesting to point out that there is a total of three locations where the peaks for the three scenarios coincide with each other, while all the other locations do not match. As it was pointed out previously, all the *repeated* or *invariant* peaks represent exactly the locations of the *damage* or *where the tractor encounters slope discontinuity*, which won't be affected by the traffic flow. Since the tractor is always 5 m ahead of the vehicle, the positions of three peaks identified by rectangles (from left to right in order) in Figure 11.15 should be interpreted as the location of the tractor "reaching the damage," of the vehicle "reaching the damage," and of the tractor "leaving the bridge," respectively. As a result, the location of the damage identified is 17.5 m from the left end of the beam, which agrees exactly with the assumed value. The above analysis demonstrated that *by allowing the tractor-trailer to travel over the bridge for more than one times, the location of the damage on the beam can be identified from the repeated or invariant peaks generated under different traffic flows.*

## 11.7 Concluding Remarks

The contact-point acceleration, rather than vehicle response, was used herein for damage detection of bridge. First, an approximate analytical solution of the contact-point acceleration was derived, followed by a procedure for computing the response in field tests. Then the algorithm for damage detection by the component analysis and Hilbert transform was outlined. Theoretically, the IAS of the driving component of the contact-point acceleration, which is of low frequency and narrow-band, was shown to contain abundant information for damage detection. With each damage modeled by a hinge-spring unit, the overall capability of the present approach in damage detection was investigated by the finite element analysis. Also studied are the effects of test vehicle speed, measurement noise and road roughness to be encountered in practice. Specifically, a multi-travel strategy was presented for capturing damages under random, multi-traffic flows.

Based on the numerical studies and the data assumed, this chapter presented four conclusions:

1) No baseline is needed for the damage detection of bridges using the present approach, which obviates the need of prior measurement of bridges to the occurrence of damages.
2) The IAS response of the contact point is sensitive to damage, even under the circumstances of tiny damages, multi damages, measurement noise and road roughness.
3) When in the presence of random traffic, the irrelevant peaks induced by ongoing vehicles in the IAS result may mingle with the relevant ones, rendering the identification of damage location difficult.
4) In the presence of ongoing traffic, the damages of the bridge can be identified from the repeated or invariant IAS peaks generated for different traffic flows by the same test vehicle moving over the bridge.

# Appendix

# Finite Element Simulation

The vehicle scanning method will be verified by numerical simulation and experiment. The vehicle–bridge interaction (VBI) element used is the one given in Eq. (2.18). In this Appendix, a detailed derivation will first be given of the element stiffness equations in Eq. (2.18) following basically the procedure presented in Yang and Yau (1997) and the corrections made by Chang et al. (2010). Then the procedure for assembling the VBI element will be described.

## A.1 Derivation of VBI Element

Consider the VBI element shown in Figure A.1, in which the vehicle is modeled as a sprung mass of magnitude $m_v$ supported by a spring of stiffness $k_v$ and a dashpot of damping $c_v$ plus a wheel mass $m_w$. The vehicle interacts with the beam element at the contact point with coordinate $x_c$, which moves as long as the vehicle moves.

The equation of motion for the vehicle is

$$m_v \ddot{q}_v + c_v \left( \dot{q}_v - \dot{q}_w \right) + k_v \left( q_v - q_w \right) = 0, \tag{A.1}$$

where $q_v$ and $q_w$ denote the vertical displacement of the vehicle body and wheel, respectively, and a dot indicates differentiation with respect to time $t$. Note that the vehicle displacement is measured from the static equilibrium position.

With reference to the free body diagram shown in Figure A.2, the vehicle's wheel is subjected to the elastic force $F_1$ and damping force $F_2$ from the top and the contact force $p_1$ from the bottom, i.e., from the bridge surface. Hence, the equation of motion for the vehicle's wheel is

$$m_w \ddot{q}_w = p_1 + F_1 + F_2, \tag{A.2}$$

where

$$F_1 = k_v \left( q_v - q_w \right), \qquad F_2 = c_v \left( \dot{q}_v - \dot{q}_w \right). \tag{A.3}$$

*Vehicle Scanning Method for Bridges*, First Edition. Yeong-Bin Yang, Judy P. Yang, Bin Zhang and Yuntian Wu.

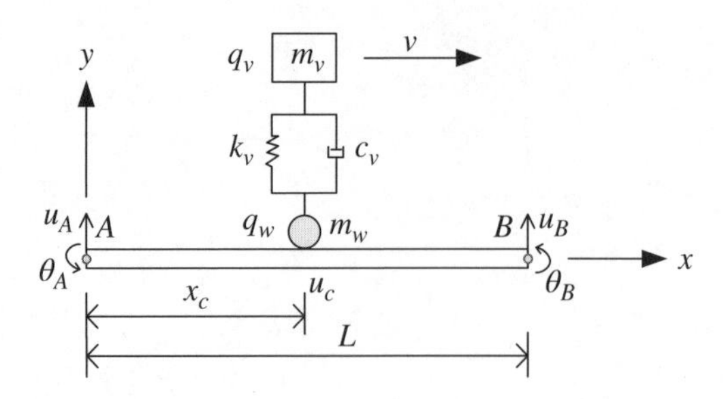


Figure A.1 Vehicle-bridge interaction (VBI) element.

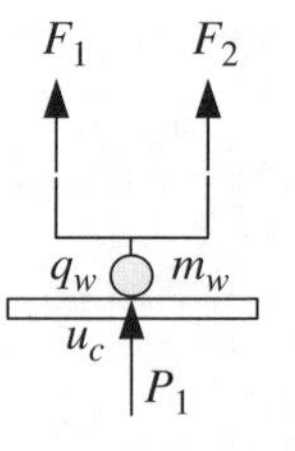


Figure A.2 Free body diagram of vehicle's wheel.

By substituting Eq. (A.3) into Eq. (A.2), one obtains the equation of motion for the vehicle's wheel as

$$m_w\ddot{q}_w = p_1 + k_v\left(q_v - q_w\right) + c_v\left(\dot{q}_v - \dot{q}_w\right), \tag{A.4}$$

in which all the forces acting on the wheel can be appreciated.

For illustration of the VBI effect, only the 2D beam element shown in Figure A.1 will be considered. For this element, the equation of motion is

$$\left[m_b\right]\left\{\ddot{u}_b\right\} + \left[c_b\right]\left\{\dot{u}_b\right\} + \left[k_b\right]\left\{u_b\right\} = -\left(p_1 + m_w g + m_v g\right)\left\{N\right\}, \tag{A.5}$$

where $[m_b]$, $[c_b]$, $[k_b]$ denote the mass, damping, stiffness matrices, respectively, $\{u_b\}$ the displacement vector of the beam element,

$$\left\{u_b\right\} = \left\{u_A \quad \theta_A \quad u_B \quad \theta_B\right\}^T \tag{A.6}$$

and $\{N\}$ is the vector of cubic Hermitian interpolation functions evaluated at the contact point with coordinate $x_c$. For the 2D beam element shown in Figure A.1, the mass matrix $[m_b]$ of the consistent form is

$$\left[m_b\right] = \frac{\bar{m}L}{420}\begin{bmatrix} 156 & 22L & 54 & -13L \\ 22L & 4L^2 & 13L & -3L^2 \\ 54 & 13L & 156 & -22L \\ -13L & -3L^2 & -22L & 4L^2 \end{bmatrix}, \tag{A.7}$$

where $\bar{m}$ denote the mass of the beam per unit length, and the stiffness matrix $[k_b]$ is

$$\left[k_b\right] = \frac{2EI}{L^3}\begin{bmatrix} 6 & 3L & -6 & 3L \\ 3L & 2L^2 & -3L & L^2 \\ -6 & -3L & 6 & -3L \\ 3L & L^2 & -3L & 2L^2 \end{bmatrix}, \tag{A.8}$$

where $EI$ denote the flexural stiffness of the beam. The damping matrix $[c_b]$ is usually not determined on the element level. Rather, it is determined on the structural level. Assumed to be of the Rayleigh type, the structural damping matrix $[C_b]$ is determined as a linear combination of the structural mass matrix $[M_b]$ and stiffness matrix $[K_b]$ (Clough and Penzien 1995), i.e.

$$[C_b]=a_0[M_b]+a_1[K_b], \tag{A.9}$$

and the coefficients are determined as

$$\begin{Bmatrix} a_0 \\ a_1 \end{Bmatrix}=2\frac{\omega_m\omega_n}{\omega_n^2-\omega_m^2}\begin{bmatrix} \omega_n & -\omega_m \\ -\omega_n^{-1} & \omega_m^{-1} \end{bmatrix}\begin{Bmatrix} \xi_m \\ \xi_n \end{Bmatrix}, \tag{A.10}$$

where $\omega_m$ and $\omega_n$ are the two frequencies of the structure used to calibrate the damping ratios, and $\xi_m$, $\xi_n$ are the damping ratios assigned for the two modes. The cubic Hermitian interpolation function $\{N\}$ evaluated at the contact point $x_c$ (see Figure A.1) is

$$\{N\}=\begin{Bmatrix} N_1 \\ N_2 \\ N_3 \\ N_4 \end{Bmatrix}=\begin{Bmatrix} 1-3\left(\frac{x_c}{L}\right)^2+2\left(\frac{x_c}{L}\right)^3 \\ x\left(1-\frac{x_c}{L}\right)^2 \\ 3\left(\frac{x_c}{L}\right)^2-2\left(\frac{x_c}{L}\right)^3 \\ \frac{x_c^2}{L}\left(\frac{x_c}{L}-1\right) \end{Bmatrix}, \tag{A.11}$$

Assuming the vehicle's wheel to remain in perfect contact with the bridge surface, one can write

$$q_w=u_c=\{N\}^T\{u_b\}, \tag{A.12}$$

Differentiating Eq. (A.12) with respect to time $t$ once and twice yields

$$\dot{q}_w=\{N\}^T\{\dot{u}_b\}+v\{N'\}^T\{u_b\}, \tag{A.13}$$

and

$$\ddot{q}_w=\{N\}^T\{\ddot{u}_b\}+2v\{N'\}^T\{\dot{u}_b\}+v^2\{N''\}^T\{u_b\}, \tag{A.14}$$

where use has been made of the following relationship:

$$\frac{d\{N\}}{dt}=\frac{d\{N\}}{dx}\cdot\frac{d_x}{dt}=v\{N'\}. \tag{A.15}$$

By substituting Eqs. (A.12) and (A.13) into Eq. (A.1), the equation of motion for the sprung mass can be expressed as

$$m_v\ddot{q}_v + c_v\dot{q}_v - c_v\{N\}^T\{\dot{u}_b\} + k_v q_v - \left(vc_v\{N'\}^T + k_v\{N\}^T\right)\{u_b\} = 0. \tag{A.16}$$

Then, by substituting Eqs. (A.12–A.14) into Eq. (A.4), the contact force between the vehicle's wheel and bridge surface can be obtained as

$$\begin{aligned} p_1 = {} & m_w\{N\}^T\{\ddot{u}_b\} - c_v\dot{q}_v + \left(2vm_w\{N'\}^T + c_v\{N\}^T\right)\{\dot{u}_b\} \\ & -k_v q_v + \left(v^2 m_w\{N''\}^T + vc_v\{N'\}^T + k_v\{N\}^T\right)\{u_b\}. \end{aligned} \tag{A.17}$$

Further, by substituting Eq. (A.17) into Eq. (A.5), the equation of motion for the beam element can be expressed as

$$\begin{aligned} & \left([m_b] + m_w\{N\}\{N\}^T\right)\{\ddot{u}_b\} - c_v\{N\}\dot{q}_v + \left([c_b] + 2vm_w\{N\}\{N'\}^T \right. \\ & \left. + c_v\{N\}\{N\}^T\right)\{\dot{u}_b\} - k_v\{N\}q_v \\ & + \left([k_b] + v^2 m_w\{N\}\{N''\}^T + vc_v\{N\}\{N'\}^T + k_v\{N\}\{N\}^T\right)\{u_b\} \\ & = -(m_w g + m_v g)\{N\} \end{aligned} \tag{A.18}$$

Finally, by combining Eq. (A.16) and Eq. (A.18) into a matrix form, one arrives at the following general equation for the VBI element:

$$\begin{aligned} & \begin{bmatrix} m_v & 0 \\ 0 & [m_b] + m_w\{N\}\{N\}^T \end{bmatrix} \begin{Bmatrix} \ddot{q}_v \\ \{\ddot{u}_b\} \end{Bmatrix} \\ & \quad + \begin{bmatrix} c_v & -c_v\{N\}^T \\ -c_v\{N\} & [c_b] + 2vm_w\{N\}\{N'\}^T + c_v\{N\}\{N\}^T \end{bmatrix} \begin{Bmatrix} \dot{q}_v \\ \{\dot{u}_b\} \end{Bmatrix} \\ & + \begin{bmatrix} k_v & -vc_v\{N'\}^T - k_v\{N\}^T \\ -k_v\{N\} & [k_b] + v^2 m_w\{N\}\{N''\}^T + vc_v\{N\}\{N'\}^T + k_v\{N\}\{N\}^T \end{bmatrix} \begin{Bmatrix} q_v \\ \{u_b\} \end{Bmatrix} \\ & \qquad = \begin{Bmatrix} 0 \\ -(m_v g + m_w g)\{N\} \end{Bmatrix}. \end{aligned} \tag{A.19}$$

For the case where the effect of vehicle's wheel can be ignored, then, the VBI element can be simplified as

$$\begin{aligned} & \begin{bmatrix} m_v & 0 \\ 0 & [m_b] \end{bmatrix} \begin{Bmatrix} \ddot{q}_v \\ \{\ddot{u}_b\} \end{Bmatrix} + \begin{bmatrix} c_v & -c_v\{N\}^T \\ -c_v\{N\} & [c_b] + c_v\{N\}\{N\}^T \end{bmatrix} \begin{Bmatrix} \dot{q}_v \\ \{\dot{u}_b\} \end{Bmatrix} \\ & \quad + \begin{bmatrix} k_v & -vc_v\{N'\}^T - k_v\{N\}^T \\ -k_v\{N\} & [k_b] + vc_v\{N\}\{N'\}^T + k_v\{N\}\{N\}^T \end{bmatrix} \begin{Bmatrix} q_v \\ \{u_b\} \end{Bmatrix} \end{aligned}$$

$$
=\begin{Bmatrix} 0 \\ -m_v g\{N\} \end{Bmatrix}. \tag{A.20}
$$

which is exactly the one given in Eq. (2.18) of Chapter 2.

## A.2 Assembly of VBI Element

As can be seen from Eq. (A.20), the VBI element is different from the ordinary beam element in that it possesses an extra degree-of-freedom (DOF) for the vehicle. In the implementation of computer codes, it is more convenient to treat the VBI element equation as the summation of two equations, one for the ordinary beam element and the other for the moving sprung mass, namely,

$$
\begin{aligned}
&\left[\begin{bmatrix} 0 & 0 \\ 0 & [m_b] \end{bmatrix}+\begin{bmatrix} m_v & 0 \\ 0 & 0 \end{bmatrix}\right]\begin{Bmatrix} \ddot{q}_v \\ \{\ddot{u}_b\} \end{Bmatrix}+\left[\begin{bmatrix} 0 & 0 \\ 0 & [c_b] \end{bmatrix}+\begin{bmatrix} c_v & -c_v\{N\}^T \\ -c_v\{N\} & c_v\{N\}\{N\}^T \end{bmatrix}\right]\begin{Bmatrix} \dot{q}_v \\ \{\dot{u}_b\} \end{Bmatrix} \\
&+\left[\begin{bmatrix} 0 & 0 \\ 0 & [k_b] \end{bmatrix}+\begin{bmatrix} k_v & -vc_v\{N'\}^T-k_v\{N\}^T \\ -k_v\{N\} & vc_v\{N\}\{N'\}^T+k_v\{N\}\{N\}^T \end{bmatrix}\right]\begin{Bmatrix} q_v \\ \{u_b\} \end{Bmatrix} \\
&=\begin{Bmatrix} 0 \\ 0 \end{Bmatrix}+\begin{Bmatrix} 0 \\ -m_v g\{N\} \end{Bmatrix},
\end{aligned} \tag{A.21}
$$

or

$$
[m_b]\{\ddot{u}_b\}+[c_b]\{\dot{u}_b\}+[k_b]\{u_b\}=\{0\} \tag{A.22}
$$

$+$

$$
\begin{aligned}
&\begin{bmatrix} m_v & 0 \\ 0 & 0 \end{bmatrix}\begin{Bmatrix} \ddot{q}_v \\ \{\ddot{u}_b\} \end{Bmatrix}+\begin{bmatrix} c_v & -c_v\{N\}^T \\ -c_v\{N\} & c_v\{N\}\{N\}^T \end{bmatrix}\begin{Bmatrix} \dot{q}_v \\ \{\dot{u}_b\} \end{Bmatrix} \\
&+\begin{bmatrix} k_v & -vc_v\{N'\}^T-k_v\{N\}^T \\ -k_v\{N\} & vc_v\{N\}\{N'\}^T+k_v\{N\}\{N\}^T \end{bmatrix}\begin{Bmatrix} q_v \\ \{u_b\} \end{Bmatrix}=\begin{Bmatrix} 0 \\ -m_v g\{N\} \end{Bmatrix}.
\end{aligned} \tag{A.23}
$$

Clearly, Eq. (A.22) is the equation of motion for the ordinary beam element, and Eq. (A.23) is the one for the moving sprung mass.

In this Appendix, only the 2D beam element is considered, of which each node has two DOFs, i.e., the vertical and rotational DOFs. For this element, the mass matrix $[m_b]$, damping matrix $[c_b]$ and stiffness matrix $[k_b]$ are $4\times 4$ matrices. If a beam is divided into $N$ elements (i.e. with $N+1$ nodes), the total number of DOFs for the beam is $2N+2$. Since the moving sprung mass brings an extra DOF, the total number of DOFs for the entire VBI system with the vehicle's passage is $2N+2+1 = 2N+3$. In order to avoid

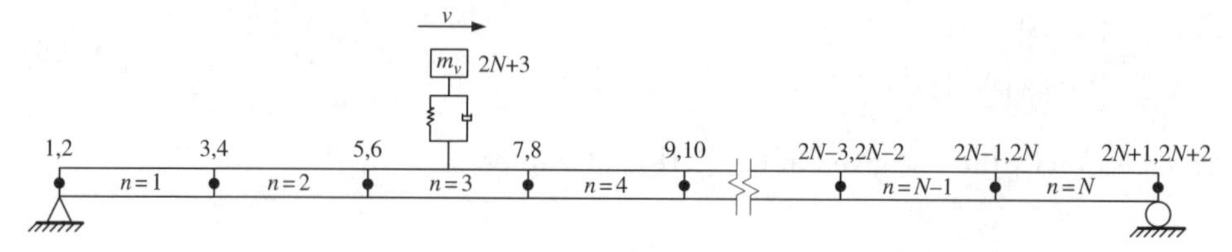


**Figure A.3** Coding of nodal degree-of-freedoms DOFs for the vehicle–bridge interaction (VBI) system.

confusion in coding the nodal DOFs of the ordinary beam elements, the vehicle's DOF will be numbered as the last DOF of the VBI system, i.e., as DOF $2N+3$ in Figure A.3.

In this section, the procedure for assembling all the elements, including the VBI element and ordinary beam elements, will be explained. As was mentioned earlier, the VBI element equation can be regarded as the summation of the associated ordinary beam element equation given in Eq. (A.22) and an extra vehicle equation given in Eq. (A.23). To account for this feature, a two-stage technique will be adopted for assembly of the elements for the VBI system. In the first stage, all the equations in the form Eq. (A.22) for the ordinary elements of the structure, including the one under the vehicle's action, will be assembled, and in the second stage, the extra vehicle equation given in Eq. (A.23) will be added. The two-stage procedure is illustrated in Figures A.4 and A.5.

In order to illustrate this two-stage assembly procedure, we will demonstrate how to construct the overall stiffness matrix for the VBI system. It is known that the same procedure applies to construction of the mass matrix of the VBI system. However, for the damping matrix of the system, distinction should be made between *vehicle' damping* ($c_v$) and *beam's damping* $[C_b]$. The assembly procedure mentioned above works only for updating vehicle's damping ($c_v$) in response to vehicle's movement, according to Eq. (A.23).

As already mentioned, the damping matrix $[C_b]$ of the beam (consisting of all the beam elements) is not determined on the element level. Rather, it is determined on the structural level based the hypothesis of Rayleigh damping, as a linear combination of the mass matrix $[M_b]$ and stiffness matrix $[K_b]$. This part of damping remains unaltered regardless of the movement of the vehicle.

With reference to Figure A.3, the nodal DOFs of the stiffness matrix for the beam element ($n$ = 3) under the vehicle action, i.e. the VBI element, are assigned the numbers 5, 6, 7, 8. For the first stage of assembly, the stiffness coefficients related to the nodal DOFs 5, 6, 7, 8 of the VBI element can be obtained from Eq. (A.8) in this Appendix for the associated ordinary beam element as follows:

$$
\begin{array}{ccccc}
\textit{DOFs} & 5 & 6 & 7 & 8 \\
5 & \dfrac{12EI}{L^3} & \dfrac{6EI}{L^2} & -\dfrac{12EI}{L^3} & \dfrac{6EI}{L^2} \\
6 & \dfrac{6EI}{L^2} & \dfrac{4EI}{L} & -\dfrac{6EI}{L^2} & \dfrac{2EI}{L} \\
7 & -\dfrac{12EI}{L^3} & -\dfrac{6EI}{L^2} & \dfrac{12EI}{L^3} & -\dfrac{6EI}{L^2} \\
8 & \dfrac{6EI}{L^2} & \dfrac{2EI}{L} & -\dfrac{6EI}{L^2} & \dfrac{4EI}{L}
\end{array}
. \tag{A.24}
$$

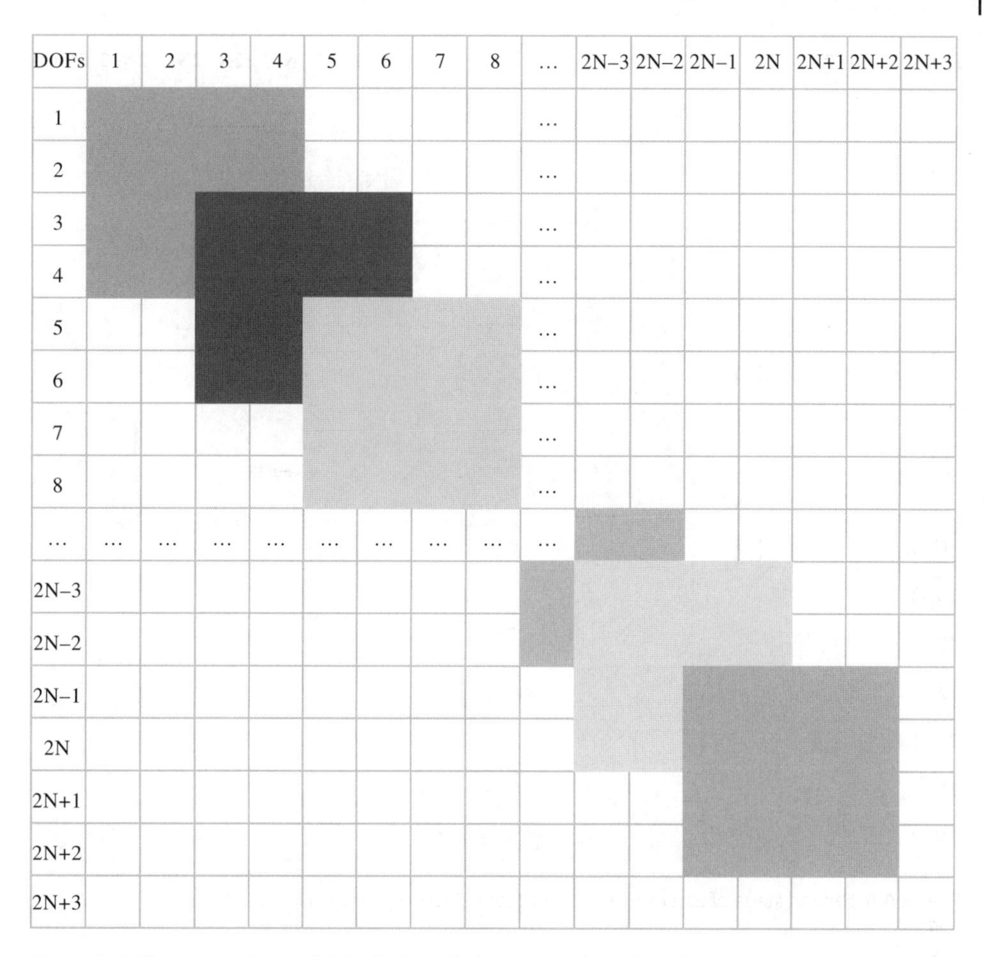

DOFs
1 2 3 4 5 6 7 8 … 2N−3 2N−2 2N−1 2N 2N+1 2N+2 2N+3
1
2
3
4
5
6
7
8
…
2N−3
2N−2
2N−1
2N
2N+1
2N+2
2N+3


**Figure A.4** First stage of assembly including all elements with Eq. (A.22).

Before the second stage of assembly can be performed, the extra equation for the VBI element in Eq. (A.23) should be rewritten with the order of the vehicle DOF ($q_v$) and beam DOFs $\{u_b\}$ reversed:

$$\begin{bmatrix} 0 & 0 \\ 0 & m_v \end{bmatrix} \begin{Bmatrix} \{\ddot{u}_b\} \\ \ddot{q}_v \end{Bmatrix} + \begin{bmatrix} c_v\{N\}\{N\}^T & -c_v\{N\} \\ -c_v\{N\}^T & c_v \end{bmatrix} \begin{Bmatrix} \{\dot{u}_b\} \\ \dot{q}_v \end{Bmatrix}$$

$$+ \begin{bmatrix} vc_v\{N\}\{N'\}^T + k_v\{N\}\{N\}^T & -k_v\{N\} \\ -vc_v\{N'\}^T - k_v\{N\}^T & k_v \end{bmatrix} \begin{Bmatrix} \{u_b\} \\ q_v \end{Bmatrix} = \begin{Bmatrix} -m_v g\{N\} \\ 0 \end{Bmatrix}. \tag{A.25}$$

Then the second stage of assembly can be performed by adding the following stiffness coefficients to the specific entries of the system stiffness matrix indicated by the following expressions:

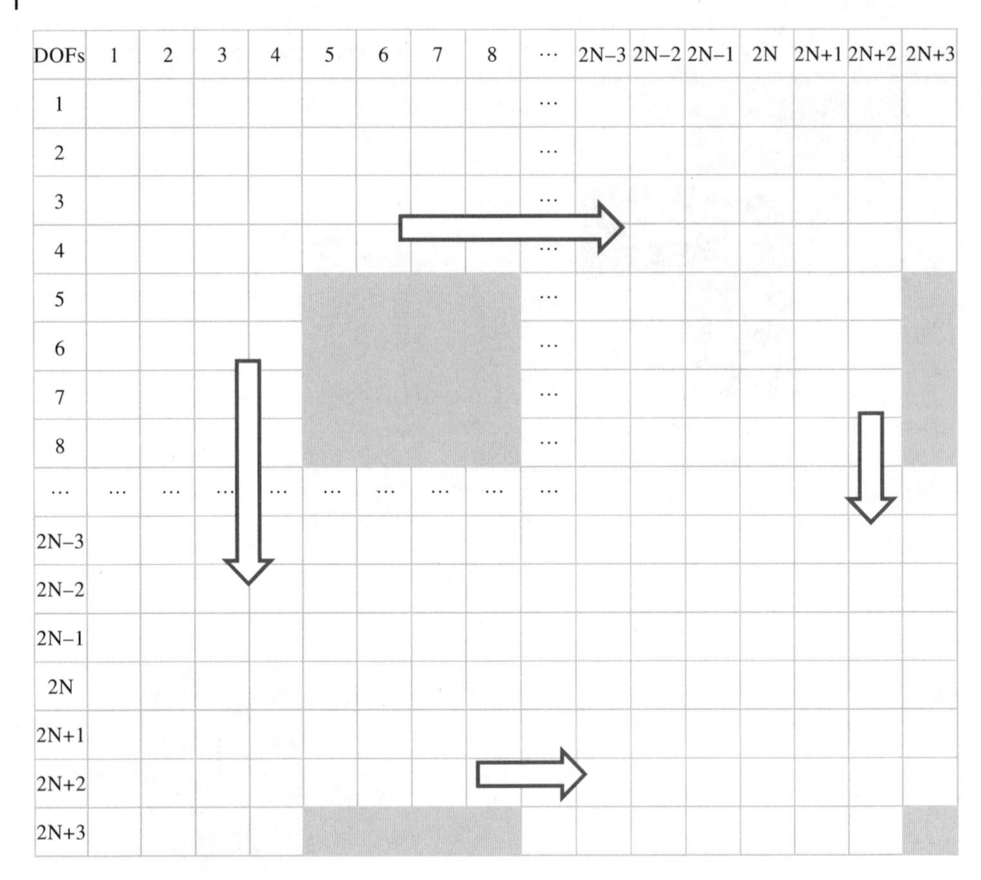


**Figure A.5** Second stage of assembly to account for vehicle equation in Eq. (A.23).

$$
\begin{array}{cccccc}
DOFs & 5 & 6 & 7 & 8 & 2N+3 \\
5 & vc_vN_1N_1^{'}+k_vN_1N_1 & vc_vN_1N_2^{'}+k_vN_1N_2 & vc_vN_1N_3^{'}+k_vN_1N_3 & vc_vN_1N_4^{'}+k_vN_1N_4 & -k_vN_1 \\
6 & vc_vN_2N_1^{'}+k_vN_2N_1 & vc_vN_2N_2^{'}+k_vN_2N_2 & vc_vN_2N_3^{'}+k_vN_2N_3 & vc_vN_2N_4^{'}+k_vN_2N_4 & -k_vN_2 \\
7 & vc_vN_3N_1^{'}+k_vN_3N_1 & vc_vN_3N_2^{'}+k_vN_3N_2 & vc_vN_3N_3^{'}+k_vN_3N_3 & vc_vN_3N_4^{'}+k_vN_3N_4 & -k_vN_3 \\
8 & vc_vN_4N_1^{'}+k_vN_4N_1 & vc_vN_4N_2^{'}+k_vN_4N_2 & vc_vN_4N_3^{'}+k_vN_4N_3 & vc_vN_4N_4^{'}+k_vN_4N_4 & -k_vN_4 \\
2N+3 & -vc_vN_1^{'}-k_vN_1 & -vc_vN_2^{'}-k_vN_2 & -vc_vN_3^{'}-k_vN_3 & -vc_vN_4^{'}-k_vN_4 & k_v
\end{array}
\tag{A.26}
$$

Note that the value of $N$ and $N'$ in Eq. (A.26) change with the acting position $x_c$ of the moving vehicle, as can be seen from Eq. (A.11). Through the above two-stage assembly procedure, the stiffness matrix of the VBI system at a certain instant $t$ can be assembled. It is easy to see that the same procedure applies to construction of the mass and damping matrices and the load vector of the system.

For the VBI problem, the contact position $x_c$ varies as the vehicle moves. As a result, the system matrix presented in Figure A.5 also changes in response to the vehicle's movement. Since the dynamic analysis of the VBI system with the vehicle moving at speed $v$ is conducted in a step-by-step manner with time increment $\Delta t$, the system matrix as schematically shown in Figure A.5 should also be updated in a step-by-step manner. The key point is to judge where the vehicle is located for the next step at time $t+\Delta t$. Take the case with the vehicle acting on element $n = 3$ at time $t$ as an example. There are two scenarios.

*Scenario 1:* The vehicle remains in the same element (e.g., the element numbered by $n = 3$ in Figure A.3) at time $t+\Delta t$. For this scenario, the only adjustment to be made is the stiffness coefficients (i.e., DOFs 5, 6, 7, 8, and $2N+3$) given in Eq. (A.26) for the moving sprung mass. All the other entries contributed by the ordinary beam elements remain unchanged. This can be easily done by deleting the previous addition at time $t$ to the corresponding DOFs of the stiffness matrix by the vehicle, and then inserting the new addition at time $t+\Delta t$ by the vehicle to the same element DOFs.

*Scenario 2:* The vehicle moves to the next element (e.g., the element numbered by $n = 4$ in Figure A.3) at time $t+\Delta t$. For this scenario, one need to delete the previous addition at time $t$ to the corresponding DOFs (i.e., DOFs 5, 6, 7, 8, and $2N+3$) of the stiffness matrix by the vehicle, and then insert the new addition at time $t+\Delta t$ by the vehicle to the element DOFs corresponding to vehicle's new acting position (i.e., DOFs 7, 8, 9, 10, and $2N+3$).

With the stiffness equation made available for the VBI system for each time step, the response of the entire VBI system within that time step can be solved by proper numerical schemes, say by Newmark's $\beta$ method (Clough and Penzien 1995). The procedure presented herein is simple and efficient in that both the vertical responses of the vehicle and supporting beam can be obtained simultaneously.

# References

Abdel-Ghaffar, A.M. and Housner, G.W. (1978). Ambient vibration tests of suspension bridge. *Journal of the Engineering Mechanics Division, ASCE* 104 (EM5): 983–999.

Abdel-Ghaffar, A.M. and Scanlan, R.H. (1985). Ambient vibration studies of Golden Gate Bridge: I. Suspended structure. *Journal of Engineering Mechanics-ASCE* 111: 462–482.

Akin, J.E. and Mofid, M. (1989). Numerical solution for response of beams with moving mass. *Journal of Structural Engineering, ASCE* 115 (1): 120–131.

Alonso, F.J., Del Castillo, J.M., and Pintado, P. (2005). Application of singular spectrum analysis to the smoothing of raw kinematic signals. *Journal of Biomechanics* 38: 1085–1092.

Altunisik, A.C., Bayraktar, A., and Özdemir, H. (2012). Seismic safety assessment of eynel highway steel bridge using ambient vibration measurements. *Smart Structures and Systems* 10 (2): 131–154.

Bandat, J.S. and Piersol, A.G. (1986). *Random Data: Analysis and Measurement Procedures*, 2e. New York: Wiley.

Bao, T. and Liu, Z. (2017). Vibrationbased bridge scour detection: a review. *Structural Control and Health Monitoring* 24 (7): e1937.

Berg, G.V. (1988). *Elements of Structural Dynamics*. Englewood Cliffs, New York: Prentice-Hall International Inc.

Biggs, J.M. (1964). *Introduction to Structural Dynamics*. New York: McGraw-Hill.

Brownjohn, J.M.W., Bocciolone, M., Curami, A. et al. (1994). Humber Bridge full-scale measurement campaigns 1990-1991. *Journal of Wind Engineering and Industrial Aerodynamics* 52: 185–218.

Brownjohn, J.M.W., Magalhaes, F., Caetano, E., and Cunha, A. (2010). Ambient vibration re-testing and operational modal analysis of the Humber Bridge. *Engineering Structures* 32: 2003–2018.

Brownjohn, J.M.W., Xia, P.Q., Hao, H., and Xia, Y. (2001). Civil structure condition assessment by FE model updating: methodology and case studies. *Finite Elements in Analysis and Design* 37 (10): 761–775.

Bu, J.Q., Law, S.S., and Zhu, X.Q. (2006). Innovative bridge condition assessment from dynamic response of a passing vehicle. *Journal of Engineering Mechanics-ASCE* 132 (12): 1372–1379.

Cantero, D., Hester, D., and Brownjohn, J. (2017). Evolution of bridge frequencies and modes of vibration during truck passage. *Engineering Structures* 152: 452–464.

Carden, E.P. and Fanning, P. (2004). Vibration-based condition monitoring: a review. *Structural Health Monitoring* 3 (4): 355–377.

*Vehicle Scanning Method for Bridges*, First Edition. Yeong-Bin Yang, Judy P. Yang, Bin Zhang and Yuntian Wu.

Casas, J.R. (1995). Full-scale dynamic testing of the Alamillo cable-stayed bridge in Sevilla (Spain). *Earthquake Engineering and Structural Dynamics* 24: 35–51.

Casciati, F. and Wu, L.J. (2013). Local positioning accuracy of laser sensors for structural health monitoring. *Structural Control and Health Monitoring* 20 (5): 728–739.

Cerda, F., Chen, S.H., Bielak, J. et al. (2014). Indirect structural health monitoring of a simplified laboratory-scale bridge model. *Smart Structures and Systems* 13 (5): 849–868.

Cerda, F., Garrett, J., Bielak, J., Rizzo, P. Barrera, J.A., Zhang, Z., Chen, S., McCann, M.T. and Kovacevic, J. (2012). Indirect structural health monitoring in bridges: scale experiments. *Proceedings of Bridge Maintenance, Safety, Management, Resilience and Sustainability*, Lago di Como, Italy (July), 346–353.

Chang, C.C., Chang, T.Y.P., and Zhang, Q.W. (2001). Ambient vibration of long-span cable-stayed bridge. *Journal of Bridge Engineering* 6 (1): 46–53.

Chang, P.C., Flatau, A., and Liu, S.C. (2003). Review paper: health monitoring of civil infrastructure. *Structural Health Monitoring* 2 (3): 257–267.

Chang, K.C., Kim, C.W., and Borjigin, S. (2014a). Variability in bridge frequency induced by a parked vehicle. *Smart Structures and Systems* 13 (5): 755–773.

Chang, K.C., Kim, C.W., and Kawatani, M. (2014b). Feasibility investigation for a bridge damage identification method through moving vehicle laboratory experiment. *Structure and Infrastructure Engineering* 10 (3): 328–345.

Chang, K.C., Wu, F.B., and Yang, Y.B. (2010). Effect of road surface roughness on indirect approach for measuring bridge frequencies from a passing vehicle. *Interaction and Multiscale Mechanics* 3 (4): 299–308.

Chang, K.C., Wu, F.B., and Yang, Y.B. (2011). Disk model for wheels moving over highway bridges with rough surfaces. *Journal of Sound and Vibration* 330 (20): 4930–4944.

Chatterjee, P.K., Datta, T.K., and Surana, C.S. (1994). Vibration of suspension bridges under vehicular movement. *Journal of Structural Engineering, ASCE* 120: 681–703.

Chen, S.H., Cerda, F., Rizzo, P. et al. (2014). Semi-supervised multiresolution classification using adaptive graph filtering with application to indirect bridge structural health monitoring. *IEEE Transactions on Signal Processing* 62: 2879–2893.

Chen, S.Y. and Xia, H. (2009). An identification method for fundamental frequency of bridge from dynamic responses due to passing vehicle. *Engineering Mechanics* 26 (8): 88–94. (In Chinese).

Chondros, T.G., Dimarogonas, A.D., and Yao, J. (1998). A continuous cracked beam vibration theory. *Journal of Sound and Vibration* 215 (1): 17–34.

Chrysostomou, C.Z., Demetriou, T., and Stassis, A. (2008). Health-monitoring and system-identification of an ancient aqueduct. *Smart Structures and Systems* 4 (2): 183–104.

Chu, K.H., Garg, V.K., and Wang, T.L. (1986). Impact in railway prestressed concrete bridges. *Journal of Structural Engineering, ASCE* 112 (5): 1036–1051.

Clough, R.W. and Penzien, J. (1995). *Dynamics of Structures*, 3e. Berkeley, California: Computers and Structures, Inc.

Conner, G.H., Stallings, J.M., McDuffie, T.L. et al. (1997). Steel bridge testing in Alabama. *Transportation Research Record* 1594: 134–139.

Cook, R.D., Malkus, D.S., Plesha, M.E., and Witt, R.J. (2002). *Concepts and Applications of Finite Element Analysis*, 4e. New York: Wiley.

Doebling, S.W., Farrar, C.R., and Prime, M.B. (1998). A summary review of vibration-based damage identification methods. *Shock and Vibration Digest* 30: 91–105.

Douglas, B.M. and Reid, W.H. (1982). Dynamic tests and system identification of bridges. *Journal of the Structural Division, ASCE* 108 (10): 2295–2312.

Dusseau, R.A. and Dusaisi, H.N. (1993). Natural frequencies of concrete bridges in the Pacific Northwest. *Transportation Research Record* 1393: 119–132.

Elhattab, A., Uddin, N., and OBrien, E. (2016). Drive-by bridge damage monitoring using bridge displacement profile difference. *Journal of Civil Structural Health Monitoring* 6 (5): 839–850.

Ewins, D.J. (2000). *Modal Testing: Theory, Practice and Application*, 2e. Letchworth, England: Research Studies Press, Ltd.

Fafard, M., Laflamme, M., Savard, M., and Bennur, M. (1998). Dynamic analysis of existing continuous bridge. *Journal of Bridge Engineering* 3 (1): 28–37.

Fan, W. and Qiao, P.Z. (2011). Vibration-based damage identification methods: a review and comparative study. *Structural Health Monitoring* 10 (1): 83–111.

Fang, S.E. and Perera, R. (2009). Power mode shapes for early damage detection in linear structures. *Journal of Sound and Vibration* 324 (1–2): 40–56.

Farrar, C.R., Doebling, S.W., and Nix, D.A. (2001). Vibration-based structural damage identification. *Philosophical Transactions of the Royal Society A* 359: 131–149.

Farrar, C.R. and James, C.H. III (1997). System identification from ambient vibration measurements on a bridge. *Journal of Sound and Vibration* 205 (1): 1–18.

Feng, D.M. and Feng, M.Q. (2016). Output-only damage detection using vehicle-induced displacement response and mode shape curvature index. *Structural Control and Health Monitoring* 23: 1088–1107.

Foda, M.A. and Abduljabbar, Z. (1998). A dynamic green function formulation for the response of a beam structure to a moving mass. *Journal of Sound and Vibration* 210 (3): 295–306.

Fryba, L. (1972). *Vibration of Solids and Structures Under Moving Loads*. Groningen, Netherlands: Noordhoff International Publishing.

Fujino, Y., Abe, M., Shibuya, H. et al. (2000). Forced and ambient vibration tests and vibration monitoring of Hakucho Suspension Bridge. *Transportation Research Record* 1696 (2): 57–63.

Fujino, Y., Kitagawa, K., Furukawa, T. and Ishii, H. (2005). Development of vehicle intelligent monitoring system (VIMS). *Proceedings of SPIE, Smart Structures and Materials 2005: Sensors and Smart Structures Technologies for Civil, Mechanical, and Aerospace Systems*, 5765 (March): 148–157.

Fujino, Y. and Siringoringo, D.M. (2011). Bridge monitoring in Japan: the needs and strategies. *Structure and Infrastructure Engineering* 7 (7–8): 597–611.

Golyandina, N., Nekrutkin, V., and Zhigljavsky, A. (2001). *Analysis of Time Series Structure – SSA and Related Techniques*. New York: Chapman and Hall/CRC.

Gomez, H.C., Fanning, P.J., Feng, M.Q., and Lee, S. (2011). Testing and long-term monitoring of a curved concrete box girder bridge. *Engineering Structures* 33 (10): 2861–2869.

Gonzalez, A. and Hester, D. (2013). An investigation into the acceleration response of a damaged beam-type structure to a moving force. *Journal of Sound and Vibration* 332 (13): 3201–3217.

Gonzalez, A., O'Brien, E.J., Li, Y.Y., and Cashell, K. (2008). The use of vehicle acceleration measurements to estimate road roughness. *Vehicle System Dynamics* 46 (6): 483–499.

Gonzalez, A., Obrien, E.J., and McGetrick, P.J. (2012). Identification of damping in a bridge using a moving instrumented vehicle. *Journal of Sound and Vibration* 331 (18): 4115–4131.

Guebailia, M., Ouelaa, N., and Guyader, J.L. (2013). Solution of the free vibration equation of a multi span bridge deck by local estimation method. *Engineering Structures* 48: 695–703.

Hahn, S.L. (1966). *Hilbert Transform in Signal Processing*. Boston: Artech House Publishers.

He, X., Moaveni, B., Conte, J.P. et al. (2009). System identification of Alfred Zampa Memorial Bridge using dynamic field test data. *Journal of Structural Engineering, ASCE* 135: 54–66.

He, W.Y. and Zhu, S.Y. (2016). Moving load-induced response of damaged beam and its application in damage localization. *Journal of Vibration and Control* 22 (16): 3601–3617.

Hester, D. and Gonzalez, A. (2015). A bridge-monitoring tool based on bridge and vehicle accelerations. *Structure and Infrastructure Engineering* 11 (5): 619–637.

Hester, D. and Gonzalez, A. (2017). A discussion on the merits and limitations of using drive-by monitoring to detect localized damage in a bridge. *Mechanical Systems and Signal Processing* 90: 234–253.

Huang, N.E., Shen, Z., and Long, S.R. (1999a). A new view of nonlinear water waves: the Hilbert spectrum. *Annual Review of Fluid Mechanics* 31: 417–457.

Huang, N.E., Shen, Z., Long, S.R. et al. (1998). The empirical mode decomposition and the Hilbert spectrum for nonlinear and non-stationary time series analysis. *Proceedings of the Royal Society: Mathematical, Physical and Engineering Sciences* 454: 903–995.

Huang, C.S., Yang, Y.B., Lu, L.Y., and Chen, C.H. (1999b). Dynamic testing and system identification of a multi-span highway bridge. *Earthquake Engineering and Structural Dynamics* 28 (8): 857–878.

Inglis, C.E. (1934). *A Mathematical Treatise on Vibration in Railway Bridges*. Cambridge: Cambridge University Press.

ISO 8608 (1995). *Mechanical Vibration-Road Surface Profiles-Reporting of Measured Data*. Geneva: International Organization for Standardization.

Jaishi, B. and Ren, W.X. (2005). Structural finite element model updating using ambient vibration test results. *Journal of Structural Engineering, ASCE* 131 (4): 617–628.

Jeffcott, H.H. (1929). On the vibration of beams under the action of moving loads. *Philosophical Magazine*, Series 7 8 (48): 66–97.

Kafle, B., Zhang, L., Mendis, P. et al. (2017). Monitoring the dynamic behavior of the Merlynston creek bridge using interferometric radar sensors and finite element modeling. *International Journal of Applied Mechanics* 9 (1): 1750003.

Kato, M. and Shimada, S. (1986). Vibration of PC bridge during failure process. *Journal of Structural Engineering, ASCE* 112 (7): 1692–1703.

Keenahan, J., and OBrien, E.J. (2014). Allowing for a rocking datum in the analysis of drive-by bridge inspections. *Civil Engineering Research in Ireland*, Belfast, UK (28–29 August 2014): 117–124.

Keenahan, J., OBrien, E.J., McGetrick, P.J., and Gonzalez, A. (2014). The use of a dynamic truck-trailer drive-by system to monitor bridge damping. *Structural Health Monitoring* 13 (2): 143–157.

Khaji, N., Shafiei, M., and Jalalpour, M. (2009). Closed-form solutions for crack detection problem of Timoshenko beams with various boundary conditions. *International Journal of Mechanical Sciences* 51 (9): 667–681.

Khorram, A., Bakhtiari-Nejad, F., and Rezaeian, M. (2012). Comparison studies between two wavelet based crack detection methods of a beam subjected to a moving load. *International Journal of Engineering Science* 51: 204–215.

Kim, C.W., Chang, K.C., Kitauchi, S., and McGetrick, P.J. (2016). A field experiment on a steel Gerber-truss bridge for damage detection utilizing vehicle-induced vibrations. *Structural Health Monitoring* 15 (2): 421–429.

Kim, C.W., Chang, K.C., McGetrick, P.J. et al. (2017). Utilizing moving vehicles as sensors for bridge condition screening-a laboratory verification. *Sensors and Materials* 29 (2): 153–163.

Kim, C.W., Isemoto, R., McGetrick, P.J. et al. (2014). Drive-by bridge inspection from three different approaches. *Smart Structures and Systems* 13 (5): 775–796.

Kim, C.W. and Kawatani, M. (2008). Pseudo-static approach for damage identification of bridges based on coupling vibration with a moving vehicle. *Structure and Infrastructure Engineering* 4 (5): 371–379.

Kim, J. and Lynch, J.P. (2012). Experimental analysis of vehicle-bridge interaction using a wireless monitoring system and a two-stage system identification technique. *Mechanical Systems and Signal Processing* 28: 3–19.

Kong, X. and Cai, C.S. (2016). Scour effect on bridge and vehicle responses under bridge-vehicle-wave interaction. *Journal of Bridge Engineering* 21 (4): 04015083.

Kong, X., Cai, C.S., Deng, L., and Zhang, W. (2017). Using dynamic responses of moving vehicles to extract bridge modal properties of a field bridge. *Journal of Bridge Engineering* 22 (6): 04017018.

Kong, X., Cai, C.S., and Kong, B. (2015). Damage detection based on transmissibility of a vehicle and bridge coupled system. *Journal of Engineering Mechanics-ASCE* 141 (1): 04014102.

Kong, X., Cai, C.S., and Kong, B. (2016). Numerically extracting bridge modal properties from dynamic responses of moving vehicles. *Journal of Engineering Mechanics-ASCE* 142 (6): 04016025.

Law, S.S., Bu, J.Q., Zhu, X.Q., and Chan, S.L. (2006). Vehicle condition surveillance on continuous bridges based on response sensitivity. *Journal of Engineering Mechanics-ASCE* 132: 78–86.

Lederman, G., Wang, Z., Bielak, J. et al. (2014). Damage quantification and localization algorithms for indirect SHM of bridges. In: *Bridge Maintenance, Safety, Management and Life Extension* (eds. A.R. Chen, D.M. Frangopol and X. Ruan), 640–647. New York: CRC Press.

Lee, H.P. (1996). Dynamic response of a beam with a moving mass. *Journal of Sound and Vibration* 191 (2): 289–294.

Lee, J.W., Kim, J.D., Yun, C.B. et al. (2002). Health-monitoring method for bridges under ordinary traffic loadings. *Journal of Sound and Vibration* 257 (2): 247–264.

Li, Z.H. and Au, F.T.K. (2014). Damage detection of a continuous bridge from response of a moving vehicle. *Shock and Vibration* 2014: 146802.

Li, Z.H. and Au, F.T.K. (2015). Damage detection of bridges using response of vehicle considering road surface roughness. *International Journal of Structural Stability and Dynamics* 15 (3): 1450057.

Li, Y., Cai, C.S., Liu, Y. et al. (2016a). Dynamic analysis of a large span specially shaped hybrid girder bridge with concrete-filled steel tube arches. *Engineering Structures* 106: 243–260.

Li, J. and Hao, H. (2015). Damage detection of shear connectors under moving loads with relative displacement measurements. *Mechanical Systems and Signal Processing* 60–61: 124–150.

Li, W.M., Jiang, Z.H., Wang, T.L., and Zhu, H.P. (2014). Optimization method based on generalized pattern search algorithm to identify bridge parameters indirectly by a passing vehicle. *Journal of Sound and Vibration* 333 (2): 364–380.

Li, H.L., Lu, Z.R., and Liu, J.K. (2016b). Identification of distributed damage in bridges from vehicle-induced dynamic responses. *Advances in Structural Engineering* 19 (6): 945–952.

Lin, C.W. and Yang, Y.B. (2005). Use of a passing vehicle to scan the bridge frequencies – an experimental verification. *Engineering Structures* 27 (13): 1865–1878.

Lu, Z.R. and Liu, J.K. (2011). Identification of both structural damages in bridge deck and vehicular parameters using measured dynamic responses. *Computers and Structures* 89: 1397–1405.

Magalhães, F., Caetano, E., Cunha, Á. et al. (2012). Ambient and free vibration tests of the Millau Viaduct: evaluation of alternative processing strategies. *Engineering Structures* 45: 372–384.

Maizuar, M., Zhang, L., Miramini, S. et al. (2017). Detecting structural damage to bridge girders using radar interferometry and computational modelling. *Structural Control and Health Monitoring* 24 (10): e1985.

Majumder, L. and Manohar, C.S. (2004). Nonlinear reduced models for beam damage detection using data on moving oscillator-beam interactions. *Computers and Structures* 82: 301–314.

Malekjafarian, A., McGetrick, P.J., and OBrien, E.J. (2015). A review of indirect bridge monitoring using passing vehicles. *Shock and Vibration* 2015: 286139.

Malekjafarian, A. and OBrien, E.J. (2014). Identification of bridge mode shapes using short time frequency domain decomposition of the responses measured in a passing vehicle. *Engineering Structures* 81: 386–397.

Malekjafarian, A. and OBrien, E.J. (2017). On the use of a passing vehicle for the estimation of bridge mode shapes. *Journal of Sound and Vibration* 397: 77–91.

Maragakis, E.M., Douglas, B.M., and Chen, Q. (2001). Full-scale field failure tests of railway bridge. *Journal of Bridge Engineering* 6 (5): 356–362.

Maragakis, E.M., Douglas, B.M., Chen, Q., and Sandirasegaram, U. (1998). Full-scale tests of a railway bridge. *Transportation Research Record* 1624: 140–147.

Mazurek, D.F. and DeWolf, J.T. (1990). Experimental study of bridge monitoring technique. *Journal of Structural Engineering, ASCE* 116 (9): 2532–2549.

McGetrick, P.J., Gonzalez, A., and O'Brien, E.J. (2009). Theoretical investigation of the use of a moving vehicle to identify bridge dynamic parameters. *Insight* 51 (8): 433–438.

McGetrick, P.J., Hester, D., and Taylor, S.E. (2017). Implementation of a driveby monitoring system for transport infrastructure utilising smartphone technology and GNSS. *Journal of Civil Structural Health Monitoring* 7 (2): 175–189.

McGetrick, P.J. and Kim, C.W. (2013). A parametric study of a drive by bridge inspection system based on the Morlet wavelet. *Key Engineering Materials* 569–570: 262–269.

McGetrick, P., and Kim, C.W. (2014a). A wavelet based drive-by bridge inspection system. *Proceedings of 7th International Conference on Bridge Maintenance Safety and Management*, Shanghai, China: 613–621.

McGetrick, P., and Kim, C.W. (2014b). An indirect bridge inspection method incorporating a wavelet-based damage indicator and pattern recognition. *Proceedings of 9th International Conference on Structural Dynamics*, Porto, Portugal (July): 2605–2612.

McGetrick, P.J., Kim, C.W., Gonzalez, A., and OBrien, E.J. (2015). Experimental validation of a drive-by stiffness identification method for bridge monitoring. *Structural Health Monitoring* 14 (4): 317–331.

McLamore, V.R., Hart, G.C., and Stubbs, I.R. (1971). Ambient vibration of two suspension bridges. *Journal of the Structural Division, ASCE* 97 (ST10): 2567–2582.

Meredith, J., Gonzalez, A., and Hester, D. (2012). Empirical mode decomposition of the acceleration response of a prismatic beam subject to a moving load to identify multiple damage locations. *Shock and Vibration* 19: 845–856.

Miyamoto, A. and Yabe, A. (2011). Bridge condition assessment based on vibration responses of passenger vehicle. *Journal of Physics: Conference Series* 305 (1): 012103.

Miyamoto, A. and Yabe, A. (2012). Development of practical health monitoring system for short- and medium-span bridges based on vibration responses of city bus. *Journal of Civil Structural Health Monitoring* 2 (1): 47–63.

Moschas, F. and Stiros, S. (2011). Measurement of the dynamic displacements and of the modal frequencies of a short-span pedestrian bridge using GPS and an accelerometer. *Engineering Structures* 33 (1): 10–17.

Nagayama, T., Reksowardojo, A.P., Su, D., and Mizutani, T. (2017). Bridge natural frequency estimation by extracting the common vibration component from the responses of two vehicles. *Engineering Structures* 150: 821–829.

Nasrellah, H.A. and Manohar, C.S. (2010). A particle filtering approach for structural system identification in vehicle-structure interaction problems. *Journal of Sound and Vibration* 329 (9): 1289–1309.

Newmark, N.M. (1959). A method of computation for structural dynamics. *Journal of the Engineering Mechanics Division, ASCE* 85 (EM3): 67–94.

Nguyen, K.V. and Tran, H.T. (2010). Multi-cracks detection of a beam-like structure based on the on-vehicle vibration signal and wavelet analysis. *Journal of Sound and Vibration* 329: 4455–4465.

Ni, Y.C. and Zhang, F.L. (2016). Bayesian operational modal analysis of a pedestrian bridge using a field test with multiple setups. *International Journal of Structural Stability and Dynamics* 16 (8): 1550052. (23 pages).

OBrien, E.J. and Keenahan, J. (2015). Drive-by damage detection in bridges using the apparent profile. *Structural Control and Health Monitoring* 22 (5): 813–825.

OBrien, E.J. and Malekjafarian, A. (2016). A mode shape-based damage detection approach using laser measurement from a vehicle crossing a simply supported bridge. *Structural Control and Health Monitoring* 23: 1273–1286.

OBrien, E.J., Malekjafarian, A., and Gonzalez, A. (2017a). Application of empirical mode decomposition to drive-by bridge damage detection. *European Journal of Mechanics - A/Solids* 61: 151–163.

OBrien, E.J., Martinez, D., Malekjafarian, A., and Sevillano, E. (2017b). Damage detection using curvatures obtained from vehicle measurements. *Journal of Civil Structural Health Monitoring* 7 (3): 333–341.

OBrien, E.J., McGetrick, P.J., and Gonzalez, A. (2014). A drive-by inspection system via vehicle moving force identification. *Smart Structures and Systems* 13 (5): 821–848.

Okauchi, I., Miyata, T., Tatsumi, M., and Kiyota, R. (1992). Dynamic field tests and studies on vibrational characteristics of long-span suspension bridges. *Structural Engineering/Earthquake Engineering, JSCE* 9 (1): 89–100.

Olsson, M. (1991). On the fundamental moving load problem. *Journal of Sound and Vibration* 145 (2): 299–307.
Oshima, Y., and Yamamoto, K. (2009). Assessment of bridge vibration based on vehicle responses using by independent component analysis, *Proceedings of the Twenty-Second KKCNN Symposium on Civil Engineering*: 99–104, Chiangmai, Thailand.
Oshima, Y., Yamamoto, K., and Sugiura, K. (2014). Damage assessment of a bridge based on mode shapes estimated by responses of passing vehicles. *Smart Structures and Systems* 13 (5): 731–753.
Pandey, A.K., Biswas, M., and Samman, M.M. (1991). Damage detection from changes in curvature mode shapes. *Journal of Sound and Vibration* 145 (2): 321–332.
Paultre, P., Proulx, J., and Talbot, M. (1995). Dynamic testing procedures for highway bridges using traffic loads. *Journal of Structural Engineering, ASCE* 121 (2): 362–376.
Peeters, B. and de Roeck, G. (1999). Reference-based stochastic subspace identification for output-only model analysis. *Mechanical Systems and Signal Processing* 13 (6): 855–878.
Pesterev, A.V., Yang, B., Bergman, L.A., and Tan, C.A. (2001). Response of elastic continuum carrying multiple moving oscillators. *Journal of Engineering Mechanics-ASCE* 127 (3): 260–265.
Qi, Z.Q. and Au, F.T.K. (2017). Identifying mode shapes of girder bridges using dynamic responses extracted from a moving vehicle under impact excitation. *International Journal of Structural Stability and Dynamics* 17 (8): 1750081.
Quirke, P., Bowe, C., OBrien, E.J. et al. (2017). Railway bridge damage detection using vehiclebased inertial measurements and apparent profile. *Engineering Structures* 153: 421–442.
Ren, W.X., Zhao, T., and Harik, I.E. (2004). Experimental and analytical modal analysis of steel arch bridge. *Journal of Structural Engineering, ASCE* 130: 1022–1031.
Roveri, N. and Carcaterra, A. (2012). Damage detection in structures under traveling loads by Hilbert-Huang transform. *Mechanical Systems and Signal Processing* 28: 128–144.
Sadiku, S. and Leipholz, H.H.E. (1987). On the dynamics of elastic systems with moving concentrated masses. *Ingenieur Archiv* 57: 223–242.
Salawu, O.S. (1997). Detection of structural damage through changes in frequency: a review. *Engineering Structures* 19 (9): 718–723.
Salawu, O.S. and Williams, C. (1995). Review of full-scale dynamic testing of bridge structures. *Engineering Structures* 17 (2): 113–121.
Siringoringo, D.M. and Fujino, Y. (2012). Estimating bridge fundamental frequency from vibration response of instrumented passing vehicle: analytical and experimental study. *Advances in Structural Engineering* 15 (3): 417–433.
Soyoz, S. and Feng, M.Q. (2009). Long-term monitoring and identification of bridge structural parameters. *Computer-Aided Civil and Infrastructure Engineering* 24 (2): 82–92.
Stanisic, M.M. (1985). On a new theory of the dynamic behaviour of the structures carrying moving masses. *Ingenieur Archiv* 55: 176–185.
Stokes, G.G. (1849). Discussion of a differential equation relating to the braking of railway bridges. *Transactions of the Cambridge Philosophical Society* 8 (5): 707–735.
Tada, H., Paris, P.C., and Irwin, G.R. (2000). *The Stress Analysis of Cracks Handbook*, 3e. Del Research Corporation.
Tan, G.H., Brameld, G.H., and Thambiratnam, D.P. (1998). Development of an analytical model for treating bridge-vehicle interaction. *Engineering Structures* 20 (1–2): 54–61.

Tan, C., Elhattab, A., and Uddin, N. (2017). "Drive-by" bridge frequency-based monitoring utilizing wavelet transform. *Journal of Civil Structural Health Monitoring* 7 (5): 615–625.

Tan, C.P. and Shore, S. (1968). Response of horizontally curved bridge to moving load. *Journal of Structural Engineering, ASCE* 94 (9): 2135–2151.

Timoshenko, S.P. (1922). On the forced vibrations of bridges. *Philosophical Magazine Series* 6 (43): 1018–1019.

Ting, E.C., Genin, J., and Ginsberg, J.H. (1974). A general algorithm for the moving mass problem. *Journal of Sound and Vibration* 33 (1): 49–58.

Tsai, H.C., Wang, C.Y., Huang, N.E. et al. (2015). Railway track inspection based on the vibration response to a scheduled train and the Hilbert-Huang transform. *Proceedings of Institute of Mechanical Engineers Part F: Journal of Rail and Rapid Transit* 229 (7): 815–829.

Urushadze, S. (2017). Theoretical predictive models of interaction between varying and moving loads and bridges for structural health monitoring. *1st Workshop for Taiwan-Czech Collaborative Research*, Taipei, Taiwan, March 27.

Ventura, C.E., Felber, A.J., and Stiemer, S.F. (1996). Determination of the dynamic characteristics of the Colquitz River Bridge by full-scale testing. *Canadian Journal of Civil Engineering* 23 (2): 536–548.

Wallin, J., Leander, J., and Karoumi, R. (2011). Strengthening of a steel railway bridge and its impact on the dynamic response to passing trains. *Engineering Structures* 33 (2): 635–646.

Wang, S.D., Bu, J.Q., and Lou, G.C. (2008). Bridge damage identification by dynamic response of passing vehicle. *Journal of Chang'an University (Natural Science Edition)* 28 (3): 63–67. (In Chineese).

Wang, N.B., He, L.X., Ren, W.X., and Huang, T.L. (2017d). Extraction of influence line through a fitting method from bridge dynamic response induced by a passing vehicle. *Engineering Structures* 151: 648–664.

Wang, H., Nagayama, T., Zhao, B., and Su, D. (2017a). Identification of moving vehicle parameters using bridge responses and estimated bridge pavement roughness. *Engineering Structures* 153: 57–70.

Wang, N.B., Ren, W.X., and Chen, Z.W. (2017c). Waveletbased automatic identification method of axle distribution information. *Structural Engineering and Mechanics* 63 (6): 761–769.

Wang, L., Zhang, Y., and Lie, S.T. (2017b). Detection of damaged supports under railway track based on frequency shift. *Journal of Sound and Vibration* 392: 142–153.

Ward, H.S. (1984). Traffic generated vibrations and bridge integrity. *Journal of Structural Engineering, ASCE* 110 (10): 2487–2498.

Wenzel, H. and Pichler, P. (2005). *Ambient Vibration Monitoring*. New York: Wiley.

Willis, R. (1849). *Appendix to the Report of the Commissioners Appointed to Inquire into the Application of Iron to Railway Structures*. London, UK: H.M. Stationary Office.

Wilson, J.C. and Liu, T. (1991). Ambient vibration measurements on a cable-stayed bridge. *Earthquake Engineering and Structural Dynamics* 20: 723–747.

Xia, H., Xu, Y.L., and Chan, T.H.T. (2000). Dynamic interaction of long suspension bridges with running trains. *Journal of Sound and Vibration* 237 (2): 263–280.

Xiang, Z., Dai, X., Zhang, Y., and Lu, Q. (2010). The tap-scan method for damage detection of bridge structures. *Interaction and Multiscale Mechanics* 3 (2): 173–191.

Xu, Y.L., Zhu, L.D., Wong, K.Y., and Chan, K.W.Y. (2000). Field measurement results of Tsing Ma Suspension Bridge during Typhoon Victor. *Structural Engineering and Mechanics* 10 (6): 545–559.

Yamamoto, K., Oshima, Y., Tanaka, A., and Mori, M. (2009). Monitoring of coupled vibration between bridge and train. *Proceedings of the Twenty-Second KKCNN Symposium on Civil Engineering*, Chiangmai, Thailand, 105–110.

Yang, Y.B. and Chang, K.C. (2009a). Extraction of bridge frequencies from the dynamic response of a passing vehicle enhanced by the EMD technique. *Journal of Sound and Vibration* 322 (4–5): 718–739.

Yang, Y.B. and Chang, K.C. (2009b). Extracting the bridge frequencies indirectly from a passing vehicle: parametric study. *Engineering Structures* 31 (10): 2448–2459.

Yang, Y.B., Chang, K.C., and Li, Y.C. (2013a). Filtering techniques for extracting bridge frequencies from a test vehicle moving over the bridge. *Engineering Structures* 48: 353–362.

Yang, Y.B., Chang, C.H., and Yau, J.D. (1999). An element for analysing vehicle-bridge systems considering vehicle's pitching effect. *International Journal for Numerical Methods in Engineering* 46: 1031–1047.

Yang, Y.B. and Chen, W.F. (2016). Extraction of bridge frequencies from a moving test vehicle by stochastic subspace identification. *Journal of Bridge Engineering* 21 (3): 04015053.

Yang, Y.B., Chen, W.F., Yu, H.W., and Chan, C.S. (2013b). Experimental study of a hand-drawn cart for measuring the bridge frequencies. *Engineering Structures* 57: 222–231.

Yang, Y.B., Cheng, M.C., and Chang, K.C. (2013c). Frequency variation in vehicle-bridge interaction systems. *International Journal of Structural Stability and Dynamics* 13 (2): 1350019.

Yang, J.P. and Lee, W.C. (2017). Damping effect of a passing vehicle for indirectly measuring bridge frequencies by EMD technique. *International Journal of Structural Stability and Dynamics* 17 (10): 1850008. (17 pages).

Yang, Y.B., Li, Y.C., and Chang, K.C. (2012a). Effect of road surface roughness on the response of a moving vehicle for identification of bridge frequencies. *Interaction and Multiscale Mechanics* 5 (4): 347–368.

Yang, Y.B., Li, Y.C., and Chang, K.C. (2012b). Using two connected vehicles to measure the frequencies of bridges with rough surface – a theoretical study. *Acta Mechanica* 223 (8): 1851–1861.

Yang, Y.B., Li, Y.C., and Chang, K.C. (2014). Constructing the mode shapes of a bridge from a passing vehicle: a theoretical study. *Smart Structures and Systems* 13 (5): 797–819.

Yang, Y.B., Liao, S.S., and Lin, B.H. (1995). Impact formulas for vehicles moving over simple and continuous beams. *Journal of Structural Engineering, ASCE* 121 (11): 1644–1650.

Yang, Y.B. and Lin, B.H. (1995). Vehicle-bridge interaction analysis by dynamic condensation method. *Journal of Structural Engineering, ASCE* 121 (11): 1636–1643.

Yang, Y.B. and Lin, C.W. (2005). Vehicle-bridge interaction dynamics and potential applications. *Journal of Sound and Vibration* 284 (1–2): 205–226.

Yang, Y.B., Lin, C.W., and Yau, J.D. (2004a). Extracting bridge frequencies from the dynamic response of a passing vehicle. *Journal of Sound and Vibration* 272: 471–493.

Yang, Y.B. and Wu, Y.S. (2001). A versatile element for analyzing vehicle-bridge interaction response. *Engineering Structures* 23: 452–469.

Yang, Y.B. and Yang, J.P. (2018). State-of-the-art review on modal identification and damage detection of bridges by moving test vehicles. *International Journal of Structural Stability and Dynamics* 18 (2): 1850025.

Yang, Y.B. and Yau, J.D. (1997). Vehicle-bridge interaction element for dynamic analysis. *Journal of Structural Engineering, ASCE* 123 (11): 1512–1518. (Errata: 124(4), p. 479).

Yang, Y.B. and Yau, J.D. (2015). Vertical and pitching resonance of train cars moving over a series of simple beams. *Journal of Sound and Vibration* 337: 135–149.

Yang, Y.B., Yau, J.D., and Hsu, L.C. (1997). Vibration of simple beams due to trains moving at high speeds. *Engineering Structures* 19 (11): 936–944.

Yang, Y.B., Yau, J.D., and Urushdaze, S. (2019). Scanning the modal coupling of slender suspension footbridges by a virtual moving vehicle. *Engineering Structures* 180: 574–585.

Yang, Y.B., Yau, J.D., and Wu, Y.S. (2004b). *Vehicle-Bridge Interaction Dynamics—with Applications to High-Speed Railways*. Singapore: World Scientific.

Yang, Y.B., Zhang, B., Qian, Y., and Wu, Y.T. (2018a). Contact-point response for modal identification of bridges by a moving vehicle. *International Journal of Structural Stability and Dynamics* 18 (5): 1850073. (24 pages).

Yang, Y.B., Zhang, B., Qian, Y., and Wu, Y.T. (2018b). Further revelation on damage detection by IAS computed from contact-point response of moving vehicle. *International Journal of Structural Stability and Dynamics* 18 (11): 1850137. (13 pages).

Yau, J.D., Yang, Y.B., and Kuo, S.R. (1999). Impact response of high speed rail bridges and riding comfort of rail cars. *Engineering Structures* 21 (9): 836–844.

Yau, J.D., Yang, J.P., and Yang, Y.B. (2017). Wave number-based technique for detecting slope discontinuity in simple beams using the moving test vehicle. *International Journal of Structural Stability and Dynamics* 17 (6): 1750060.

Yin, S.H. (2016). Vibration of a simple beam subjected to a moving sprung mass with initial velocity and constant acceleration. *International Journal of Structural Stability and Dynamics* 16 (3): 1450109.

Yin, S.H. and Tang, C.Y. (2011). Identifying cable tension loss and deck damage in a cable-stayed bridge using a moving vehicle. *Journal of Vibration and Acoustics* 133 (2): 021007.

Zhang, Y., Lie, S.T., and Xiang, Z.H. (2013). Damage detection method based on operating deflection shape curvature extracted from dynamic response of a passing vehicle. *Mechanical Systems and Signal Processing* 35: 238–254.

Zhang, B., Qian, Y., Wu, Y.T., and Yang, Y.B. (2018). An effective means for damage detection of bridges using the contact-point response of a moving test vehicle. *Journal of Sound and Vibration* 419: 158–172.

Zhang, Y., Wang, L., and Xiang, Z.H. (2012). Damage detection by mode shape squares extracted from a passing vehicle. *Journal of Sound and Vibration* 331 (2): 291–307.

Zhu, X.Q. and Law, S.S. (2006). Wavelet-based crack identification of bridge beam from operational deflection time history. *International Journal of Solids and Structures* 43 (7): 2299–2317.

Zhu, X.Q. and Law, S.S. (2015). Structural health monitoring based on vehicle-bridge interaction: accomplishments and challenges. *Advances in Structural Engineering* 18 (12): 1999–2015.

Zhu, X.Q. and Law, S.S. (2016). Recent developments in inverse problems of vehicle-bridge interaction dynamics. *Journal of Civil Structural Health Monitoring* 6 (1): 107–128.

Zhu, X.Q., Law, S.S., Huang, L., and Zhu, S.Y. (2018). Damage identification of supporting structures with a moving sensory system. *Journal of Sound and Vibration* 415: 111–127.

Zienkiewicz, O.C. and Taylor, R.L. (2005). *The Finite Element Method for Solid and Structural Mechanics*, 6e. Burlington: Elsevier Butterworth-Heinemann.

# Author Index



*Vehicle Scanning Method for Bridges*, First Edition. Yeong-Bin Yang, Judy P. Yang, Bin Zhang and Yuntian Wu.

# Subject Index



*Vehicle Scanning Method for Bridges*, First Edition. Yeong-Bin Yang, Judy P. Yang, Bin Zhang and Yuntian Wu.